Analog Circuits and Systems for Voltage-Mode and Current-Mode Sensor Interfacing Applications

ANALOG CIRCUITS AND SIGNAL PROCESSING SERIES

Consulting Editor*: Mohammed Ismail. *Ohio State University

For further volumes:
http://www.springer.com/series/7381

Andrea De Marcellis • Giuseppe Ferri

Analog Circuits and Systems for Voltage-Mode and Current-Mode Sensor Interfacing Applications

Springer

Andrea De Marcellis
Electrical and Information Engineering
Department
University of L'Aquila
via G. Gronchi 18
67100 L'Aquila
Italy
andrea.demarcellis@univaq.it

Giuseppe Ferri
Electrical and Information Engineering
Department
University of L'Aquila
via G. Gronchi 18
67100 L'Aquila
Italy
giuseppe.ferri@univaq.it

ISBN 978-90-481-9827-6 e-ISBN 978-90-481-9828-3
DOI 10.1007/978-90-481-9828-3
Springer Dordrecht Heidelberg London New York

Library of Congress Control Number: 2011931893

Printed on acid-free paper

Springer is part of Springer Science+Business Media (www.springer.com)

Preface

This book proposes recent scientific results concerning the research of novel electronic integrated circuits and system solutions for sensor interfacing, many of which developed by the authors, utilizing the deep experience in analog microelectronics of the research team from University of L'Aquila both in sensor field and in Low Voltage Low Power analog integrated circuit design with Voltage-Mode and Current-Mode approaches. In particular, this monograph describes and discusses a number of analog interfaces, suitable for resistive, capacitive and temperature sensors, some of which developed by the authors also in a standard *CMOS* integrated technology (AMS 0.35 μm).

The book is organized as follows.

After a fast "excursus" on physical and chemical sensors (Chap. 1) and a state of art analysis of the main resistive, capacitive and temperature sensors and their related basic analog interfaces (Chap. 2), novel and improved solutions of Low Voltage Low Power analog circuits and systems, designed both in Voltage-Mode (Chap. 3) and in Current-Mode (Chap. 4) approaches, suitable for portable sensor interfacing applications, will be described and investigated. Then, the lock-in technique will be considered (Chap. 5) with the aim to improve the sensor system characteristics. In the Appendices, the Second Generation Current Conveyor theory and applications, together with some novel design implementations at transistor level, as well as the noise and offset compensation techniques for the design of high-accuracy instrumentation voltage amplifiers, will be also described.

More in detail, concerning resistive sensors, the book describes the main design aspects and different circuit solutions of the first analog front-ends, performing resistance-to-voltage (for small measurand variations) and resistance-to-period or frequency (for wide variation ranges) conversions, both in Voltage-Mode and in Current-Mode approaches and using AC as well as DC excitation voltages for the sensors; also, it has been proved that the analog lock-in amplifier can be employed for enhancing resistive sensor system sensitivity and resolution.

Regarding capacitive sensors, both the Voltage-Mode and the Current-Mode approaches have been utilized to develop suitable interface systems converting the capacitance change of the sensing element into a voltage or a frequency variation.

Moreover, temperature sensors and their interfaces have been described. They have proved to be necessary in many sensor systems, since their characteristics are strictly related to operating thermal conditions. In this sense, electronic heater circuits for temperature control are shown.

We want to mention the fact that after an accurate design by means of a suitable simulation software, as ORCAD PSpice and CADENCE Virtuoso-Affirma, some of the described circuits (in particular, those developed by the authors of this book) have been implemented through prototype boards, with commercial discrete components, so to characterize and validate the new ideas, studying also other possible improvements. The final step has been, in some cases, the fabrication of the integrated circuit on-chip, in a standard *CMOS* technology, which follows the implementation of the circuit layout.

This book originated from the Ph.D. final dissertation of the first author and wants to give an overview of Voltage-Mode and Current-Mode analog sensor interfaces. In our opinion, it can be useful for analog electronic circuit designers, as well as for sensor companies, but can be also utilized as reference text book in advanced graduate or Ph.D. courses covering these topics. In this sense, the presented interfaces can be easily fabricated both as prototype boards, for a fast characterization (in this sense, they can be simply implemented by students and technicians), and as integrated circuits, also using modern design techniques (well known to specialist analog microelectronic students and designers).

We hope that this book will be interested and useful for readers at the same level of which it has been exciting and difficult to write it.

Furthermore, we want to address some acknowledgements. In particular, we want to thank Prof. Arnaldo D'Amico (University of Roma Tor Vergata) to have been an invaluable reference in all our working and scientific research activities. Then, we thank all the people with whom we have collaborated and discussed, at different levels, in particular, in an alphabetic order, Carlo Cantalini, Alessandro Depari, Claudia Di Carlo, Corrado Di Natale, Christian Falconi, Ferdinando Feliciangeli, Alessandra Flammini, Fabrizio Mancini, Paolo Mantenuto, Daniele Marioli, Eugenio Martinelli, Roberto Paolesse, Andrea Pelliccione, Stefano Ricci, Emiliano Sisinni, Vincenzo Stornelli and all the students who helped us to develop, simulate and test some of the described circuits.

Finally, we would like especially to thank our families for their continuous support and encouragement in every our activity and daily life.

University of L'Aquila, 2011 *Andrea De Marcellis*
Giuseppe Ferri

Contents

Introduction

Modern silicon Very Large Scale Integration (*VLSI*) Complementary Metal-Oxide Semiconductor (*CMOS*) technologies can place and interconnect several million transistors on a single Integrated Circuit (*IC*) having sizes approximately lower than 100 mm^2. These integrated technologies have evolved over a long period of time, starting with only few transistors per *IC*, then doubling them about every 18–24 months (according to the well-known Moore's Law), towards the present high densities (about 1–2 billions of transistors, considering recently developed commercial microprocessors). In parallel with the technology evolution, also Computer-Aided Design (*CAD*) and electronic design automation (*EDA*) tools have been developed with the aim to help *IC* designers. Through the use of these tools, design teams have employed very "experienced" designers completely embedded in the same tool management. Therefore, *IC* functionalities, together with the *CAD*/*EDA* tools which guide the design towards the *IC* fabrication, have made available the actual technology to system designers.

All these facilities allow to detect and quantify the bigger part of natural phenomena related to the energy transformation of the parameters, through the use of sensors (i.e., sensing elements), their electronic interfaces and suitable instrumentation and measurement systems.

In fact, recent progresses in physics, chemistry, electronics, material science, bottom/up and top/down technologies have allowed the integration of high performance and low-cost low-size systems, achieving the so-called System-on-Chip (*SoC*), for a variety of applications (i.e., sensor interfacing, signal processing and signal conditioning systems, medical and biological instrumentations, Micro-Electronic-Mechanical System (*MEMS*), etc.). In particular, one of the main aims of actual sensor research is the design of full integrated electronic systems, formed by the sensor, its first analog interface and the processing circuitry, possibly in a miniaturized microelectronic environment (i.e., microsystem). Furthermore, the capability to minimize the sensor element to a nano-scale level and the integration of the sensor itself with the electronic circuit by micromachining and silicon technology, respectively, have opened also new opportunities for electronic interfaces. In this case, a suitable sensor front-end has to be able also to adapt itself to different kinds of both sensors and measurands, through appropriate electronic circuits, and to improve signal processing by apposite circuit design.

Obviously, the first stage of a sensor interface has to be analog, because of the analog nature of the signal coming from sensor. Moreover, analog signal processing offers a high functional density and the capability to directly interface the analog real world of sensors. Furthermore, an Analog-to-Digital (A/D) conversion of the analog output signal is always possible, so as to improve the quality of data display. In this case, owing to the sensor nature, no particular speed constraints are generally necessary; therefore, traditional low-cost and commercial A/D converters can be quite good for a lot of purposes.

Nowadays, fully analog or mixed analog/digital electronic circuits are becoming more and more important for sensors, because the chip-scale integration can be utilized for combining, on the same chip, existing standard *IC* processes, the sensing elements and the processing electronics so to fabricate the so-called smart sensors. This is exalted by the fact that actually the same materials (silicon, polysilicon, aluminium, dielectrics, metal-oxides, etc.) are used to fabricate the majority of sensors, such as, for example, resistive chemical gas sensors based on Metal-Oxide (*MOX*) and silicon-based capacitive pressure sensors, and their front-ends. In this way, standard *CMOS* has been proved to be the main sensor technology, because is able to match the reduction of technological costs with the design of new attractive integrated electronic interface solutions showing low supply voltages and reduced power consumption characteristics.

Starting from these considerations, actually there are basic performances which have to be achieved in *IC* design for the first analog sensor interface: high sensitivity and resolution, high dynamic range, good linearity and high precision, good accuracy, low input noise and offset, long-term temperature stability, reduced silicon area, low effect of parasitic capacitances, calibration and compensation of the transducer characteristics, etc.. These characteristics have to be satisfied by suitable integrated electronic circuits whose typology depends on both the nature of the measurand and the amount of its variation. These interfaces, if designed also with Low Voltage (*LV*) and Low Power (*LP*) characteristics, can be utilized in portable, remote and wireless electronic systems for domestic, industrial, biomedical, automotive and consumer applications, where a great need of reliable and miniature sensor systems has recently grown.

Considering both *LV* and *LP* techniques, the Current-Mode (*CM*) approach, which utilizes the information provided by a current signal instead of voltage as in Voltage-Mode (*VM*), can become, in some cases, a good alternative solution. The main active basic block in *CM* approach is the Second Generation Current Conveyor (*CCII*) which, in different applications, can represent a possible alternative to the traditional Operational Amplifier (*OA*), typically employed in *VM* circuits.

Finally, in the design of a complete integrated sensing system, the capability to operate at environment temperature, as well as at higher temperatures, with also a high linearity, is generally required. This kind of integrated front-end is often formed by a sensor heater (which fixes the employed sensor temperature at a suitable operating point), a proper electronic circuit, converting non-electrical value of the sensing element into a electrical parameter which can be easily utilized by the next stage, and a signal processing unit (typically of digital kind). Therefore, since this complete system sometimes has to be able to reveal a large number of sensing element variation decades, a suitable and accurate design of the analog front-end is mandatory. This is why in this book the main sensor interfacing techniques and front-end circuits, as discrete element prototype boards and, when possible, as integrated architectures, will be described.

Chapter 1
Physical and Chemical Sensors

In this chapter we give an introduction and classification on some examples of physical sensors (devices placed at the input of an instrumentation system that quantitatively measures a physical parameter, for example pressure, displacement or temperature) and chemical sensors (devices which are part of an instrumentation system that determines, typically, the concentration of a chemical substance, such as a toxic gas or oxygen), describing their working principles and main characteristic parameters.

1.1 Sensors and Transducers: Principles, Classifications and Characteristics

The sensor represents the first and main element in measurement and control systems. It is the sensing element, in a revelation equipment, which reacts to the either physical or chemical phenomenon to be detected. It makes use of suitable transduction components so to convert a physical or chemical characteristic into a parameter of different nature, more suitable for the next elaboration through an electronic system. The use of computer-compatible sensors has closely followed the advances in circuit and system design and the advent of the microprocessor. Together with the always-present need for sensors in science and medicine, the demand for sensors in automated manufacturing and processing has rapidly grown. In addition, small and cheap sensors have become important in a large number of consumer products, from children toys to dishwashers and automobiles. Then, the process automation, the fabrication of auto-calibrated devices, the control of the operating condition of a system are only other possible applications of sensors [1–9]. The need of novel sensors and related electronic interfaces showing reduced dimensions and, possibly, *LV LP* characteristics (that is the capability of working with reduced supply voltages and showing a low power consumption) is in a continuous growth since wireless detectors and devices have emerged and moved

A. De Marcellis and G. Ferri, *Analog Circuits and Systems for Voltage-Mode and Current-Mode Sensor Interfacing Applications*, Analog Circuits and Signal Processing, DOI 10.1007/978-90-481-9828-3_1, © Springer Science+Business Media B.V. 2011

towards commercialization. The possibility for a wide number of devices that give accurate, remote and quick access information about their environment has started to spread. Application areas include health care (verification of the environmental conditions during transport or in storage of diapers, bandages, etc.), food monitoring (food quality during transport, storage and sales) and environmental monitoring (meteorology, road safety, indoor climate, detection of toxic and dangerous gases, etc.). Therefore, one of the important requirements in these researches is the development of *LV LP* and low-cost sensors [10–26]. In particular, the expansion of miniaturized integrated circuits and the advances in microelectronic technologies have made more and more important the design of analog interfaces suitable for the read-out and the processing of signals coming from sensors: in this way, both sensor and electronic circuitry for its interfacing, which have to be developed in a suitable integrated technology (e.g., standard *CMOS*), can be also combined into only one chip, implementing the so-called "Smart Sensor" [27–49].

This Chapter introduces basic definitions and features of sensors, together with some their possible classifications, and illustrates them with some typical examples.

There are many terms which are often used as synonymous for the word "sensor" such as, for an example, transducer, meter, detector, gauge, actuator, etc.. For precision sake, transducers convert signals from an energy domain into signals in a different energy domain. In particular, sensors may be defined as systems which convert signals from non-electrical domains into electrical ones. Actuators are the complementary class of systems which convert electrical signals into non-electrical ones. Concerning transducers, the most widely used definition is that which has been applied to electrical transducers by the Instrument Society of America: *"the transducer is a device which provides a usable output in response to a specified measurand"*. A "usable output" generally refers to an optical, electrical or mechanical signal. In the context of electrical engineering, however, it refers to an electrical output signal. On the other hand, sensors are physical devices which transfer information from different energy domains, such as chemical, optical, mechanical, thermal, magnetic, electrical into an electrical one, providing a broad variety of electrical signals, which are normally of analog kind. In this sense, the "measurand" is defined as the physical, chemical (or biological) property or condition to be measured [1–9].

Sometimes sensors are classified as ***direct*** and ***indirect*** sensors according if one or more than one transduction mechanism is used, respectively. For example, a mercury thermometer is an indirect sensor since it produces a change in volume of mercury in response to a temperature change via thermal expansion, but the output is a mechanical displacement and not an electrical signal, then another transduction mechanism is required. This thermometer is a sensor because humans can read the change in mercury height using their eyes as a second transducing element. On the other hand, in order to produce an electrical output, the height of the mercury has to be converted to an electrical signal; this could be accomplished using a capacitive effect, as an example [1–9].

Fig. 1.1 depicts a simple sensor block diagram identifying the measurand according to the type of input signal and the primary and secondary transduction

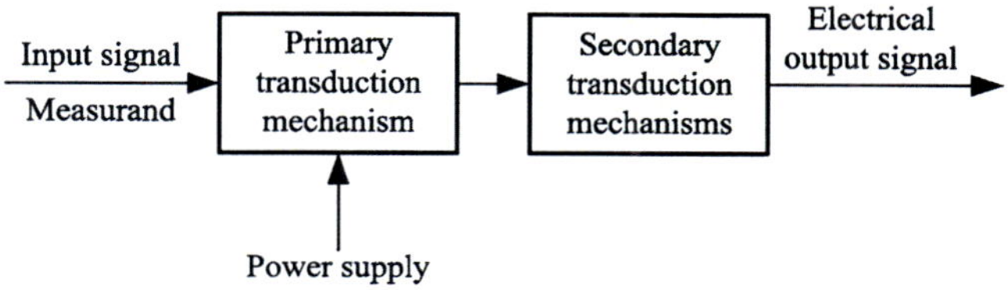

Fig. 1.1 A typical sensor block diagram

mechanisms which give the readable electrical output signal. Classification of sensors can be done according to different approaches. In the following we will show some of these possible points of view [1–9].

In Table 1.1 we report a detailed description of the more commonly employed transduction mechanisms (in particular, primary and secondary signals can be: mechanical, thermal, electrical, magnetic, radiant, chemical, etc.). Many of the effects listed in this Table will be shown in detail in this and next Chapters [1–9].

In order to choose a particular sensor for a given application, there are many factors to be considered. These factors (or specifications) can be divided into two main categories: ***environmental*** factors and ***economic*** factors, as listed in Table 1.2 together with their main characteristics. Most of the environmental factors determine also the packaging of the sensor. The term packaging stands for the encapsulation or insulation which provides protection and isolation and the input/output leads or connections and cabling. The economic factors determine the type of manufacturing and materials used in the sensor and to some extent the quality of the materials (with respect to lifetime). For example, a very expensive sensor may be employed if it is used repeatedly or for very long time periods. On the other hand, a not reusable sensor, often desired in many medical applications, is really inexpensive [1–9].

Another characterization of the sensors regards the type of non-electrical stimulus to be measured; in this sense, we can mention four main families of sensors: sensors for ***mechanical*** phenomenon, sensors for ***hydraulic*** phenomenon, sensors for ***environmental*** phenomenon and sensors for ***electromagnetic*** phenomenon [1–9].

On the other hand, sensors are most often classified simply according to the type of measurand; in particular, there are mainly physical and chemical (or biological) sensors. More in detail:

- ***Physical*** measurands mainly sense temperature, strain, force, pressure, displacement, position, velocity, acceleration, optical radiation, sound, flow rate, humidity, viscosity, electromagnetic fields, etc..
- ***Chemical*** measurands generally detect ion concentration, chemical composition, rate of reactions, reduction-oxidation potentials, gas concentration, etc..

Moreover, with respect to electronic circuits that have to be integrated on the same chip as first analog front-end, sensors are normally divided into two main groups as reported in Table 1.3: ***active*** sensors, which directly produce an output current

Table 1.1 Physical and chemical transduction principles

Primary signal	Secondary signal: Mechanical	Thermal	Electrical	Magnetic	Radiant	Chemical
Mechanical	(Fluid) Mechanical and acoustic effects (e.g., diaphragm, gravity balance, echo sounder)	Friction effects (e.g., friction calorimeter); Cooling effects (e.g., thermal flow meters)	Piezoelectricity; Piezoresistivity; Resistive, capacitive and inductive effects	Magneto-mechanical effects (e.g., piezomagnetic effect)	Photoelastic systems (e.g., stress-induced birefringence); Interferometers; Sagnac effect; Doppler effect	–
Thermal	Thermal expansion (bimetal strip, liquid-in-glass and gas thermometers, resonant frequency); Radiometer effect (light mill)	–	Seebeck effect; Thermoresistance; Pyroelectricity; Thermal (Johnson) noise	–	Thermooptical effects (e.g., in liquid crystals); Radiant emission	Reaction activation (e.g., thermal dissociation)
Electrical	Electrokinetic and electromechanical effects (e.g., piezoelectricity, electrometer, Ampere's law)	Joule (resistive) heating; Peltier effect	Charge collectors; Langmuir probe	Biot-Savart's law	Electrooptical effects (e.g., Kerr effect); Pockel's effect; Electroluminescence	Electrolysis; Electromigration
Magnetic	Magnetomechanical effects (e.g., magnetorestriction, magnetometer)	Thermomagnetic effects (e.g., Righi-Leduc effect); Galvanomagnetic effects (e.g., Ettingshausen effect)	Thermomagnetic effects (e.g., Ettingshausen-Nernst effect); Galvanomagnetic effects (e.g., Hall effect, magnetoresistance)	–	Magnetooptical effects (e.g., Faraday effect); Cotton-Mouton effect	–

Radiant	Radiation pressure	Bolometer thermopile	Photoelectric effects (e.g., photovoltaic effect, photoconductive effect)	–	Photorefractive effects; Optical bistability	Photosynthesis; Dissociation
Chemical	Hygrometer; Electrodeposition cell; Photoacoustic effect	Calorimeter; Thermal conductivity cell	Potentiometry; Conductimetry; Amperometry; Volta effect; Flame ionization; Gas-sensitive field effect	Nuclear magnetic resonance	(Emission and absorption) spectroscopy; Chemiluminescence	–

Table 1.2 Main factors in sensor applications

Environmental factors	Economic factors	Sensor characteristics
Temperature range	Cost	Sensitivity
Humidity effects	Availability	Range
Corrosion	Lifetime	Stability
Size		Repeatability
Overrange protection		Linearity
Susceptibility to EM interferences		Error
Ruggedness		Response time
Power consumption		Frequency response
Self-test capability		

Table 1.3 Another possible sensor classification: active and passive sensors and their typical electrical outputs

Main group	Type of sensor	Type of signal	Typical range
Active sensors	Thermopiles, pyroelectric, piezoelectric	Voltage	μV – mV
	Pyroelectric, magnetic	Current	μA – mA
Passive sensors	Humidity, gas, pressure	Capacitance	fF – μF
	Piezoelectric	Charge	fC – pC
	Pressure, chemical, gas	Resistance	kΩ–GΩ

or voltage but require an external power source in order to give a usable output signal, and ***passive*** sensors, which directly modify their internal parameters if an external phenomenon occurs. In the first case, either resistive or capacitive bridges can be interfaced to signal processing and conditioning circuitry such as low noise voltage or current amplifiers. In the second case, the basic parameters of the passive sensors, such as capacitance and resistance, can be measured (according to their variation range) either directly or through some suitable circuits such as oscillators, bridges, charge amplifiers and switched-capacitors based converters [1–10].

Finally, we want to mention that other sensor classifications depend on: how they are fabricated, what is the sensing element, at what physical and/or chemical phenomenon they are able to react, how "electrically" they respond, etc.. In this sense, three main types of sensors will be considered: ***resistive***, ***capacitive*** and ***temperature*** sensors. Since the analog electronic interface especially depends on the kind of sensor and the amount of its variation, this classification seems to be the better and most useful for sensor interface designers, so it will be that mainly adopted in this book.

In the next Paragraphs we will describe firstly the main sensor parameters and then the fundamentals and the working principles of some different kinds of sensors (classifying them with respect to the physical or chemical transduction mechanisms which they show), in a non-exhaustive way. Moreover, in the next Chapters we will present these and other kind of sensors, considering their electrical outputs (type of generated signals by means of different transduction mechanisms), describing also in detail the main analog front-end circuits and interfacing techniques.

1.2 Sensor Main Parameters

In sensor analysis and characterization, it is opportune to evaluate the performances given by the sensor also under different operating conditions. In this sense, the following sensor characteristics can be identified:

- ***static*** characteristics, which describe the performances and the environmental conditions for null or very slow variations of the phenomenon which has to be revealed;
- ***dynamic*** characteristics, which show the performances of a sensor when the phenomenon which has to be detected suffers extensive variations during the observation time;
- ***environmental*** characteristics, which refer to the sensor performances after the exposure (static environmental characteristics) or during the exposure (dynamic environmental characteristics) to specific external conditions (such as temperature, bumps or vibrations).

The main sensor parameters, which have to be considered for evaluating the goodness of a sensor, are the following:

- ***sensitivity***: it is the ratio between the output electrical variation and the input non-electrical parameter variation (measurand variation). It represents the relationship (transfer function) between the output electrical signal and the input non-electrical signal. A sensor will result to be very sensitive when, for the same phenomenon variation to be measured, the electrical signal shows a larger variation. Generally, sensitivity value depends on the operating point of the sensor system, except in the case of direct proportionality between measurand and output value; in this case, it shows a constant value for any working condition.
- ***resolution***: it is the ratio between the output noise level and the sensor sensitivity. It is the minimum detectable non-electrical parameter value under the condition of unitary Signal-to-Noise Ratio (*SNR*). On the other hand, it is defined as the smallest variation of the non-electrical information appreciable from the sensor and which provides a detectable output variation. Least variations of the input non-electrical information below the value of the resolution do not cause valuable variations of output generated signal.

Resolution is definitively the most important sensor characteristic; numerically speaking, it must be minimized. In fact, a system with a very low resolution value is typically mentioned as a "High-Resolution System". Sensitivity and resolution have to be the best possible and must be evaluated in the typical variation range of the non-electrical parameter where, possibly, have to be constant or linear: it means that their value does not depend on the working operating point. These two parameters can be determined also after the interfacing of the "basic" sensor with the first analog front-end that, typically, improves their value [7, 8, 10].

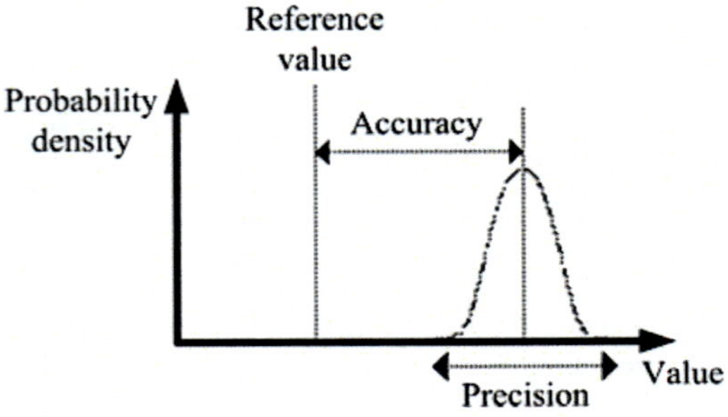

Fig. 1.2 Accuracy and precision definitions and their relationship

Other significant sensor parameters are the following:

- ***Linearity***: proportionality between input and output signals. This parameter is related to the sensor response curve, which correlates the output signal of the sensor to the measurand parameter. Generally, for small measurand variation, linearity is always ensured.
- ***Repeatability***: capability to provide the same performances after a number of utilizations, that is to reproduce output readings for the same value of measurand, when applied consecutively and under the same conditions.
- ***Accuracy***: agreement of the measured values with a standard reference (i.e., ideal characteristic). On the other hand, accuracy is the degree of closeness of a measured or calculated quantity to its reference (expected) value. Accuracy is closely related to precision, also called reproducibility. As a consequence, accuracy is related to percentage relative error between ideal and measured value, as shown in Fig. 1.2.
- ***Precision***: capability to replicate output signals with similar values, for different and repeated measurements, when the same input signal is applied. The precision can be also intended as the degree to which repeated measurements or calculations show the same or similar results. Precision can be considered as the repeatability in the same measurement conditions.
- ***Reproducibility***: it is the repeatability obtained under different measurement conditions (e.g., in different times and/or places).
- ***Stability***: time-invariability of the main sensor characteristics, that is the capability of a sensor to provide the same characteristics over a relatively long period of time.
- ***Hysteresis***: difference among the output signal values, generated by the sensor in correspondence of the same non-electrical input signal range, achieved a first time for increasing values and a second time for decreasing values of the input signal.
- ***Processing speed***: it defines the speed of the generated output signal to reach its final value starting from the instant when the input signal suffers a variation (in this case, a time constant can be also defined).
- ***Noise***: output unwanted signal, produced when the input signal or its variation (to be revealed) is null.

- ***Drift***: it is the (slow and statistically unpredictable) temporal variation of sensor characteristics, due to aging and/or other effects related to sensing materials.
- ***Selectivity***: the presence of different sensitivities to various measurands, sometimes, avoids a useful detection of the sensor answer. Selectivity, or cross-sensitivity, is the capability of the sensor system to maximize only the sensitivity to the desired measurand and to reduce that related to the other chemical or physical parameters that are unavoidably present.

Moreover, output signals coming from sensors, typically, have the following characteristics: low-level values, relatively slow sensing parameter variations and the need of initial calibration for long-term drift (it means they generally can be considered time-variant). For these reasons, in order reduce measuring errors, the use or the design of suitable low-noise low-offset analog interfaces with low parasitic transistors and impedances is essential. In this sense, another important feature to be considered is the electrical impedance of the sensor, which determines the frequency measurement range.

Finally, we want to underline that a sensor is suitable only if all its main parameters are tightly specified for a given range of measurand and time of operation. For example, a highly sensitive device is not useful if its output signal drifts greatly during the measurement time and the data obtained is not reliable if the measurement is not repeatable. Moreover, selectivity and linearity can often be compensated using either additional independent sensor inputs or signal conditioning circuits. In fact, most sensor responses are related to their working temperature, since most transducing effects are temperature-dependent.

1.3 Piezoelectric, Ferroelectric, Electret and Pyroelectric Sensors

The root of the word "piezo" means pressure; hence, the original meaning of the word piezoelectric implied "pressure electricity" (the generation of electric field through an applied pressure). However, this definition ignores the fact that these materials are reversible, allowing the generation of a mechanical movement by applying an electric field. The prefix "ferro" refers to the permanent nature of the electric polarization in analogy with the magnetization in the magnetic case. Even though the root of the word means iron, it does not imply the presence of this material. Then, the "electret" term comes from the words "electrostatic" and "magnet"; in particular, it is formed by "electr", from "electricity", and "et", from "magnet". An electret material generates internal and external electric fields and is the electrostatic equivalent of a permanent magnet [1–9, 50–67].

Among these sensors, examples of the classes of materials and applications are given in Table 1.4, from which it is evident that many materials exhibit electric phenomena which can be attributed to piezoelectric, ferroelectric and electret

Table 1.4 Electret, ferroelectric, piezoelectric and electrostrictive materials classification

Type	Material class	Example	Applications
Electret	Organic	Waxes	No recent
Electret	Organic	Fluorine based	Microphones
Ferroelectric	Organic	PVF2	No known
Ferroelectric	Organic	Liquid crystals	Displays
Ferroelectric	Ceramic	PZT thin film	NV-memory
Piezoelectric	Organic	PVF2	Transducers
Piezoelectric	Ceramic	PZT	Transducers
Piezoelectric	Ceramic	PLZT	Optical
Piezoelectric	Single crystal	Quartz	Freq. control
Piezoelectric	Single crystal	$LiNbO_3$	SAW devices
Electrostrictive	Ceramic	PMN	Actuators

materials. Here we will discuss the basic concepts in the use of these materials, highlight their applications and describe the constraints that limit their utilization [1–9].

Piezoelectric and ferroelectric materials derive their properties from a combination of structural and electrical properties. As the name implies, both types of materials have electric attributes. A large number of ferroelectric materials are also piezoelectric; however, the contrary is not true. Ferroelectric materials show permanent charge dipoles which arise from asymmetries in the crystal structure. The electric field due to these dipoles can be observed externally to the material when opportune conditions are satisfied (ordered material and absence of charge on the surfaces). Ferroelectrics react to the external fields with a polarization hysteresis and can retain this polarization permanently owing to the thermodynamic equilibrium. Alternatively, some materials consist of large numbers of unit cells; the manifestation of the individual charged groups is, consequently, an internal and an external electric field that arise when the material is stressed. The interaction of an external electric field with a charged group causes a displacement of some atoms in the group, so a macroscopic displacement of the material surfaces. This motion is called piezoelectric effect, that is the conversion of an applied field into a displacement. On the other hand, piezoelectric materials exhibit an external electric field when a stress is applied to it and a charge flow proportional to the strain is observed when a closed circuit is attached to electrodes on the material surface. In ferroelectric materials a crystalline asymmetry exists and allows electric dipoles to form. In symmetrical structures the dipoles are absent and the internal field disappears. All ferroelectric and piezoelectric materials have phase transitions at which the material changes its crystalline symmetry. For example, in these materials there is a change of the symmetry when the temperature is increased. The temperature at which the material spontaneously changes its crystalline phases orsymmetry is called the Curie temperature [1–9].

Electret material is a stable dielectric material that has a permanent electrostatic charge or oriented dipole polarization, which, due to the high resistance of the material, does not decay for hundreds of years. It is similar to ferroelectric one but charges are macroscopically separated and thus are not structural. In some cases, the net charge in the electrets is not zero, for instance when an implantation process is used to embed the charge. Real-charge electrets contain either positive or negative excess charges or both, while oriented-dipole electrets contain oriented dipoles. Moreover, there is a similarity between electrets and the dielectric layer used in capacitors. The difference is that dielectrics in capacitors show an induced polarization that is only transient, dependent on the potential applied on the dielectric, while dielectrics with electret properties exhibit permanent charge storage. Electrets are commonly made by first melting a suitable dielectric material such as a plastic or wax that contains polar molecules and then allowing it to re-solidify in a powerful electrostatic field. The polar molecules of the dielectric align themselves to the direction of the electrostatic field, producing a permanent electrostatic bias. Electret materials are quite common in nature: for example, quartz and other forms of silicon dioxide are naturally occurring electrets, as well as most electrets are made from synthetic polymers (e.g., fluoropolymers, polypropylene, etc.). Although electrets are often characterized as solid (dielectric) materials, a less restrictive view encompasses both solid and liquid systems. Rigid particles or macroscopic surfaces that retain permanent charge or oriented dipoles are rightly termed "solid electrets", while "liquid electrets," on the other hand, are formed by inserting charge in the form of electrons, ions, nanometer size micelles or charged colloidal particles into a liquid or onto a liquid-gas or liquid-solid interface. The electret material can be manipulated with external electrostatic fields. With some liquid electrets (e.g., a polymer above its glass transition temperature), unique interface morphologies can be "frozen in" by cooling. The permanent internal or external electric fields, created by electret materials, can be exploited in various applications. Therefore, electrets have recently found commercial and technical interests. For example, they are used in copy machines, microphones, in some types of air filters, for electrostatic collection of dust particles and in ion chambers for measuring ionizing radiation or radon [1–9, 50–53].

As shown in Table 1.4, among these sensors there are three dominant classes of materials: organics, ceramics and single crystals. All these classes have important applications of their piezoelectric properties. In order to exploit the ferroelectric property, recently a strong effort has been devoted to produce thin films of *PZT* (common name for piezoelectric materials of the lead (Pb) zirconate titanate family) on various substrates of silicon-based memory chips for non-volatile storage. In these devices, data is retained without an external power as positive and negative polarization. The polarization is the amount of charge associated with the dipolar or free charge in a ferroelectric or an electret, respectively; it corresponds to the external charge which must be supplied to the material to produce a polarized state from a random state (twice that amount is necessary to reverse the polarization). The statement is rigorously true if all movable charges in the material are reoriented (i.e., saturation can be achieved). Organic materials have not been used for their ferroelectric properties. Liquid crystals in display applications are used for their ability to rotate the plane of polarization of light and not their ferroelectric attribute. Materials are acted on by forces (stresses) and the resulting deformations are called strains. An example of a strain due to a force applied to the material is the change of dimension, in parallel and perpendicularly to the applied force (e.g., *PZT* converts electrical fields into mechanical displacements and vice versa) [1–9].

Historically, the material which was used earliest for its piezoelectric properties was the single-crystal quartz. Crude sonar devices were built by Langevin using quartz transducers, but the most important application was, and still is, the frequency control. Crystal oscillators are today at the heart of every clock that does not derive its frequency reference from the AC power line. They are also used in every colour television set and personal computer, as well as in cellular phones. In these applications at least one (or more) "quartz crystal" controls frequency or time. This explains the label "quartz" which appears on many clocks and watches. The use of quartz resonators for frequency control relies on another unique property. Not only the material is piezoelectric (which allows to excite mechanical vibrations), but has also a very high mechanical quality factor Q ($Q > 10^5$, considering that the typical Q for *PZT* is about between 10^2 and 10^3). The actual value depends also on the mounting details, whether the crystal is in a vacuum, and on other details. The Q factor is a measurement of the rate of decay and thus of the mechanical losses of an excitation with no external drive. A high Q leads to a very sharp resonance and thus to a tight frequency control. To this purpose, it has been possible to find suitable orientations of quartz cuts which reduce the influence of temperature on the vibration frequency. Ceramic materials of the *PZT* family have also found increasingly important applications. The piezoelectric but not the ferroelectric property of these materials is used in transducer applications. *PZT* has higher efficiency than quartz crystal. Probably the most important applications of *PZT* today are based on ultrasonic echo ranging [1–9].

There is another class of ceramic materials which has recently become important. The *PMN* (lead [Pb], Magnesium Niobate, typically doped with $\approx$10% lead titanate) is an electrostrictive material that can be used in those applications where the absence of hysteresis is important. Electrostrictive materials exhibit a strain which is quadratic as a function of the applied field; producing a displacement requires an internal polarization. Since the latter polarization is induced by the applied field and is not permanent, as it is in the ferroelectric materials, electrostrictive materials have essentially no hysteresis, but, unlike *PZT*, are not reversible. In fact, *PZT* will change shape on application of a field and generate a field when a strain is induced, while electrostrictive materials only change shape on application of a field and, therefore, cannot be used as receivers. *PZT* has inherently large hysteresis because of the domain nature of the polarization [1–9].

Concerning the organic electrets, they have important applications in self-polarized condenser microphones where the required electric bias field in the gap is generated by the diaphragm material rather than by an external power supply. More in detail, an electret microphone is a type of condenser microphone, which eliminates the need of an added external power supply by using a permanently-charged material [1–9].

Pyroelectricity is closely related to piezoelectricity and ferroelectricity via the symmetry properties of the crystals. In fact, the pyroelectric effect appears in each material which shows a polar symmetry axis. A temperature change on a pyroelectric material induces a current to flow in an external circuit, dependent on the electrode area of the material, on the pyroelectric coefficient (related to the specific material) and on the rate of temperature change. Pyroelectric devices detect changes in temperature in sensitive materials, so they are detectors of supplied energy. It can be seen that the pyroelectric current is proportional to the rate of change of the material characteristics and that, in order to obtain a measurable signal, it is necessary to modulate the source of energy. As energy detectors, they are most frequently applied to the detection of incident electromagnetic energy, particularly in the infrared wavebands. More in detail, concerning the pyroelectric sensor, the pyroelectric effect is a property of few ferroelectric crystals, such as $Sr_{1-x}Ba_xNb_2O_6$ and $LiNbO_3$, having a spontaneous electric polarization which can be measured as a voltage level at the material terminations. However, the internal charges distribution, at a constant temperature, is neutralized by both the free electrons and the surface charges, providing a null external voltage level. If the temperature rapidly changes, the internal dipole moments vary generating a transient voltage signal. Therefore, by means of the pyroelectric effect, it is possible to implement detectors of modulated radiation working at ambient temperature. Generally, the pyroelectric detectors are capacitors having metallic electrodes applied on the opposite surfaces of a temperature-sensible ferroelectric crystal [1–9].

The modulated incident radiation on the detector surface generates a temperature variation ΔT which induces a charge variation ΔQ on the external electrodes expressed by the following relation:

$$\Delta Q = p \cdot A \cdot \Delta T \tag{1.1}$$

being p the pyroelectric coefficient of the material and A the area of the detector. Then, the generated "photo-current" is proportional to the temperature variation rate, as described by the following expression:

$$i(t) = \frac{dQ}{dt} = p \cdot A \cdot \frac{dT}{dt} \tag{1.2}$$

In practice, pyroelectrics are polar dielectric materials showing their internal dipole moments as temperature dependent; this leads to a change in the charge balance at the surface of the material which can be detected as either a potential difference or as a charge flowing in an external circuit [7, 9, 68–75]. Typically, a pyroelectric detector consists of a thin layer of a pyroelectric material, cut perpendicularly to its polar axis; it shows electrodes fabricated with a conducting material such as an evaporated metal and connected to a low-noise, high-input impedance amplifier, such as a junction field-effect transistor (*JFET*) or a metal-oxide-semiconductor field-effect transistor (*MOSFET*), as shown in Fig. 1.3 [1–9].

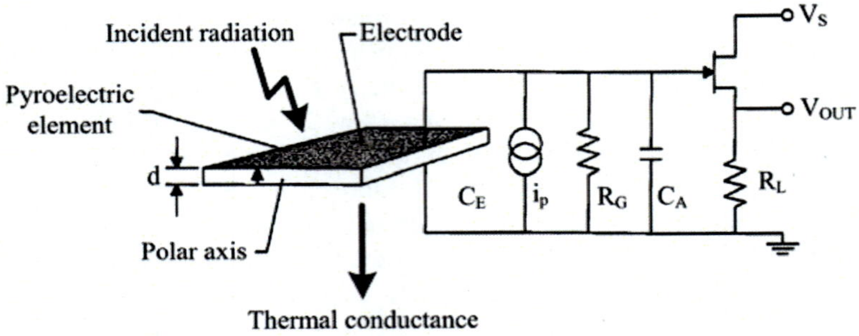

Fig. 1.3 Pyroelectric detector with FET amplifier

Recently, the pyroelectric effect is mainly utilized for the fabrication of infrared radiation detectors. These devices are employed as "people detectors" for intruder alarms and energy conservation systems, fire and flame detectors, spectroscopic gas analyzers, especially looking for pollutants from car exhausts, and, more recently, for thermal imaging. Such thermal imagers can be used for night vision and, by exploiting the smoke-penetrating properties of long-wavelength infrared radiation, in devices to assist fire-fighters in smoke-filled spaces. The major advantages of these devices, when compared to the infrared detectors that exploit narrow band-gap semiconductors, are that no cooling is necessary and that they are cheap and consume a reduced power [1–9].

1.4 Magnetic Field Sensors

Typically, a magnetic field, having a time-variable intensity, generates in a solid an electrical current by means of the well-known electromagnetic induction law. All the electromagnetic phenomenon characteristics, so also those related to the magnetic field, are summarized in the four Maxwell equations which describe the natural point source form of the electrical field and the force line circulation as regards the magnetic field. The latter, due to its characteristic to have closed force lines and to interact with electrical currents, provides, in particular, the possibility to implement sensors for the proximity revelation. In this sense, it is possible to measure the intensity of a magnetic field between a source and a detector. In general, magnetic field sensors are devices whose characteristics change as a function of an external magnetic field. They are mainly based on Hall effect which is a consequence of the Lorentz force on the charges in a semiconductor crossed by a magnetic field [76,77].

More in detail, a Hall effect sensor measures the magnetic field B by applying a known constant current I and revealing the voltage V (Hall voltage) orthogonal to the same current, as shown in Fig. 1.4.

The measured Hall voltage is proportional directly to the magnetic field to be detected, the applied current and inversely to charge carrier concentration and

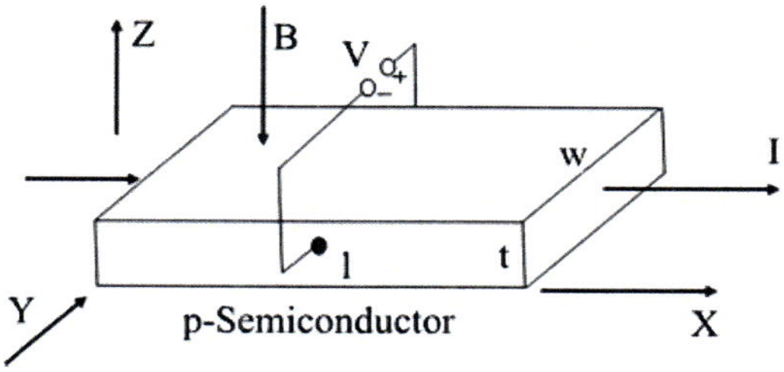

Fig. 1.4 Hall effect sensor: a principle scheme

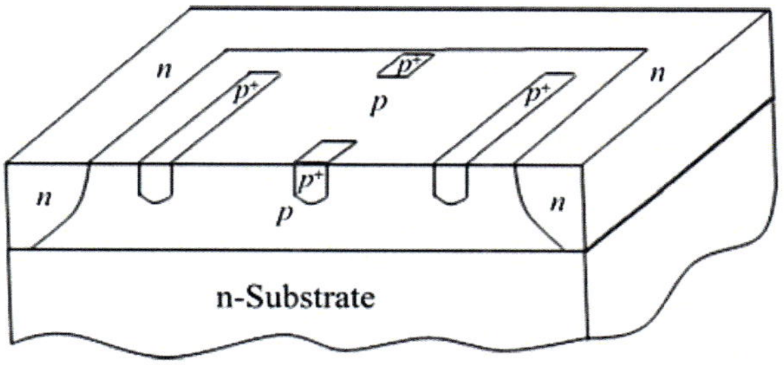

Fig. 1.5 An example of an integrated Hall effect sensor

thickness t (the Hall field direction, for the same current and the magnetic field intensity, depends on the kind of the charge carriers, i.e., electrons or holes). This simple relationship is valid if the device length is much higher than its width and if the voltage electrodes are perfectly aligned (otherwise a voltage offset arises).

Since the sensitivity of a Hall effect sensor, for constant current and magnetic field, is inversely proportional to both the charge carrier density (so semiconductor-based sensors show very high sensitivities) and the sensor material sizes (i.e., its thickness), this sensor is typically based on a thin film of lightly doped semiconductor (e.g., *InSb*, *InAs*, *GaAs*, *Si* or *Ge*), deposited on an insulator material. It shows a regular shape where four electric contacts (orthogonally mounted and in opposition two by two) have been implemented, two of which crossed by the constant current, while the other two utilized to measure the generated Hall voltage. The Hall sensor, whose integrated version can be fabricated with modern microelectronic technologies, can be used also to detect if in a semiconductor the conductivity is dominated by electrons or holes. Fig. 1.5 shows the scheme of a typical integrated Hall effect sensor: it is composed by a semiconductor-based bar having four contacts in correspondence of the four orthogonal faces. The current is injected through the longitudinally extended contacts so to maximize the sensor transversal dimension.

Hall effect can be efficiently revealed also employing the *MOSFET* structure, in particular its channel: in this case the sensor is called *MagFET* and shows high sensitivities to the magnetic field. Moreover, if the device length is reduced, it is better to measure the resistance variation in the semiconductor, instead of the Hall effect voltage; in this case it is called magneto-resistive sensor. Finally, we want to mention the Fluxgate Magnetometer, a flux sensor, based on electrical coils, which exploits the interaction between magnetic field B and magnetic induction H, allowing to perform zero measurements with better resolutions than in the case of Hall effect based sensors. Another kind of magnetometer is that based on superconductors (materials that, at a very low temperatures, e.g., 4 K, show a strong reduction of their electric resistivity): in this sense, the *SQUID* (Superconductive QUantum Interference Device) is utilized, being implemented by a superconductor material ring showing the highest sensitivity to the magnetic field. In this device, a magnetic flux induces an electrical current (instead of voltage) whose value is proportional to the intensity of the magnetic field H to be detected [7].

1.5 Optical Radiation Sensors

The intensity and frequency of optical radiation are parameters of growing interest and utility in consumer products such as video camera and home security systems and in optical communication systems. The conversion of optical energy to electronic signals can be accomplished by several mechanisms (see radiant to electronic transduction in Table 1.1), but the most commonly used is the photo-generation of carriers in semiconductors and the most often-used devices are the *p-n* junction and the avalanche photodiodes. The construction of these devices is very similar to the diodes used in electronic circuits as rectifiers.

The diode operates in reverse bias so a very little current normally flows. When the light is incident on the structure and is absorbed in the semiconductor, energetic electrons are produced. These electrons flow thanks to the electric field sustained internally across the junction, so producing a current which is externally measurable through a suitable electronic circuit. The current magnitude is proportional to light intensity and also depends on the light frequency (or on its wave length). Fig. 1.6 shows the effects of different incident optical intensities on the diode current as a function of its voltage in a *p-n* junction. Note that for zero applied voltage, a negative current flows when the junction is illuminated; therefore, this device can also be considered a source of power (i.e., a solar cell) [9].

More in detail, the photoconductivity consists of an electrical conductivity variation produced by an electromagnetic irradiation. The corresponding signal can be revealed either through the voltage variation in a load resistor in-series with the detector, as shown in Fig. 1.7, or by the evaluation of the current variation into the same device. Typically, the voltage detection is obtained through a load resistor whose value is equal to the "dark resistance" related to the considered material [7].

The exposition of the detector to the light involves an additional current (the photocurrent) generated by charge movements produced through both the radiation and the applied voltage. Referring to Fig. 1.7, the generated voltage signal can be expressed by:

$$V_{OUT} = R_L \cdot (I_{dark} + I_{light}) = R_L \cdot \frac{V_{IN}}{R_L + R_D} + R_L \cdot I_{light} \qquad (1.3)$$

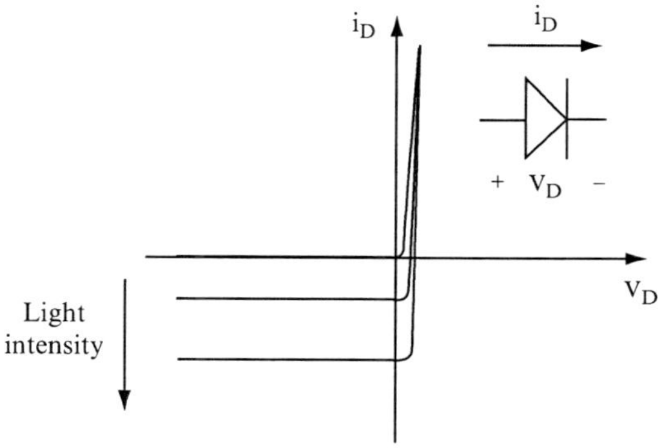

Fig. 1.6 Sketch of the variation of current versus voltage characteristics of a *p-n* photodiode with different incident light intensities

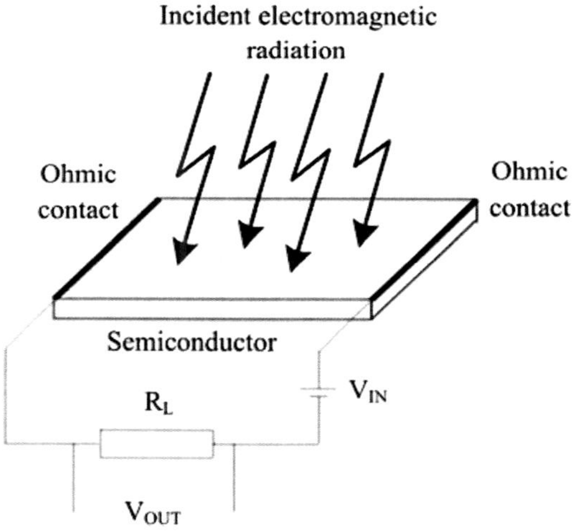

Fig. 1.7 Generic photoconductor biasing scheme

being I_{dark} the dark current, I_{light} the additional current generated during the light exposition, V_{IN} the biasing voltage, R_L the load resistance and R_D the "dark resistance". Therefore, the photoconductivity can be associated to the conductivity variation of a resistor.

In the photovoltaic effect, even if the physical phenomena are the same of those shown in the photoconduction, the device is able to generate a signal without any external biasing source, thanks to the irradiation effect. In practice, the junction-based photodetectors are often utilized with an inverted biasing and the generated photosignal results to be a current rather than a voltage signal. It is important to notice that, with a direct biasing, the diode current is due to the diffusion and, therefore, is slightly influenced by the drift current. On the contrary, when the diode is biased in inverse mode, its current is dominated by the drift caused by the built-in electrical field.

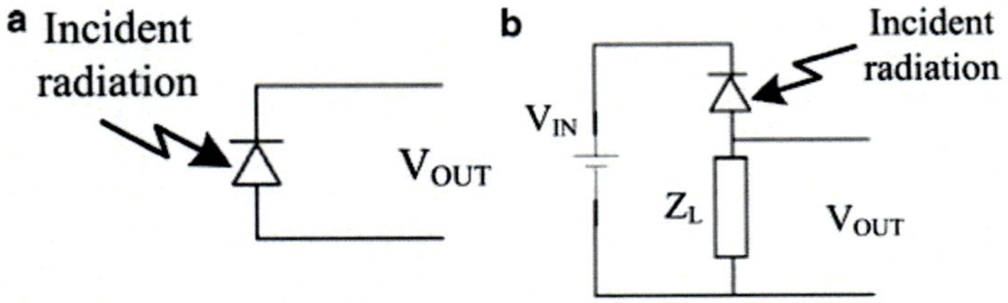

Fig. 1.8 Typical measurement configurations for photodiode devices: (**a**) an incident radiation generates a photo-current (photovoltaic mode); (**b**) the generated photo-current, depending on the incident radiation, flows into a load resistor providing an output voltage signal (photoconductive mode)

The photoconductivity effect is detectable in a homogeneous semiconductor, while the photovoltaic one can be observed in a semiconductor device excited by a built-in electrical field. The more diffused device for these aims is the junction diode, whose measurement configurations are shown in Fig. 1.8, even if more complicated structures can be also utilized, such as avalanche diodes, Schottky diodes, eterojunction devices, etc.. Generally, the photovoltaic detectors are faster than the photoconductor ones fabricated with similar materials [7].

Resuming, the photodiode can be biased and so utilized under two different modalities: photovoltaic mode and photoconductive mode. In the first configuration, the diode is characterized by a slow time response since the generated charges have to charge the diode capacitor so to produce a detectable voltage signal (in this way, a signal delay, as in an *RC*-cell, is achieved). On the contrary, in the photoconductive configuration, the diode needs an inverse biasing, so the current which flows into the device is converted into a voltage through a resistor. The main advantage of this operating mode is in the fact that the utilized biasing reduces the diode internal capacitor, increasing the spatial charge region. Therefore, in this case, a faster time response, with respect to that achieved in the photovoltaic mode, is obtained. Unfortunately, the diode constant biasing causes also a leakage current which can affect the radiation measurement. In Fig. 1.9 two possible read-out electronic circuits (based on an *OA*) related to the two different operating modes for the photodiode have been reported [7].

1.6 Displacement and Force Sensors

Many types of forces are sensed through the displacements they create. For example, the force due to acceleration of a mass at the end of a spring will cause the same spring to stretch and the mass m to move. Its displacement from the zero acceleration position is governed by the force F generated by the acceleration a (through the well-known law $F = m \cdot a$) and the restoring force of the spring. Another example

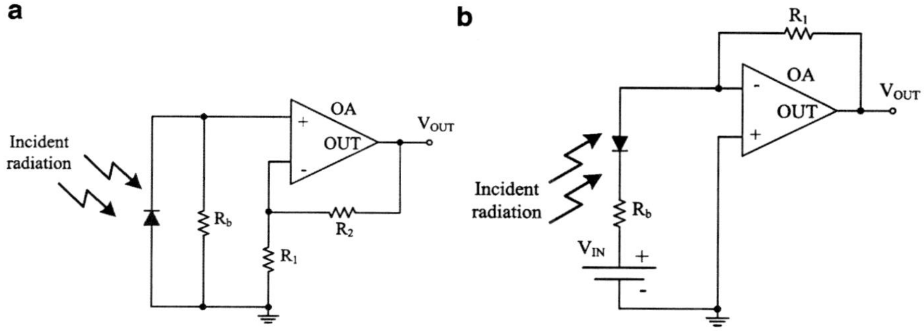

Fig. 1.9 Examples of possible photodiode connections: (**a**) photovoltaic mode; (**b**) photoconductive mode

is the displacement of the centre of a deformable membrane due to a pressure difference across it. Both these examples utilize a multiple transduction mechanism to produce an electronic output: a primary mechanism which converts the force to the displacement (mechanical to mechanical conversion) and then a secondary mechanism to convert the displacement to an electrical signal (mechanical to electrical conversion).

Generally, the displacement can be evaluated through the measurement of an associated capacitance. For example, the capacitance C associated with a gap which is changing in length is given by $C = \text{area} \times \text{dielectric constant}/\text{gap length}$. The gap must be very small when compared to the surface area of the capacitor, since most dielectric constants are in the order of 0.1 pF/cm and, with modern detection methods, capacitance is readily resolvable to about some fF. This is because measurement leads and contacts create parasitic capacitances in the same order of magnitude. If the capacitance is measured by an integrated circuit, fabricated on the same chip, capacitances as small as a few tens (or hundreds) of fF can be also revealed and measured. Displacement is also commonly measured by the movement of a ferromagnetic core inside of an inductor coil. The displacement produces a change in inductance which can be measured by placing the inductor in an oscillator circuit and measuring the oscillation frequency variation.

The most commonly used force sensor is the strain gauge. It consists of metal wires which are stretched by the application of an external force. The resistance of the wire changes as it undergoes strain, i.e., a change in length, since the resistance of a wire is $R = \text{resistivity} \times \text{length}/\text{cross-sectional area}$. The wire resistivity is a bulk property of the metal which is a constant for a constant temperature. For example, a strain gauge can be used to measure acceleration by attaching both the ends of the wire to a cantilever beam, with one end at the attached beam end and the other kept free. The typical strain gauge equation is the following:

$$\frac{\Delta R}{R} = K\frac{\Delta l}{l} \tag{1.4}$$

where $\Delta l/l$ is the relative deformation and K is the gauge factor, typical of the specific material used for the sensor (e.g., a metallic wire). The cantilever beam free end moves in response to an applied force, such as the force due to acceleration which produces strain in the wire and a subsequent change in resistance.

Semiconductors are known to exhibit piezoresistivity, that is a change in resistance, with a high sensitivity, in response to strain which involves a large change in resistivity in addition to a variation in the linear dimension. Moreover, it is important to consider that also a sonar uses the conversion of electrical signals to mechanical displacements as well as the reverse transducer property, which is the generation of electrical signals in response to a stress wave (medical diagnostic ultrasound and non-destructive testing system devices rely on this property). In this case, some actuators have also been developed, but their drawback is the small displacement which can be obtained (required voltages are typically hundreds of Volts and the displacements are only a few hundred Angstroms) [7, 9].

1.7 Ion-Selective Electrodes Based Sensors

In order to perform an electrical measurement of ions contained into a liquid solution, it is important to consider what happens at the interface area between a solid conductor and a liquid. This phenomenon is very similar to what occurs at the interface area between two semiconductors or a semiconductor and a metal. In fact, also in this case, there are species which migrate from a region into another, because of the electrochemical potential difference. The migration spontaneously occurs so to provide the equilibrium of the electrochemical potential, defined as the amount of a term depending on the activities (more simply the concentrations for diluted solutions, defined as chemical potential) and of an electrical potential. At the beginning of the phenomenon, there is a difference of the ionic species concentration between the solid and the liquid, which creates an imbalance between the electrochemical potentials both in the species and in the solid. In order to balance the chemical potential, at the conclusion of the migration, an electrical potential is created and its value results to be proportional to the logarithm of the ionic species activity.

Therefore, exploiting this principle, it is possible to implement the so-called Ion Selective Electrodes (*ISE*): when these electrodes are dipped into a solution, they assume a potential which is a function of its concentration. On the other hand, the *ISE*, as the name implies, allows to measure the concentration of a specific ion in a solution of many ions. To accomplish this, a membrane generates selectively an electrical potential (more commonly named Nernst potential) which is dependent on the concentration of the ion of interest. This is usually an equilibrium potential and develops across the interface of the membrane with the solution. It is generated by the initial net flow of ions (charge) across the membrane in response to a concentration gradient, and, then, the diffusion force is balanced by the generated

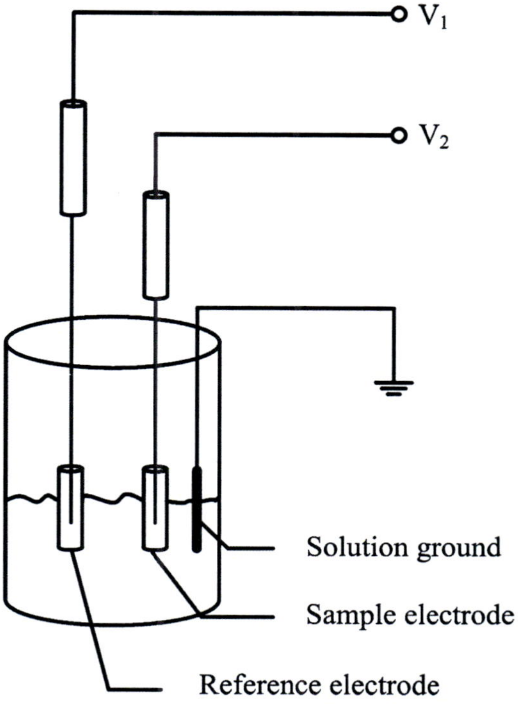

Fig. 1.10 Schematic view of an *ISE*-based system.

electric force so that an equilibrium is established. More in detail, an *ISE* consists of a glass tube with the ion-selective membrane closing the end of the tube which is immersed into the test solution. Fig. 1.10 shows a simple representation of a *ISE*-based system. The Nernst potential is measured by making an electrical contact to each side of the membrane. This is done by placing both a fixed concentration of conductive filling solution inside of the tube and a wire into the solution. The other side of the membrane is contacted by a reference electrode placed inside of the same solution under test [7, 9].

The reference electrode is constructed in the same manner as the *ISE* but has a porous membrane which creates a liquid junction between its inner filling solution and the test solution. This junction is designed to have a potential which is invariant with changes in concentration of any ion in the test solution. The reference electrode, the solution under test and the *ISE* form an electrochemical cell. The reference electrode potential acts like the ground reference in electric circuits and the *ISE* potential is measured between the two wires emerging from the related two electrodes.

The *ISE*-based system gives a behaviour similar to the so-called built-in potential of a *p*-*n* junction diode. The ion-selective membrane acts to ensure that the generated potential is dependent mostly on the ion of interest and is insensitive to

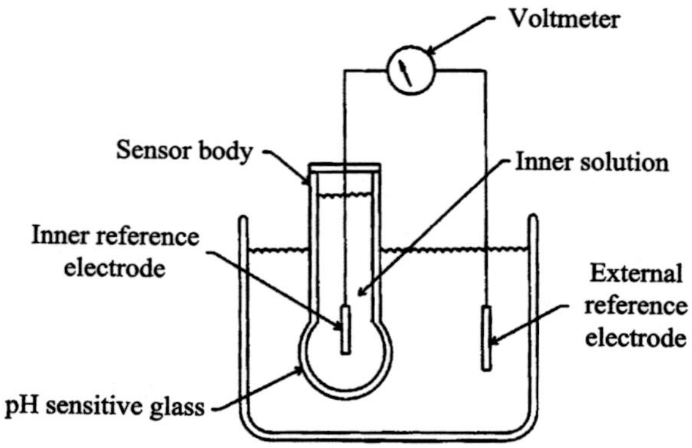

Fig. 1.11 A glass pH electrode chemical sensor

the other ions in solution. This is done by enhancing the exchange rate of the ions of interest across the membrane; therefore, the species generate and maintain the potential.

The most familiar *ISE* is the pH electrode. In this sensor, generally, the membrane is a sodium glass which shows a high exchange rate for H^+ ions. In this case, the generated Nernst potential is dependent on both the H^+ concentration and the solution operating temperature. Considering that the acidity or alkalinity of a solution is characterized by its pH (which represents the activity of the hydrogen ions in the solution), one pH unit change corresponds to a tenfold change in the molar concentration of H^+ and to about tens of mV change in the Nernst potential at room temperature.

The glass pH electrode, which is frequently used in several laboratories, is illustrated in Fig. 1.11. This sensor works only in an aqueous environment. It consists of an inner chamber containing an electrolytic solution of a known pH value and an outer solution with an unknown pH to be measured. The membrane consists of a specially formulated glass that will allow only hydrogen ions to pass in both the directions. If the concentration of hydrogen ions in the external solution is greater than that in the internal solution, there will be a gradient forcing hydrogen ions to diffuse through the membrane into the internal solution. This will cause the internal solution to have a positive charge greater than the external solution so that an electrical potential and, hence, an electric field will be generated across the membrane. This field will counteract the diffusion of hydrogen ions due to the concentration difference, so an equilibrium state will be established. The potential across the membrane at this equilibrium condition is related to the hydrogen ion concentration difference between inner and outer solutions. Thus, the potential measured across the glass membrane is proportional to the pH of the solution under study. It is not practical to measure the potential across the membrane directly, so reference electrodes (elements that can be used to measure electrical potential of an electrolytic solution) are utilized to contact the solution on each side of the membrane and to measure the potential difference across it. The reference electrodes and the glass membrane are incorporated into the structure known as a glass pH electrode. Other *ISEs* have the same type of response, but specific to a different kind of ions, depending on the choice of the membrane [1–9].

1.8 Gas Chromatograph and Gas Sensors

Molecules in gases have thermal conductivities depending on their masses. Therefore, a pure gas can be identified by its thermal conductivity. One way to determine the composition of a gas is the use of a gas chromatograph that first separates the gas into its components and then measures the thermal conductivity of each of them. The gas flows through a long narrow column, which is packed with an adsorbant solid (for gas–solid chromatography) wherein the gases are separated according to the retentive properties of the packing material for each gas. As the individual gases exit the end of the tube one at a time, they flow over a heated wire. The amount of heat transferred to the gas depends on its thermal conductivity. The gas temperature is measured by a short distance downstream and compared to a known gas flowing in a separate sensing tube. The temperature is related to the amount of heat transferred and can be used to determine the thermal conductivity according to thermodynamic theory and empirical data. Generally, this kind of sensor requires two transductions: (1) chemical to thermal; (2) thermal to electrical [1–9].

Typical gas sensors are based on *MOX* materials [78–109]. Technologically, the metal oxides are compatible with the microelectronic fabrication techniques, therefore *MOX*-based sensors are integrable on a single chip together with their electronic circuitry. These sensors show a high sensitivity which allows to detect many chemical species, generally having a concentration in the order of few ppm or lower (in some cases, also in the ppb range). Unfortunately, they suffer from some problems, such as the very low selectivity and the high power consumption. Moreover, they typically work at high temperatures, in the order of hundreds of °C. Electrically, the transition metal oxides are semiconductors, typically of n-type, and, experimentally, it can be observed that, under the presence of many different kinds of gases and vapours, the conductivity of these materials varies in a specific range, depending on different parameters. These phenomena occur at high operating temperatures, ranging typically from 100°C to 600°C, according to the kind of considered oxide. In addition, as well known, *MOX* gas sensors exhibit resistance values varying over a wide range and the main factors determining such a large value distribution include: manufacturing materials (e.g., tin oxide, titanium dioxide, etc.), fabrication techniques (e.g., thin and thick films, nanowires, etc.), excitation parameters (e.g., power supply voltage, operating temperature, etc.) and, of course, gas exposure, especially if high sensitivity sensors are used.

Among the gases to which a *MOX*-based sensor can be sensible, we mention the urban pollutants produced by combustion processes, such as the carbon monoxide (CO) and the nitrogen dioxide (NO_2). The most important and utilized metal oxide

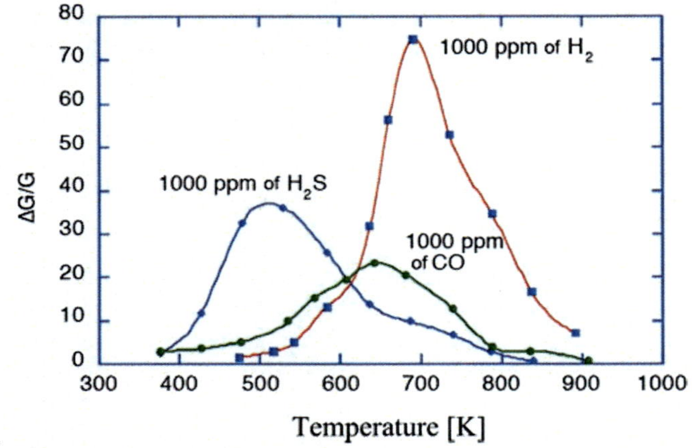

Fig. 1.12 Typical relative conductivity variation vs. the operating temperature of a SnO_2-based gas sensor, for different gas species

is the tin oxide (SnO_2) which represents the best material suitable for conductivity-variation based chemical sensors. Other materials that can be utilized for the fabrication of *MOX*-based sensors are, for example, ZnO, In_2O_3, WO_3, Fe_2O_3, Ga_2O_3, etc.. Concerning the toxic gas revelation, the CO results to be an important environmental pollutant which is generated during the combustion processes, when the oxygen quantity, present in the air, is not sufficient to properly complete the same combustion (the product of a correct combustion is the carbon dioxide, CO_2). CO toxicity is due to the fact that it affects the oxygen transport process in the human body [96].

It is important to underline that the operating temperature has an important role as regarding both the sensitivity and the selectivity of this kind of gas sensors. More in detail, considering a particular sensible material, for each specific gas an optimal working temperature for the same sensor exists.

In Fig. 1.12, a typical relative conductivity variation, as a function of the operating temperature of a SnO_2-based gas sensor, is reported, showing a different and temperature-dependent sensor selectivity for the same quantity (1000 ppm) of H_2S, CO and H_2 gases [7].

In order to fabricate a gas sensor based on conductivity variation, depending on the *MOX* properties, it is mandatory, first of all, to deposit the same sensing material on an insulating substrate, having different electrodes which constitute the necessary electrical contacts. Moreover, the deposited *MOX* has to be kept at a constant temperature of about few hundreds of Kelvin. Fig. 1.13, as an example, shows a photo of a *MOX*-based gas sensor fabricated on an aluminium oxide substrate [7]. In this case, the sensor is mounted on a standard microelectronic support and is provided of four terminals: two of them are necessary for the measurement of the sensing element (thin film MOX) conductivity, while the others two are

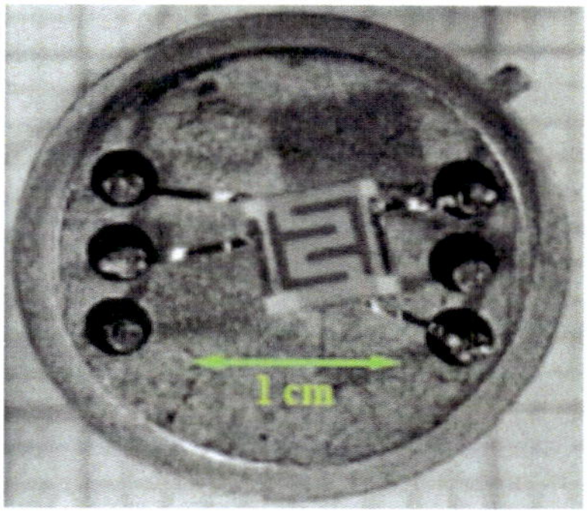

Fig. 1.13 An example of a *MOX*-based gas sensor fabricated on an aluminium oxide substrate

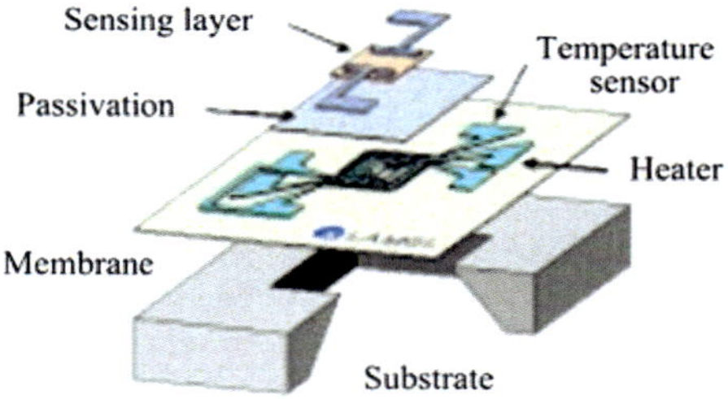

Fig. 1.14 An example of the constructive scheme of a *MOX*-based gas sensor

required to supply the heating element (metal filament). In addition, in Fig. 1.14 a possible constructive scheme of a conductivity-variation based gas microsensor is reported [7].

Recently, the demand of thin film gas sensor systems has increased, because fabrication processes have to be optimized to be faster, safer and to extend the tool life. Concerning the integration of sensor systems, it is important to remember that the size must be as small as possible or in a shape that can be easily integrated. The aim is to build up a sensor system that can be used in a large variety of applications.

Thin films based on TiO_2 can be customized for their use as gas sensors, self-cleaning surfaces, as biomaterials for orthopaedic and oral implants, for photocatalytic decomposition of organic compounds in the air, such as formaldehyde and nicotine fume, but also toxic gases such as CO, CO_2, CH_4, NO_X, ozone and also microorganisms. These *MOX* films show high stability, sensitivity, selectivity and reversibility under low-temperature conditions for NO_2, O_3 and H_2S. Gas sensors based on TiO_2 are applicable as low-cost and *LP* sensor devices for miniaturized gas monitoring [83, 84].

In order to perform a direct measurement for limited hydrocarbon (*HC*) components in the exhaust, it was proposed to detect them directly through other kinds of resistive sensors based on *MOX* such as gallium-oxide (Ga_2O_3) or doped strontium-titanate ($SrTiO_3$). Since the resistance of these materials also depends on the oxygen concentration of the exhaust, a two-sensor-setup was introduced, with one sensor being catalytically activated, whereas the other one remained non-activated [110, 111]. In some of these cases, the sensor can be completely manufactured in a ceramic multilayer and thick-film technology and is suitable as O_2 resistive gas sensor [112].

Although many different sensor nanostructures, such as those based on SnO_2, ZnO, In_2O_3, and TiO_2, have been investigated for their gas sensing properties, researchers have recently paid greater attention to SnO_2 nanowires-based sensor. Presently, different synthesizing methods have been reported for producing SnO_2 nanowires such as hydrothermal methods, thermal decomposition of precursor powders *Sn*, *SnO*, SnO_2 followed by vapour–solid or vapour–liquid–solid growth. Even if the synthesis of SnO_2 nanowires by the thermal decomposition using *SnO* as a source material is often utilized, it is rather difficult to get SnO_2 nanowires based on this procedure since the synthesis of these devices is strongly dependent on the synthesis apparatus. Thus, novel, simple and reproducible procedures and methods have been developed to easily fabricate SnO_2 nanowires for gas sensing applications [86–101].

1.9 Humidity Sensors

The term humidity refers to the water vapour content in air or other gases; its measurement can be expressed in different terms and units. The three commonly used are absolute humidity, dew point and relative humidity (*RH*), whose definitions are provided in the following. The absolute humidity is the ratio of the mass of water vapour to the volume of air or gas and it is commonly expressed in grams per cubic meter. It can be calculated from known *RH* temperature, or wet bulb, or can be measured directly. Refinements in thermistor technology have led to the development of a thermal conductivity principle that permits absolute humidity measurements at high temperatures (>200°C) even in a polluted environment. In this case, the detection system typically uses two thermistors in a bridge configuration. The dew point, expressed in °C or °F, is the temperature (depending on the pressure) at which a gas begins to condense into a liquid. Chilled mirror hygrometers have reliably made dew point measurements since the early 1960s, but the development of stable thin film capacitive sensors, in the 1980s, actually allows measurements of dew points, as low as −40°F at a reduced cost. *RH* refers to the ratio (stated as a percent) of the moisture content of air compared to the saturated moisture level at the same temperature and pressure. *RH* was derived from measuring a physical change that moisture absorption caused in some different materials such as silk, human hair, nylon, etc.. Later, most mechanical methods have been replaced by electronic *RH* sensors due to their greater accuracy, dependability and lower costs. Recently, specialized polymer-based resistive and laser-trimmed capacitive sensors with monolithic signal conditioners for *RH* measurements have been also introduced. The most important specifications for a humidity sensor are: accuracy, repeatability, interchangeability, long-term stability, ability to recover from condensation, resistance to chemical and physical contaminants, device size, packaging, cost effectiveness, durability for use in different environments, etc. [113].

Absolute humidity sensors are very durable, operate at temperatures up to about 300°C and have a good endurance to chemical vapours by means of the inert materials used for their construction (i.e., glass, semiconductor material for the thermistors, high-temperature plastics, aluminium). An interesting feature of thermal conductivity sensors is that they respond to any gas that has thermal properties different from those of dry nitrogen (this will affect the measurements). Absolute humidity sensors are commonly used in appliances such as cloth dryers and both microwave and steam-injected ovens, while industrial applications include kilns for drying wood, machinery for drying textiles, paper and chemical solids, pharmaceutical production, cooking and food dehydration. Since one of the by-products of combustion and fuel cell operation is water vapour, a particular interest has been shown in using absolute humidity sensors to monitor the efficiency of those reactions. In general, absolute humidity sensors provide a resolution, at temperatures higher than about 100°C, greater than those shown by capacitive and resistive sensors and may be used in applications where these sensors would not survive. Furthermore, the typical accuracy of an absolute humidity sensor is about $3\,g/m^3$ that corresponds to about ±5%RH at 40°C and ±0.5%RH at 100°C [113].

Generally, in order to determine air *RH*, the more utilized sensors employ a capacitive measurement technique. The sensor element is built out of a film capacitor on different substrates (glass, ceramic, etc.). The dielectric is a polymer which absorbs or releases water proportional to the relative environmental humidity and thus changes the value of the capacitor, which can be measured directly by an on-board electronic circuit. Capacitive, resistive and thermal conductivity sensing technologies for humidity evaluation offer each distinct advantages (see also next Chapter). In particular, capacitive sensors provide wide *RH* range and condensation tolerance and, if laser trimmed, are interchangeable. Resistive sensors are also interchangeable, usable for remote locations and cost effective. Thermal conductivity sensors perform well in corrosive environments and at high temperatures. Therefore, for most applications, the environmental conditions dictate the choice of the suitable humidity sensor [113–118].

Recently a new generation of integrated, digital and calibrated sensors, which combine humidity and temperature detection, using *CMOS* "micro-machined" chip technology, has been also introduced in the market (e.g., SHT1x, SHT7x and SHT2x series by SENSIRION) [118, 119]. These new products represent a single chip relative humidity and temperature multi sensor module with a calibrated digital output which allows a simple and quick system integration. By combining *CMOS* and sensor technologies, highly integrated and extremely small humidity sensors have been achieved. These devices include two calibrated microsensors, for relative humidity and temperature detection, which are followed by a suitable processing circuitry on the same chip. The temperature and the humidity sensors together form a single unit, which enables a precise determination of the dew point without incurring errors due to temperature gradients between the two sensor elements. The integration provides improved signal quality and insensitivity to external disturbances (EMC). Other advantages include very short response times (4 s at lie), high precision (±2% to ±5% according to configuration), low power consumption (<3 μA standby), small footprint (7×5×3 mm) and the sensor chip can be connected directly to any microprocessor system by means of the digital 2-wire interface. These digital humidity sensors are suitable in the automation, control and building HVAC markets; they could be used in high precision "smart" transmitters, data logging applications, automotive markets, etc. [118, 119].

1.10 Biosensors and Biomedical Sensors

Biosensors are devices for interfacing an instrumentation equipment with a biological system such as a biological specimen or an entire organism. The device serves the function of detecting and measuring a property of the biologic system. However, they are also used in industrial applications, e.g., the monitoring and control of fermentation reactions. On the other hand, biosensors are not exactly biomedical sensors (i.e., blood pressure sensors or electrocardiogram electrodes), even if many of them are used in biomedical applications. Biosensors are of a special interest because of the very high selectivity of biological reactions. More in detail, biological measurands are biologically-produced substances, such as antibodies, glucose, hormones and enzymes. A familiar commercial biosensor is the in-home pregnancy test sensor, which detects the presence of human growth factor in urine. This device is a non-electrical sensor since the information is given by a colour change easily detectable by the human eyes [1–9].

In biomedical applications the sensor and its instrumentation system are of fundamental importance, because the sensor can affect the measurand and the latter can affect the sensor performance. Therefore, biomedical sensors must be designed to minimize their interaction with the biologic host. It is important that the presence of the sensor does not affect the variable being measured in the proximity of the sensor as a result of the interaction between the sensor and the biologic system. If the sensor is placed into a living organism, that organism will probably recognize the sensor as a foreign body and react to it. This may in fact change the quantity being sensed in the proximity of the sensor so that the measurement reflects the foreign body reaction rather than a central characteristic of the host. Similarly, the biological system can affect the performance of the sensor. The foreign body reaction might cause the host to attempt to break down the materials of the sensor as a way to remove it. This may, in fact, degrade the sensor package so that the sensor can no longer perform in an adequate manner. Moreover, sensors implanted in the body are not accessible for calibration. Thus, such sensors must be extremely stable so that frequent calibrations are not possible [1–9].

There are many types of physical/chemical sensors used in biomedical measurement instrumentation. The following list shows some general categories of sensors, in particular a possible classification of physical/chemical biomedical sensors: electrochemical (amperometric, potentiometric, etc.), optical (colorimetric, emission and adsorption spectroscopy, fluorescence, chemiluminescence, etc.), thermal methods (calorimetry, thermoconductivity, etc.), nuclear magnetic resonance. Electrochemical and optical sensors are most frequently used for biomedical measurements both *in-vivo* and *in-vitro* [1–9].

Biomedical sensors can be also classified according to how they are used with respect to the biologic system: non-invasive (non-contacting, body surface) and invasive (indwelling, implanted). In particular, the non-invasive sensor is an interface device of an instrumentation system that measures a physiologic variable from an organism without interrupting the integrity of that organism. Sensors of radiant heat, sound energy coming from an organism, skin surface thermometers and biopotential electrodes placed on the skin are examples of non-invasive sensors which can be placed on the body surface. On the contrary, indwelling sensors are those which can be placed into a natural body cavity that communicates with the outside. These kind of sensors includes oral-rectal thermometers, intrauterine pressure transducers, stomach pH sensors, etc.. The most invasive sensors are typically those that need to be surgically placed and involve some tissue damage associated with their installation (i.e., a needle electrode for picking up electromyographic signals directly from muscles, a blood pressure sensor placed in an artery, vein, or in the heart itself, a blood flow transducer positioned on a major artery) [1–9].

Other parameters to be detected through biomedical sensors are liquid pressure and flow. In particular, the measurement of blood pressure and blood flow in humans and other animals remains an important problem in biomedical sensing. Direct blood pressure measurement refers to the evaluation of the blood pressure using a sensor that is in contact with the blood being measured or contacts it through an intermediate fluid such as a physiologic saline solution. Direct blood pressure sensors are typically invasive. Indirect blood pressure measurement involves a sensor that does not actually contact the blood. The most familiar indirect blood pressure measurement is the sphygmomanometer cuff that is used in most medical examinations and is a non-invasive instrument. On the contrary, until recently, the primary sensor used for direct blood pressure measurement was the unbonded strain gauge pressure transducer. The basic principle of this device is that a differential pressure seen across a diaphragm will cause it to deflect. This deflection is then measured by a displacement transducer implemented by wires whose electrical resistance increases when they are stretched. However, in biomedical applications, pressure is generally referenced to atmospheric pressure, therefore, the pressure in the chamber must be maintained at atmospheric level. In recent years, semiconductor technology has been applied to the design of novel pressure transducers: the entire sensor can be fabricated and sold inexpensively so that disposable, single-use devices can be made [1–9].

A subgroup of the chemical sensors that sense the presence and the concentration of biochemical materials in the host is known as bioanalytical sensors. In particular, they are a special case of chemical sensors for determining the amount of a biochemical substance. This type of sensor usually makes use of one of the following types of biochemical reactions: enzyme-substrate, antigen-antibody or ligand-receptor. The advantage of using these reactions in a sensor is that they are highly specific for a particular biological molecule and sensors with high sensitivity can be developed considering these reactions. A bioanalytical sensor structure consists of two main parts: the first contains one component of the biological sensing reaction such as the enzyme or the antibody and the second detects whether the biological reaction has taken place. This second part of a bioanalytical sensor is made up of either a physical or a chemical sensor that serves as the detector of the biological reaction. These sensors can be used either on a biological specimen taken from the host and tested in a laboratory or for "*in-vivo*" measurements both as non-invasive and invasive sensors, the latter being the most frequently used [9, 120, 121].

References

1. W. Gopel, J. Hesse, J.H. Zemel (eds.), *Fundamentals and General Aspects, Sensors: A Comprehensive Survey* (Wiley VCH, Germany, 1996). ISBN 3527293299
2. S. Middelhock, S.A. Audet, P. French, *Silicon Sensors* (Academic, London, 2000)
3. S.D. Senturia, *Microsystem Design* (Kluwer, Boston, 2001). ISBN 9780792372462
4. R. Pall'as-Areny, J.G. Webster, *Sensors and Signal Conditioning*, 2nd edn. (Wiley Interscience, New York, 2001). ISBN 0471332321
5. J. Fraden, *Handbook of Modern Sensors: Physics, Design and Applications*, 3rd edn. (Springer, New York, 2003). ISBN 1441964657
6. S.M. Sze, Kwok K. Ng, *Physics of Semiconductor Devices*, 3rd edn. (Wiley, New York, 2007). ISBN 9780471143239
7. A. D'Amico, C. Di Natale, *Introduzione ai sensori* (Aracne, Roma, 2008). ISBN 9788854816633
8. A. DAmico, C. Di Natale, A contribution on some basic definitions of sensors properties. IEEE Sensors J. **1**(3), 183–190 (2001)
9. R.C. Dorf, *The Electrical Engineering Handbook* (CRC Press LLC, Boca Raton, 2000). ISBN 0849385741
10. G. Ferri, N.C. Guerrini, *Low-Voltage Low-Power CMOS Integrated Architectures for Sensor Interfaces*, Electronics World, Nov 2002, Vol. 108, pp. 12–20
11. L. Zhao, E.M. Yeatman, Inherently digital micro capacitive tilt sensor for low power motion detection, in *Proceedings of Eurosensors*, Dresden, Sept 2008, pp. 621–624
12. Y.H. Hsueh, J.H. Lin, Low power integrated capacitive pressure microsensor design, in *Proceedings of Eurosensors* 2008, Dresden, Sept 2008, pp. 284–287
13. C. Hierold, B. Clasbrummel, D. Behrend, T. Scheiter, M. Steger, K. Oppermann, H. Kapels, E. Landgraf, D. Wenzel, D. Etzrodt, Low power integrated pressure sensor system for medical applications. Sensor Actuat A **73**(1–2), 58–67 (1999)
14. L. Löfgren, B. Löfving, T. Pettersson, B. Ottosson, S. Haasl, C. Rusu, K. Persson, O. Vermesan, N. Pesonen, P. Enoksson, Low-power humidity sensor, in *Proceedings of Eurosensors*, Dresden, Sept 2008, pp. 231–234
15. C. Falconi, E. Martinelli, C. Di Natale, A. DAmico, F. Maloberti, P. Malcovati, A. Baschirotto, V. Stornelli, G. Ferri, Electronic Interfaces. Sensor Actuat B **121**, 295–329 (2007)
16. J. Huijsing, Integrated smart sensors. Sensor Actuat A **30**, 167–174 (1992)
17. M. Landwehr, H. Grätz, A low-power, low-area, delay-line based CMOS temperature sensor, in *Proceedings of Eurosensors*, Dresden, Sept 2008, pp. 1392–1394
18. T.G. Constandinou, J. Georgiou, C. Toumazou, Micropower front-end interface for differential capacitive sensor systems. IET Electron. Lett. **44**(7), 470–472 (2008)
19. S.S.W. Chan, P.C.H. Chan, A resistance-variation tolerant constant-power heating circuit for integrated sensor applications. IEEE J. Solid-St. Circ. **34**(4), 432–437 (1999)

20. M. Grassi, P. Malcovati, A.Baschirotto, A high-precision wide-range front-end for resistive gas sensor arrays, in *Proceedings of Eurosensors*, Rome, Sept 2004
21. G. Ferri, N. Guerrrini, V. Stornelli, C. Catalani, A novel CMOS temperature control system for resistive gas sensor array, *Proceedings of ECCTD*, Cork, 2005, pp. 351–354
22. A. Gerosa, A. Maniero, A. Neviani, A fully integrated two-channel A/D interface for the acquisition of cardiac signals in implantable pacemakers. IEEE J. Solid-St. Circ. **39**(7), 1083–1093 (2004)
23. S. Pennisi, High-performance and simple CMOS interface circuit for differential capacitive sensors. IEEE T. Circuits II **52**(6), 322–326 (2005)
24. J.R. Kaienburg, M. Huonker, R. Schellin, Surface micromachined bridge configurations for accurate angle measurements, in *IEEE Internationl Conference on Microelectromechanical Systems*, Miyazaki, Jan 2000, pp. 120–125
25. M. Cicioni, L. Bissi, P. Placidi, A. Shehu, A. Scorzoni, E. Cozzani, I. Elmi, S. Zampolli, G.C. Cardinali, Interface circuit for an ultra low power gas sensor, in *IEEE Instrumentation and Measurement Technology Conference*, Singapore, May 2009, pp. 254–258
26. L. Bissi, M. Cicioni, P. Placidi, S. Zampolli, I. Elmi, A. Scorzoni, A programmable interface circuit for an ultralow power gas sensor. IEEE T. Instrum. Meas. **99**, 1–8 (2010)
27. H. Baltes, A. Häberli, P. Malcovati, F. Maloberti, Smart sensor interfaces DOI: dx.doi.org. Proc. IEEE Int. Symp. Circ. Syst. **4**, 380–383 (2006). Atlanta
28. M.A.P. Pertijs, K.A.A. Makinwa, J.H. Huijsing, A CMOS smart temperature sensor with a 3s inaccuracy of $\pm 0.1°C$ from $-55°C$ to $125°C$. IEEE J. Solid-St. Circ. **40**(12), 2805–2815 (2005)
29. M.A.P. Pertijs, A.L. Aita, K.A.A. Makinwa, J.H. Huijsing, Voltage calibration of smart temperature sensors, Proc. of IEEE Sensors, Lecce, Oct 2008, pp. 756–759
30. M.A.P. Pertijs, G.C.M. Meijer, J.H. Huijsing, Precision temperature measurement using CMOS substrate pnp transistors. IEEE Sens. J. **4**(3), 294–300 (2004)
31. M. Malfatti, M. Peronzoni, N. Viarani, A. Simoni, L. Lorenzelli, A. Baschirotto, A complete front-end system read-out and temperature control for resistive gas sensor array. IEEE Eur. Conf. Circ. Theor. Des. **3**, 31–34 (2005)
32. H. Baltes, CMOS as sensor technology. Sensor Actuat A **37–38**, 51–56 (1993)
33. H. Baltes, D. Moser, E. Lenggenhager, O. Brand, D. Jaeggi, Thermomechanical microtransducers by CMOS and micromachining micromechanical sensors, in *Actuators and Systems*, DSC-32, (ASME, New York, 1991), pp. 61–75
34. H. Baltes, O. Brand, CMOS-based microsensors and packaging. Sensor Actuat A **92**(1–3), 1–9 (2001)
35. A. Baschirotto, P. Malcovati, Technology-driven alternatives for smart sensor interfaces, in *Sensors Update*, ed. by H. Baltes, G. Fedder, J. Korvink (Wiley-VCH, Weinhein, 2003), Vol. 13, pp. 45–81
36. G.C.M. Meijer, G. Wang, F. Fruett, Temperature sensors and voltage references implemented in CMOS technology. IEEE Sens. J. **1**(3), 225–234 (2001)
37. C. Falconi, J. Huijsing, Curvature correction of bandgap references for low cost integrated sensors systems, in *Proceedings of Eurosensors*, Prague, 2003
38. G. Wang, G.C.M. Meijer, The temperature characteristics of bipolar transistors fabricated in CMOS technology. Sensor Actuat A **87**, 81–89 (2000)
39. C. Zhang, T. Yin, Q. Wu, H. Yang, A large dynamic range CMOS readout circuit for MEMS vibratory gyroscope, in *Proceedings of IEEE Sensors*, Lecce, Oct 2008, pp. 1123–1126
40. W.L. Liu, S.J. Chen, C.H. Shen, Sensitivity improvement of thermal conduction CMOS based accelerometer, in *Proceedings of IEEE Sensors*, Lecce, Oct 2008, pp. 407–410
41. C. Liang Dai, M. Chen Liu, Nanoparticle SnO_2 gas sensor with circuit and microheater on chip fabricated using CMOS-MEMS technique, in *IEEE Nano-Micro Engineered and Molecular Systems and Conference*, Thailand, 2007, pp. 959–963
42. C.T. Ko, S.H. Tseng, M.S.C. Lu, A CMOS micromachined capacitive tactile sensor with high frequency output. IEEE J. Microelectromech Syst. **15**(6), 1708–1714 (2006)

43. S.H. Tseng, P.C. Wu, Y.Z. Juang, M.S.C. Lu, A CMOS MEMS thermal sensor with high frequency output, in *Proceedings of IEEE Sensors*, Lecce, Oct 2008, pp. 387–390
44. C.L. Dai, Y.W. Tai, P.H. Kao, Modeling and fabrication of micro FET pressure sensor with circuits. Sensors **7**, 3386–3398 (2007)
45. G.M. Lazzerini, M. Dei, P. Bruschi, M. Piotto, VHDL-AMS modeling of an integrated gas flow sensor readout channel with pressure compensation, in *Proceedings of PRIME 2007*, Bordeaux, 2007, pp. 141–144
46. A. Lombardi, M. Grassi, L. Bruno, P. Malcovati, A. Baschirotto, A fully integrated interface circuit for 1.5°C accuracy temperature control and 130-dB dynamic-range read-out of MOX gas sensors, in *34th European Solid-State Circuits Conference*, Sept 2008, pp. 78–81
47. A. Lombardi, L. Bruno, M. Grassi, P. Malcovati, S. Capone, L. Francioso, P. Siciliano, A. Baschirotto, Integrated read-out and temperature control interface with digital I/O for a gas-sensing system based on a SnO_2 microhotplate thin film gas sensor, in *Proceedings of IEEE Sensors*, Lecce, Oct 2008, pp. 596–599
48. M. Piotto, M. Dei, P. Bruschi, An interface circuit for thermal gas flow meters with compensation of pressure effects, in *Proceedings of Eurosensors*, Dresden, Sept 2008, pp. 433–436
49. M. Grassi, P. Malcovati, A. Baschirotto, A 141-dB dynamic range CMOS gas-sensor interface circuit without calibration with 16-bit digital output word. IEEE J. Solid-St. Circ. **42**, 1543–1554 (2007)
50. Internet resource: http://en.wikipedia.org/wiki/Electret
51. Internet resource: http://www.electrets.org/html/overview.html
52. D.F. Da Silva, D. Acosta-Avalos, Light dependent resistance as a sensor in spectroscopy setups using pulsed light and compared with electret microphones. Sensors **6**, 514–525 (2006)
53. X. Qiu, Patterned piezo-, pyro- and ferroelectricity of poled polymer electrets. J. Appl. Phys. **108**(1), 011101-011101–19 (2010)
54. S. Kon, K. Oldham, R. Horowitz, Piezoresistive and Piezoelectric MEMS Strain Sensors for Vibration Detection, in *Proceedings of SPIE*, part 2, N. 65292V, 2007
55. M. Pohanka, O. Pavliš, P. Skládal, Rapid characterization of monoclonal antibodies using the piezoelectric immunosensor. Sensors **7**, 341–353 (2007)
56. M. Stobiecka, J.M. Cieśla, B. Janowska, B. Tudek, H. Radecka, Piezoelectric sensor for determination of genetically modified Soybean Roundup Ready in samples not amplified by PCR. Sensors **7**, 1462–1479 (2007)
57. M. Pohanka, F. Treml, M. Hubálek, H. Band'ouchová, M. Beklová, J. Pikula, Piezoelectric biosensor for a simple serological diagnosis of tularemia in infected European Brown Hares. Sensors **7**, 2825–2834 (2007)
58. S. Noimanee, T. Tunkasiri, K. Siriwitayakorn, J. Tantrakoon, Design considerations for aural vital signs using PZT piezoelectric ceramics sensor based on the computerization method. Sensors **7**, 3192–3208 (2007)
59. D. Ortega, J.S. Garitaonandia, C. Barrera-Solano, M. Domínguez, Ferromagnetic resonance of nanocomposites based on iron oxides. Sensor Lett. **5**, 69–72 (2007)
60. V.A. Chernenko, S. Besseghini, P. Müllner, G. Kostorz, J. Schreuer, M. Krupa, Ferromagnetic shape memory materials: Underlying physics and practical importance. Sensor Lett. **5**, 229–233 (2007)
61. A. Platil, J. Tomek, P. Kaspar, Characterization of ferromagnetic powders for magnetopneumography and other applications. Sensor Lett. **5**, 311–314 (2007)
62. J. Guyonnet, H. Bea, P. Paruch, Lateral piezoelectric response across ferroelectric domain walls in thin films. J. Appl. Phys. **108**(4), 042002-042002–11 (2010)
63. K.P. Jayachandran, J.M. Guedes, H.C. Rodrigues, Optimal configuration of microstructure in ferroelectric materials by stochastic optimization. J. Appl. Phys. **108**(2), 024101–10 (2010)
64. G.A. Salvatore, L. Lattanzio, D. Bouvet, I. Stolichnov, N. Setter, A.M. Ionescu, Ferroelectric transistors with improved characteristics at high temperature. Appl. Phys. Lett. **97**(5), 053503-053503–3 (2010)

65. SB. Lang, J.C. Lashley, K.A. Modic, R.A. Fisher, W.M. Zhu, Z.G. Ye, Specific heat of a ferroelectric PZT ceramic at the morphotropic phase boundary, in *15th IEEE Mediterranean Electrotechnical Conference*, Valletta, April 2010, pp. 23–25
66. X.W. Dong, S. Dong, K.F. Wang, J.G. Wan, J.M. Liu, Enhancement of ferroelectricity in Cr-doped $HO_2Ti_2O_7$. Appl. Phys. Lett. **96**(24), 242904-242904–3 (2010)
67. A. Cano, D. Jimenez, Multidomain ferroelectricity as a limiting factor for voltage amplification in ferroelectric field-effect transistors. Appl. Phys. Lett. **97**(13), 133509-133509–3 (2010)
68. W.L. Lew, J.A. Ole Farmer, M.C. Crowe, C.T. Campbell, Improved pyroelectric detectors for single crystal adsorption calorimetry from 100 to 350 K. Rev. Sci. Instrum. **81**(2), 024102-024102–9 (2010)
69. M. Schossig, V. Norkus, G. Gerlach, Infrared responsivity of pyroelectric detectors with nanostructured NiCr thin-film absorber. IEEE Sens. J. **10**(10), 1564–1565 (2010)
70. P. Zappi, E. Farella, L. Benini, Tracking motion direction and distance with pyroelectric IR sensors. IEEE Sens. J. **10**(9), 1486–1494 (2010)
71. Q. Hao, F. Hu, Y. Xiao, Multiple human tracking and identification with wireless distributed pyroelectric sensor systems. IEEE Syst. J. **3**(4), 428–439 (2009)
72. W. Tornow, S.M. Lynam, S.M. Shafroth, Substantial increase in acceleration potential of pyroelectric crystals. J. Appl. Phys. **107**(6), 063302-063302–4 (2010)
73. J. Wooldridge, J.F. Blackburn, N.L. McCartney, M. Stewart, P. Weaver, M.G. Cain, Small-scale piezoelectric devices: Pyroelectric contributions to the piezoelectric response. J. Appl. Phys. **107**(10), 10118-104118–6 (2010)
74. A.N. Morozovska, E.A. Eliseev, G.S. Svechnikov, S.V. Kalinin, Pyroelectric response of ferroelectric nanowires: Size effect and electric energy harvesting. J. Appl. Phys. **108**(4), 042009-042009–6 (2010)
75. J. Zhang, M.W. Cole, S.P. Alpay, Pyroelectric properties of barium strontium titanate films: Effect of thermal stresses. J. Appl. Phys. **108**(5), 054103-054103–7 (2010)
76. S. Gundogdu, O. Sahin, E.M.I. effects of cathodic protection on electromagnetic flowmeters. Sensors **7**, 75–83 (2007)
77. C. Israel, S. Kar-Narayan, N.D. Mathur, Eliminating the temperature dependence of the response of magnetoelectric magnetic-field sensors. IEEE Sens. J. **10**(5), 914–917 (2010)
78. J.G. Lu, P. Chang, Z. Fan, Quasi-one-dimensional metal oxide materials. Synthesis, properties and applications. Mater. Sci. Eng. **52**, 49–91 (2006)
79. I. Sayago, M.C. Horrillo, S. Baluk, M. Aleixandre, M.J. Fernandez, L. Ares, M. Garcia, J.P. Santos, J. Gutierrez, Detection of toxic gases by a tin oxide multisensor. IEEE Sens. J. **2**, 387–393 (2002)
80. T. Sahm, L. Mädler, A. Gurlo, N. Barsan, U. Weimar, A. Roessler, S. E. Pratsinis, High performance porous metal oxide sensors via single step fabrication, in *Proceedings of Eurosensors*, Barcelona, Sept 2005
81. A. Depari, G. Faglia, A. Flammini, A. Fort, M. Mugnaini, A. Ponzoni, E. Sisinni, S. Rocchi, V. Vignoli, CO detection by MOX sensors exploiting their dynamic behavior, in *Proceedings of Eurosensors*, Dresden, Oct 2008, pp. 1070–1073
82. A. Fort, M. B. Serrano-Santos, R. Spinicci, N. Ulivieri, V. Vignoli, Electronic noses based on metal oxide gas sensors: the problem of selectivity enhancement, *Proceedings of IEEE Instrumentation and Measurement Technology Conference* - IMTC 2004, Como, May 2004, pp. 599–604
83. G. Sberveglieri, E. Comini, G. Faglia, M.Z. Atashbar, W. Wlodarski, Titanium dioxide thin films prepared for alcohol microsensor applications. Sensor Actuat B **66**(1–3), 139–141 (2000)
84. A. Delan, A. Karuppasamy, E. Schultheiß, Gas sensing properties of pure and doped (n, c) TiO_2 thin films grown by pulsed DC magnetron sputtering, in *Proceedings of Eurosensors*, Dresden, Sept 2008, pp. 844–847

85. S.V. Kalinin, J. Shin, S. Jesse, D. Geohegan, A.P. Baddorf, Y. Lilach, M. Moskovits, A. Kolmakov, Electronic transport imaging in a multiwire SnO_2 chemical field-effect transistor device. J. Appl. Phys. **98**, 044503–8 (2005)
86. Y.X. Chen, L.J. Campbell, W.L. Zhou, Self-catalytic branch growth of SnO_2 nanowire junctions. J. Cryst. Growth **270**, 505–510 (2004)
87. D. Calestani, M. Zha, G. Salviati, L. Lazzarini, L. Zanotti, E. Comini, G. Sberveglieri, Nucleation and growth of SnO_2 nanowires. J. Cryst. Growth **275**, 2083–2087 (2005)
88. M.R. Yang, S.Y. Chu, R.C. Chang, Synthesis and study of the SnO_2 nanowires growth. Sensor Actuat B **122**, 269–273 (2007)
89. J.K. Jian, X.L. Chen, W.J. Wang, L. Dai, Y.P. Xu, Growth and morphologies of large-scale SnO_2 nanowires, nanobelts and nanodendrites. Appl. Phys. **76**, 291–294 (2003)
90. D.F. Zhang, L.D. Sun, G. Xu, C.H. Yan, Sizecontrollable one-dimensional SnO_2 nanocrystals: Synthesis, growth mechanism and gas sensing property. Phys. Chem. Chem. Phys. **8**, 4874–4880 (2006)
91. Y.J. Chen, X.Y. Xue, Y.G. Wang, T.H. Wang, Synthesis and ethanol sensing characteristics of single crystalline SnO_2 nanorods. Appl. Phys. Lett. **87**, 233503–3 (2005)
92. S. Kumar, S. Rajaraman, R.A. Gerhardt, Z.L. Wang, P.J. Hesketh, Tin oxide nanosensor fabrication using AC dielectrophoretic manipulation of nanobelts. Electrochim. Acta **51**, 943–951 (2005)
93. A. Kolmakov, D.O. Klenov, Y. Lilach, S. Stemmer, M. Moskovits, Enhanced gas sensing by individual SnO_2 nanowires and nanobelts functionalized with Pd catalyst particles. Nano Lett. **5**, 667–673 (2005)
94. Q. Wan, T.H. Wang, Single-crystalline Sb-doped SnO_2 nanowires: synthesis and gas sensor application. Chem. Commun. **14**, 3841–3843 (2005)
95. N.S. Ramgir, I.S. Mull, K.P. Vijayamohanan, A room temperature nitric oxide sensor actualized from rudoped SnO_2 nanowires. Sensor Actuat B **107**, 708–715 (2005)
96. L.H. Qian, K. Wang, Y. Li, H.T. Fang, Q.H. Lu, X.L. Ma, CO sensor based on Au-decorated SnO_2 nanobelt. Mater. Chem. Phys. **100**, 82–84 (2006)
97. E. Comini, G. Faglia, G. Sberveglieri, D. Calestani, L. Zanotti, M. Zha, Tin oxide nanobelts electrical and sensing properties. Sensor Actuat B **111–112**, 2–6 (2005)
98. Y.J. Choi, I.S. Hwang, J.G. Park, K.J. Choi, J.H. Park, J.H. Lee, Novel fabrication of an SnO_2 nanowire gas sensor with high sensitivity. Nanotechnology **19**, 095508–4 (2008)
99. N. Van Hieu, N. Duc Chien, Highly reproducible synthesis of very large-scale tin oxide nanowires used for screen-printed gas sensor, in *Proceedings of Eurosensors*, Dresden, Sept 2008, pp. 1270–1273
100. A. Ponzoni, C. Baratto, S. Bianchi, E. Comini, M. Ferroni, M. Pardo, M. Vezzoli, A. Vomiero, G. Faglia, G. Sberveglieri, Metal oxide nanowire and thin-film-based gas sensors for chemical warfare simulants detection. IEEE Sens. J. **8**(6), 735–742 (2008)
101. G. Sberveglieri, C. Baratto, E. Comini, G. Faglia, M. Ferroni, A. Vomiero, Single crystalline metal oxide nano-wires/tubes: controlled growth for sensitive gas sensor devices, 2nd IEEE *International Conference on Nano/Micro Engineered and Molecular Systems*, Bangkok, Thailand. Jan 2007, pp. 227–229
102. S. Zhang, X. Xia, C. Xie, S. Cai, H. Li, D. Zeng, A method of feature extraction on recovery curves for fast recognition application with metal oxide gas sensor array. IEEE Sens. J. **9**(12), 1705–1710 (2009)
103. S. Bicelli, A. Depari, G. Faglia, A. Flammini, A. Fort, M. Mugnaini, A. Ponzoni, V. Vignoli, S. Rocchi, Model and experimental characterization of the dynamic behavior of low-power carbon monoxide MOX sensors operated with pulsed temperature profiles. IEEE Trans. Instrum. Meas. **58**(5), 1324–1332 (2009)
104. M. Messina, F. Franze, N. Speciale, E. Cozzani, A. Roncaglia, Thermofluid analysis of ultra low power hotplates for a MOX gas sensing device. IEEE Sens. J. **9**(5), 504–511 (2009)
105. A. Hackner, A. Habauzit, G. Muller, E. Comini, G. Faglia, G. Sberveglieri, Surface ionization gas detection on platinum and metal oxide surfaces. IEEE Sens. J. **9**(12), 1727–1733 (2009)

106. J. Courbat, D. Briand, L. Yue, S. Raible, N.F. De Rooij, Ultra-low power metal-oxide gas sensor on plastic foil, *International Conference on Solid-State Sensors, Actuators and Microsystems, Transducers*, Denver, June 2009, pp. 584–587
107. S. Bicelli, A. Depari, G. Faglia, A. Flammini, A. Fort, M. Mugnaini, A. Ponzoni, V. Vignoli, Model and experimental characterization of dynamic behaviour of low power Carbon Monoxide MOX sensors with pulsed temperature profile, in *Proceedings of IEEE Instrumentation and Measurement Technology Conference* - IMTC, Victoria, May 2008, pp. 1413–1418
108. I. Elmi, S. Zampolli, E. Cozzani, M. Passini, G. Pizzochero, G. C. Cardinali, M. Severi, Ultra low power MOX sensors with ppb-level VOC detection capabilities, *Proceedings of IEEE Sensors*, Oct 2007, pp. 170–173
109. S. Bicelli, A. Flammini, A. Depari, D. Marioli, A. Ponzoni, G. Sberveglieri, A. Taroni, Low-power carbon monoxide MOX sensors for wireless distributed sensor networks, in *Proceedings of IEEE Instrumentation and Measurement Technology Conference*, May 2007, pp. 1–5
110. J. Frank, M. Fleischer, H. Meixner, Gas-sensitive electrical properties of pure and doped semiconducting Ga_2O_3 thick films. Sensor Actuat B **48**(1–3), 318–321 (1998)
111. K. Sahner, M. Fleischer, E. Magori, H. Meixner, J. Deerberg, R. Moos, HC-sensor for exhaust gases based on semiconducting doped $SrTiO_3$ for on-board diagnosis. Sensor Actuat B **114**(2), 861–868 (2006)
112. D. Biskupski, K. Wiesner, R. Moos, M. Fleischer, Automotive exhaust gas sensor based on a combination of electrochemical pumping cell and resistive gas sensor, *Proceedings of Eurosensors*, Dresden, Sept 2008, pp. 1288–1289
113. Internet resource: http://www.sensorsmag.com/sensors/humidity-moisture/choosing-a-humidity-sensor-a-review-three-technologies-840
114. A. Oprea, N. Barsan, U. Weimar, M. L. Bauersfeld, D. Ebling, Capacitive humidity sensors on flexible RFID labels, *Solid-State Sensors, Actuators and Microsystems International Conference*, Transducers, Lyon, 2007, pp. 2039–2042
115. M. Hernaez, C.R. Zamarreño, I. Del Villar, F.J. Arregui, I.R. Matias, Optical fiber humidity sensor based on lossy mode resonances. Int. J. Smart Sens. Intell. Syst. **2**(4), 653–660 (2009)
116. Z. Chen, C. Lu, Humidity sensors: A review of materials and mechanisms. Sensor Lett. **3**, 274–295 (2005)
117. C.Y. Lee, G.B. Lee, Humidity sensors: A review. Sensor Lett. **3**, 1–15 (2005)
118. Internet resource: http://www.sensorland.com/HowPage047.html
119. Internet resource: http://www.sensirion.com/en/01_humidity_sensors/00_humidity_sensors.htm
120. M. Havelková, T. Randák, V. Žlábek, J. Krijt, H. Kroupová, J. Pulkrabová, Z. Svobodová, Biochemical markers for assessing aquatic contamination. Sensors **7**, 2599–2611 (2007)
121. M. Strlič, I.K. Cigić, J. Kolar, G. de Bruin, B. Pihlar, Non-destructive evaluation of historical paper based on pH estimation from VOC emissions. Sensors **7**, 3136–3145

Chapter 2
Resistive, Capacitive and Temperature Sensor Interfacing Overview

In this chapter we consider the most important sensor typologies, according to their electrical behaviour, describing the fundamentals of electronic interfaces which are essential components in sensor systems for the detection and the quantification of a physical or chemical measurand. In particular, we will describe, in a deeper detail, some kinds of sensors presented in Chapter 1, together with other sensors, in terms of their characteristic electrical parameters and responses. Moreover, we will give some generalities on the main measurement techniques and describe the simplest analog electronic read-out circuits for the interfacing of resistive, capacitive and temperature sensors.

2.1 Resistive Sensors

Resistive sensors convert the variation of a non-electrical phenomenon (physical or chemical) into a variation of a resistance. This effect can be evidenced in a number of physical and chemical events that the sensor can reveal. For an example, variations of the environmental conditions in proximity of the sensor can modify their conductivity [1, 2]. In this sense, in the scientific world, several resistive chemical sensors, in particular gas sensors, have been already developed, also in complete systems, for environmental monitoring. Unfortunately, many of them are characterized by large dimensions, high power consumption and high costs and their interfaces are not always optimized for the specific sensor so, recently, micromachined resistive gas sensors have been developed, also with very small sizes; they reach the operating temperature (about 300–400°C) in few tens of milliseconds, exploiting a heater/thermometer embedded in the sensor itself, so reducing power consumption [3, 4]. In array-based sensor systems, composed by different sensors sensing different measurands with different sensitivities and selectivities, feature extraction techniques can be also used to pull out information also from the transient sensor response [5].

A. De Marcellis and G. Ferri, *Analog Circuits and Systems for Voltage-Mode and Current-Mode Sensor Interfacing Applications*, Analog Circuits and Signal Processing, DOI 10.1007/978-90-481-9828-3_2, © Springer Science+Business Media B.V. 2011

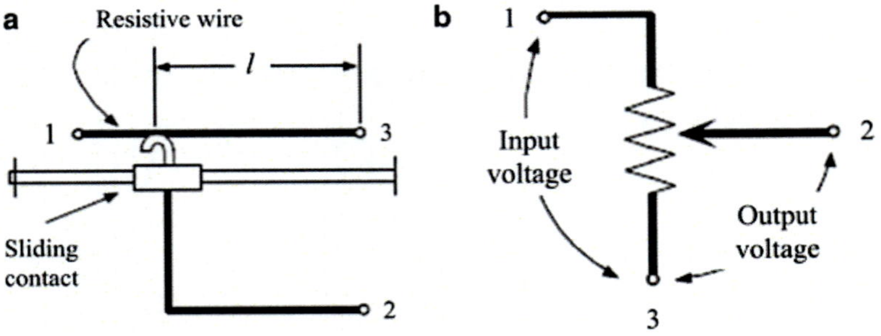

Fig. 2.1 A basic implementation of a potentiometric resistive sensor based on wire: (**a**) block scheme; (**b**) its equivalent circuit

In general, in the resistive sensing mechanisms, the measurand directly or indirectly changes the electrical characteristic (i.e., a resistance) of the sensing element, through the variation of either its material or geometry. In particular, it is well known that for a simple uniform conductor (of uniform area), the resistance can be expressed by the following Eq. 2.1:

$$R = \frac{\rho l}{A} [\Omega] \tag{2.1}$$

being ρ the resistivity, l the length and A the constant cross-sectional area through which the current flows. For typical conductors, the resistivity values in units of $\Omega \cdot \mathrm{mm}^2/\mathrm{m}$ are: Aluminium = 0.0278, pure Iron = 0.1, Constantan = 0.48, Copper = 0.0172, Gold = 0.0222, Tungsten = 0.059, Manganese = 0.423, Nickel = 0.087. In this case, resistance R is varied either by a geometric (A, l) or a material change (ρ) in the resistive element and can be measured either directly (e.g., an ohm-meter) or through a suitable signal conditioning circuit (e.g., a simple voltage divider). A kind of resistive sensor which exploits these effects is the so-called potentiometric resistive sensor. Fig. 2.1 shows an example of a basic implementation of this device together with its equivalent circuit. Other examples of potentiometric resistive sensors are reported in Fig. 2.2 (linear) and Fig. 2.3 (angular). In particular, that shown in Fig. 2.3b depicts an angular potentiometric resistive sensor whose typical electrical output response, as a function of the input mechanical parameter, is almost linear. Based on different transduction mechanisms, other resistive sensors which can be considered are the thermistors (temperature-sensitive semiconductor devices) and the light-dependent resistors (or photo-resistors), which react to the presence of the light. Moreover, resistive sensors can be employed for the level detection of liquid substances, as shown in the example reported in Fig. 2.4 [6].

A different transduction mechanism, which can be exploited in the implementation of resistive sensors, is represented by the piezoresistive effect. Piezoresistivity is a linear coupling between mechanical stress and electrical resistivity. Piezoresistance measurements can provide valuable insights concerning the conduction

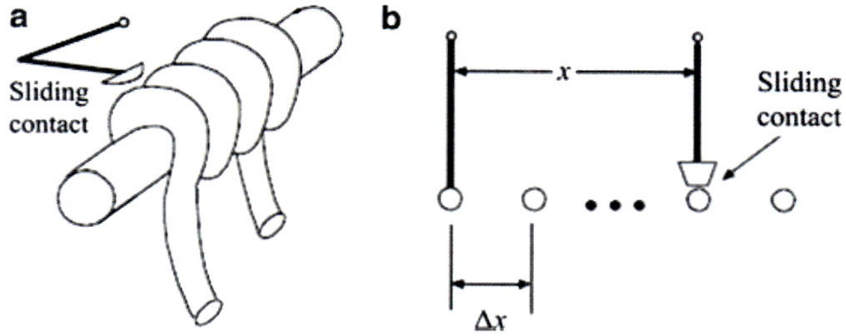

Fig. 2.2 Example of a linear potentiometric resistive sensor: (**a**) block scheme; (**b**) its operating principle (Δx defines the sensor resolution)

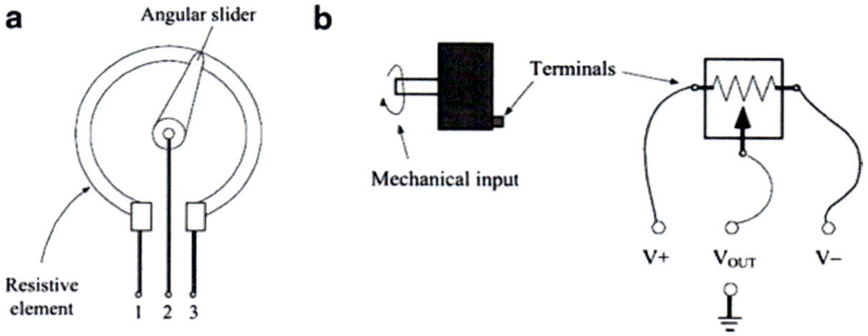

Fig. 2.3 Example of an angular potentiometric resistive sensor: (**a**) block scheme, where the resistive (or conductive) element can be wire-wound, cermet, conductive plastic, etc.; (**b**) its basic schematic representation

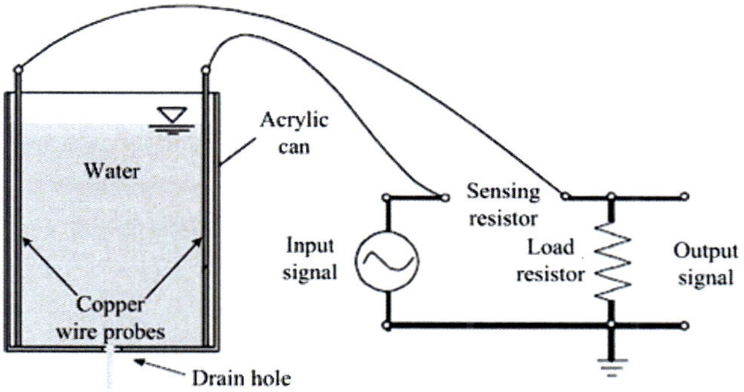

Fig. 2.4 An example of a resistive sensor for liquid level detection

mechanisms in solids, as barrier tunnelling in thick film resistors. Piezoresistive sensors based on nanostructured thin metal layer sputtered on elastic non-conductive polymer film have been recently proposed in the literature [7, 8]. Piezoresistivity has also been investigated in compound semiconductors, thin metal films,

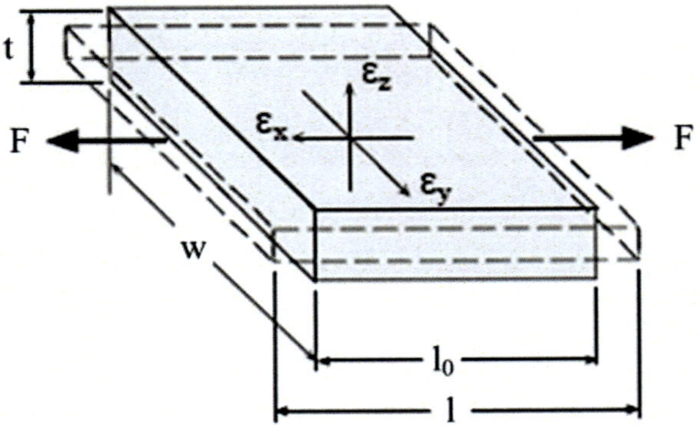

Fig. 2.5 Schematic block of piezoresistive effect

polycrystalline silicon and germanium thin films, heterogeneous solids and high Curie temperature superconductors. Several commercially available sensors (for pressure, acceleration, vibration, etc.) have been fabricated from piezoresistive materials. In order to design a highly accurate and sensitive sensor system, suitable signal conditioning electronic circuits must be provided both to compensate temperature drifts of the sensor offset and to increase its sensitivity and resolution.

More in detail, referring to Fig. 2.5, when the resistive material is elongated or compressed due to a mechanical input, the changes in the electrical conductive characteristics represent the piezoresistive effect (see also Chap. 1).

It has been shown that the resistance of copper and iron wire changes once the wires have been subjected to mechanical strain [9]. Regarding the piezoresistivity and more in general, we know that for a conductor of uniform area, the resistance is expressed by Eq. 2.2. Under strain, the change in R is related to the possible variation of the involved parameters (l, A and ρ) as follows:

$$dR = \frac{\partial R}{\partial l}dl + \frac{\partial R}{\partial A}dA + \frac{\partial R}{\partial \rho}d\rho, \tag{2.2}$$

which, from Eq. 2.1 becomes

$$dR = \frac{\rho}{A}dl - \frac{\rho l}{A^2}dA + \frac{l}{A}d\rho. \tag{2.3}$$

The fractional change of R is of more interest, so we find that:

$$\frac{dR}{R} = \frac{dl}{l} - \frac{dA}{A} + \frac{d\rho}{\rho}, \tag{2.4}$$

being dl/l the fractional change in length, dA/A that in area and $d\rho/\rho$ that in resistivity. The latter variation ($d\rho/\rho$) corresponds to the piezoresistive effect.

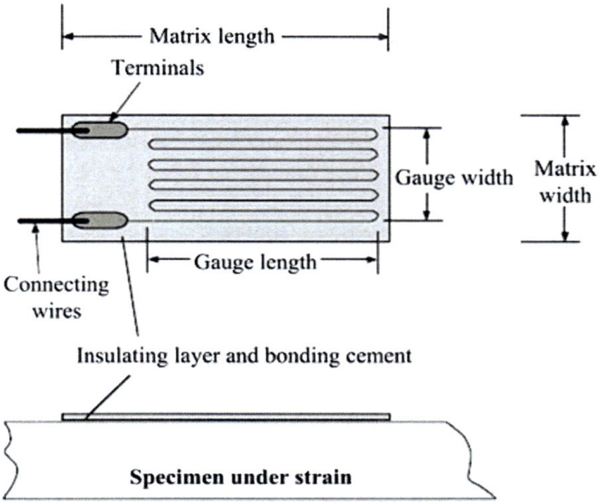

Fig. 2.6 Resistive strain gauge

Therefore, a strain of the sensing element involves both geometric (l and A) and material (ρ) characteristic variations corresponding to a resistance change. Thus, we can exploit resistance changes as a sensing mechanism and either material or geometry changes represent a physical sensing.

As direct consequence of this theory, the strain gauge, shown in Fig. 2.6, can be considered another kind of piezoresistive sensor. A measure of the sensitivity of a resistive strain gauge is given by the gauge factor G, which is defined as the ratio between fractional change in resistance and fractional change in strain. Typical values of G are: 80% Ni + 20% Cr, $G = 2$; 45% Ni + 55% Cu, $G = 2$; 100% Pt, $G = 4.8$; 95% Pt + 5%Ir, $G = 5.1$. In the semiconductors, G is the highest value since it typically ranges from 70 to 135. However, they show the following disadvantages: output not linear with strain, high dependence on temperature, lower strain limits and higher costs than metallic type, etc. [6].

However, nowadays, the most diffused and utilized resistive sensors are the chemoresistive devices, formed by *MOX* thin films for gas detection (see also Chap. 1). These sensors behave electrically as simple resistors, since they are based on direct analyte reaction and charge process between the gas molecules and the *MOX* surface, which cause an electrical resistance variation of the gas-sensing element. In particular, *MOX* sensors are constituted by sensitive materials which change their resistance as a consequence of physisorption, chemisorption and/or catalytic reactions of the particular measuring reagent gas and the surface of the sensing material [10–15]. Therefore, electrical conductivity of semiconducting metal-oxides can change under the influence of target gases, so this kind of sensor is defined as resistive gas sensor. In fact, according to their nature, n-type materials like SnO_2 and WO_3 increase their resistance when interacting with oxidizing gases, such as NO_2 and O_3, whereas p-type materials like NiO and CoO respond inversely.

Considering the state of the art of the manufacturing, the sensor resistance value may also vary across several decades, being the combination of three variable components: the nominal baseline, the deviation from this nominal baseline (due to ageing and working temperature) and the resistance variation due to gas concentration. Since each contribution is of the order of one to two decades, a wide input range is demanded. The sensor resistance measurement accuracy required to read-out gas concentrations in the order of tens of parts per million, or lower, is in the order of few percent and, of course, depends on the sensor resistance variation due to gas concentration. This feature involves, as a consequence, a suitable design of the first analog interface [16]. Typically, *MOX* sensors are used over a temperature range from 200°C to 450°C while the base line resistance (e.g., the resistance in dry air) may typically range between hundreds of kΩ up to tens of GΩ, according to the operating temperature, the intrinsic resistivity of the materials and the preparation conditions. In the literature, different sensitive materials belonging to transition metal-oxide like WO_3, NiO and CoO have been prepared by physical thermal evaporation and sol-gel synthesis in thin film form on Si/Si_3N_4 substrates provided with platinum finger type electrodes. The preparation conditions can be selected so to have different film resistances ranging from 100 kΩ up to 10GΩ.

As an example of commercial resistive gas sensor, we mention the Japanese company Figaro Engineering that represents a leader in the worldwide market for the production of solid-state gas sensors based on the SnO_2 technology. These sensors are characterized by the TGS abbreviation: Taguchi Gas Sensor. They are typically utilized for the methane gas monitoring in the domestic ambient or as sensors for the air quality detection in the motor vehicles. The more utilized TGS sensors are fabricated so to have a pellet shape, produced through SnO_2 dusts which are pressed at high temperatures (the process is called "syntherization"). They are produced on a cylindrical substrate having the sensor contacts mounted on its external surface, while into the internal volume there is the heating filament, as shown in Fig. 2.7 [17]. Actually, the Figaro Engineering produces different kind of sensors developed through the newest microfabrication technologies, such as thin film and thick film sensors. The typical responses of a TGS sensor (e.g., TGS826 model) have been reported in Fig. 2.8, where sensitivity characteristics of the same sensor, provided by the producer, are depicted. More in detail, the sensing element of this sensor is a *MOX* semiconductor which has low conductivity in clean air. In the presence of a detectable gas, in particular ammonia, the sensor conductivity increases depending

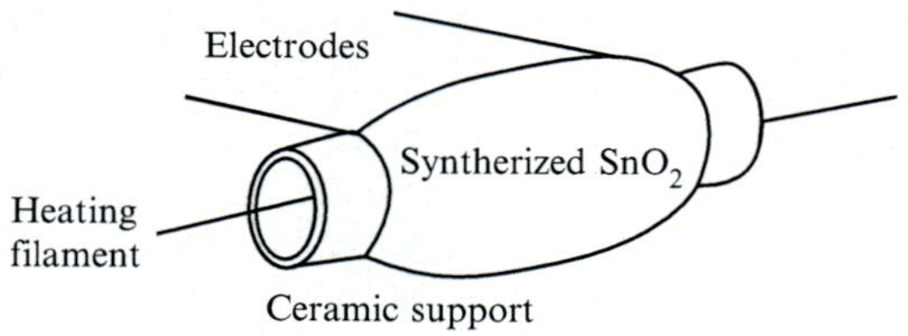

Fig. 2.7 Examples of a Figaro gas sensor: constructive scheme of a TGS "syntherized" sensor

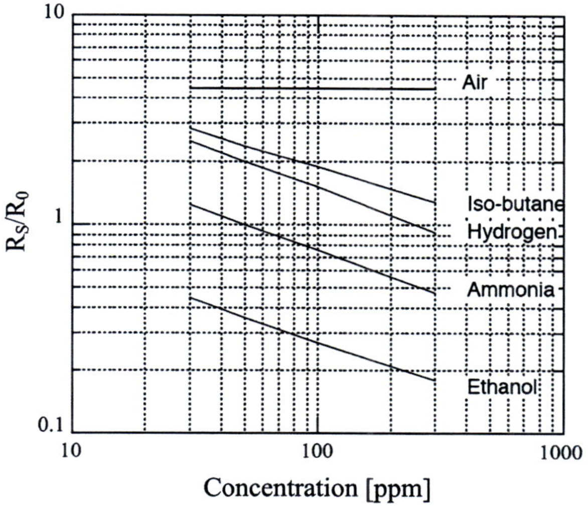

Fig. 2.8 Response curves of the TGS826: sensitivity characteristics to the exposure at Iso-Butane, Hydrogen, Ammonia and Ethanol, for different gas concentrations

on the gas concentration in the air. The sensor response assumes a behaviour similar to the law of the power: a linear trend for a double logarithmic scale. The sensor characteristic is expressed as the ratio between the sensor resistance, as a function of the gas concentration, and a reference resistance value, corresponding to the resistance shown by the sensor when it is exposed at 50 ppm of ammonia (this is the sensor baseline value R_0 which is about 20 kΩ, at 20°C and 65% *RH*) [18].

Starting from the sensor characteristics of Fig. 2.8, it is possible to calculate the analytical behavior of the sensor resistance R_S, with respect to a specific gas concentration. In fact, considering the two logarithmic scales, the relationship between the ratio R_S/R_0 and the concentration c is given by:

$$\log \frac{R_S}{R_0} = \log K + n \cdot \log c \tag{2.5}$$

being *log(K)* the curve intercept and n its angular coefficient. By choosing two different points in the curves, it is possible to easily calculate *log(K)* and n values. In this way, the relationship between the sensor resistance *Rs* and the gas concentration c can be expressed by the following simple equation:

$$R_S = R_0 \cdot K \cdot c^n. \tag{2.6}$$

In recent years, there have been extensive efforts in the synthesis, characterization and application of a new generation of semiconductor metal-oxide (*SMOX*) nanostructures such as nanowires, nanorods, nanotubes, in particular those carbon-based. They have been proposed in literature as novel materials for resistive gas sensing applications [3, 19]. In fact, under certain preparation conditions, these materials turn from metallic to semiconductor responses, where the electrical resistance in dry air may range from about tens of kΩ up to hundreds of GΩ. Moreover, such structures with a high aspect ratio (i.e., size confinement in two coordinates) offer better crystallinity, higher integration density and lower power consumption, demonstrating superior sensitivity to surface chemical processes due to the large surface-to-volume ratio and small diameter comparable to the Debye length (a measure of the field penetration into the bulk). Concerning these *SMOX* nanostructures, as an example, the current–voltage (I-V) characteristic curves, in air at different temperatures, of SnO_2 resistive gas sensors based on nanowires show a good ohmic behaviour. In particular, the sensor resistance value varies from a few MΩ up to hundreds of MΩ. This points out that not only metal–semiconductor junction between the *Au* contact layer and SnO_2 nanowires but also the semiconductor–semiconductor junction between the SnO_2 nanowires are ohmic. The ohmic behaviour is very important for the sensing properties because the sensitivity of the gas sensor device is affected by contact resistances. The I-V characteristics at different temperatures (generally up to 400°C) show that there are no differences in the I-V curves, suggesting a good reliability of the resistive gas sensors. Therefore, single-crystalline SnO_2 nanowires, fabricated on silicon and alumina substrates, can be used as resistive gas sensor devices, considering that their sensitivity and selectivity can be improved further also by surface catalytic doping or plasma treatment.

As simple and very advantageous resistive gas sensor commercial application, it is possible to mention the quick and accurate identification of food freshness and off-flavours to both winemakers and wine merchants, where the use of electronic noses (a combination of sensor arrays and pattern recognition methods) has revealed to be better than more complicated and expensive traditional methods, as gas chromatography and mass spectroscopy [20, 21].

Moreover, we want to mention resistive humidity sensors that measure the change in electrical impedance of a hygroscopic material such as a conductive polymer, salt or treated substrate. Typically, they are based on an interdigitated or bifilar winding. After deposition of a hydroscopic polymer coating, their resistance changes inversely with humidity. The sensor impedance change is typically an inverse exponential relationship to humidity, as shown in Fig. 2.9; however, this non-linear response can be linearized by a suitable signal conditioner.

Resistive humidity sensors usually consist of noble metal electrodes either deposited on a substrate by photoresist techniques or wire-wound electrodes on a plastic or glass cylinder. The substrate is coated with a salt or a conductive polymer. When it is dissolved or suspended in a liquid binder, it represents a vehicle for the sensor coating. Alternatively, the substrate may be treated with activating chemicals such as acids. The sensor absorbs the water vapour and ionic functional groups are dissociated, resulting in an increase of electrical conductivity. The response time for most resistive humidity sensors ranges from 10 to 30 s for about 60%

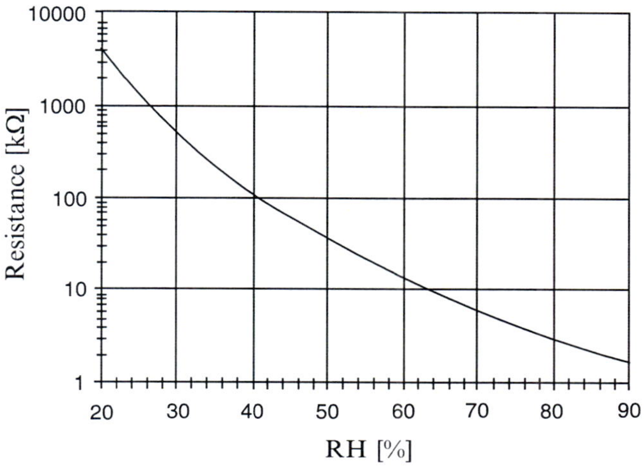

Fig. 2.9 The typical response of a resistive humidity sensor, at 25°C

step change, while the impedance range of typical resistive elements varies from about 1 kΩ to 10MΩ, considering that their nominal operating temperature ranges from about −40°C to 100°C. Most resistive humidity sensors are excited through a symmetrical AC voltage with no DC bias to prevent material polarization; the typical nominal excitation frequency ranges from about 30 Hz to 10 kHz. The resulting current flow is converted and rectified to a DC voltage signal for additional scaling, amplification, linearization or A/D conversion. Unfortunately, the resistive humidity sensor shows parasitic capacitive effects. However, an advantage of resistive *RH* sensors is their interchangeability, usually within ±2% *RH*, which allows the electronic signal conditioning circuitry to be calibrated by a resistor at a fixed *RH* point. This eliminates the need for humidity calibration standards, so they are generally field replaceable. Resistive humidity sensors have significant temperature dependencies when installed in an environment with large (>10°) temperature fluctuations. In this case, simultaneous temperature compensation have to be incorporated. Nevertheless, the small size, low cost, interchangeability and long-term stability make these resistive *RH* sensors suitable for their use in control and display products for industrial, commercial and residential applications [22].

As a final remark, we want also to underline that the mentioned sensors are, generally, not purely resistive. In this sense, the AC impedance spectroscopy is a method that provides knowledge on the different s ensor part contributions (surface, bulk, contacts, etc.). Therefore, more investigations on resistive sensors have to be performed so to attribute the different elements of the equivalent circuit to sensing resistive layer components and, consequently, to develop a suitable interface circuit.

2.2 Capacitive Sensors

Capacitive sensors, through a suitable transduction element, generally convert a physical parameter (or the change of its value) into a capacitance C (or into its variation ΔC). A basic capacitor can be constituted by two tiled metal plates, separated by a dielectric material. This structure provides a capacitance whose value can be expressed by the following well-known equation:

$$C = \varepsilon \frac{S}{d} [F] \tag{2.7}$$

being ε the dielectrical constant, S the metal plat area and d the distance between the two metal plates. Therefore, as shown in Fig. 2.10, either a variation of the distance d between the two metal plats (electrodes) of the capacitor (or the variation of their overlapping area), due to the movement of at least one of them, or a variation of the dielectric material (i.e., its permittivity), produces a capacitance variation. If we consider that the capacitance C can be a sensor, it is possible to evaluate the effect of the physical (or chemical) phenomenon which occurs, just revealing its variations through suitable electronic circuits. Generally speaking, the value of a capacitor based on plane and in parallel faces can be utilized as a transduction element of the relative position between two faces (or plates). Considering a structure having plane and in-parallel faces, there are two independent modalities to measure the displacement related either to the lateral movement of the electrodes or to their vertical separation, as shown in Fig. 2.11 and Fig. 2.12, respectively [17].

In the case shown in Fig. 2.11, the lateral displacement x of an electrode, with respect to the other, determines a capacitance value, caused by the capacitor area variation, as follows:

$$C = \varepsilon \cdot \varepsilon_0 \cdot \frac{W}{d} \cdot (L - x). \tag{2.8}$$

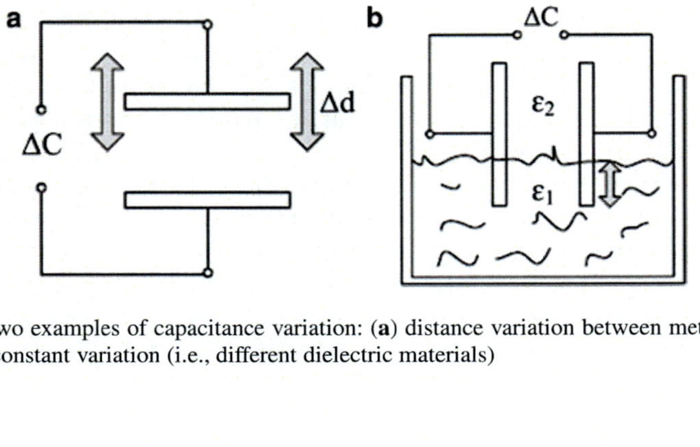

Fig. 2.10 Two examples of capacitance variation: (**a**) distance variation between metal plats; (**b**) dielectrical constant variation (i.e., different dielectric materials)

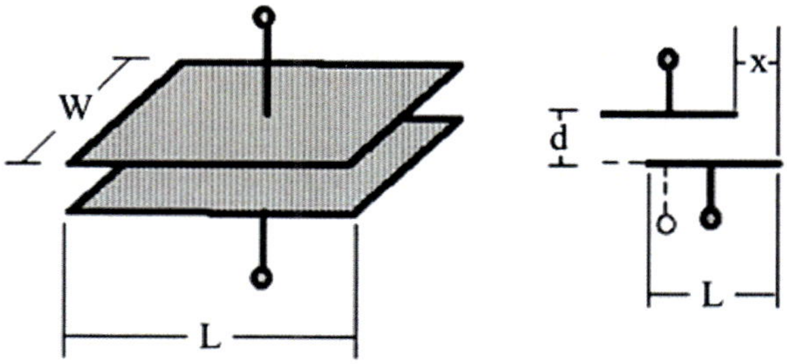

Fig. 2.11 Example of electrodes lateral displacement

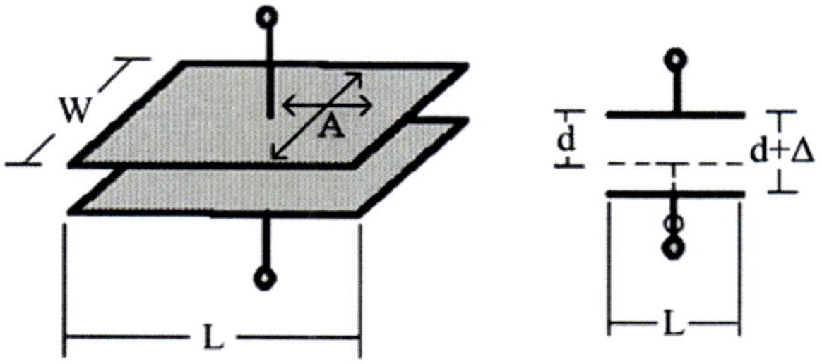

Fig. 2.12 Example of electrodes vertical displacement

In this case, for simplicity, it has been considered that only the area corresponding to the overlapped electrodes determines the capacitance (obviously, this hypothesis is not completely correct, therefore, Eq. 2.8 can be considered an approximation).

On the other hand, referring to Fig. 2.12, the measurement approach has to consider the distance variation between the two electrodes of a capacitor. Therefore, as long as the distance between the electrodes varies of a certain quantity guaranteeing that all the force lines are contained within the same electrodes, the capacitance as a function of the electrodes distance can be expressed as follows:

$$C = \varepsilon \cdot \varepsilon_0 \cdot \frac{W \cdot L}{d + \Delta}. \tag{2.9}$$

This last approach, even if shows a non-linear relation, is, generally, the more utilized in practice since in the first case (lateral displacement) the relative analytical expression, given by Eq. 2.8, has a very limited validity range. In order to evaluate the capacitive transductor linearity, it is opportune to develop Eq. 2.9 as a Taylor series with respect to d_0 (rest value), considering d the electrode distance and $\Delta = d - d_0$ the amount of the electrode displacement. Through a simple calculation, if $\Delta \ll d$, the resulting relation can be considered linear. As an example, referring to the micromechanical applications, d is in the order of few μm, so, maintaining the linearity characteristic, it is possible to measure displacement variations in the order of about 1% of distance d which corresponds to tens of nm.

An alternative measurement method considers the differential configuration, based on the use of two capacitors. It allows both to increase the sensitivity of the device and, in particular, to extend the linearity range of the same capacitive sensor.

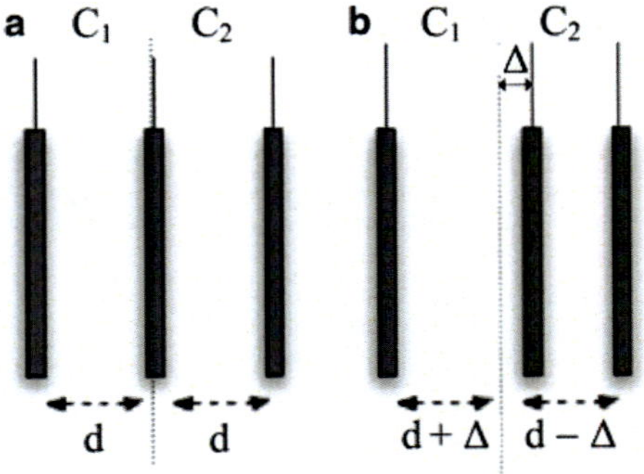

Fig. 2.13 Example of a differential capacitive sensor as a position transductor: (**a**) the equilibrium condition ($\Delta = 0$); (**b**) the Δ displacement effect with respect to the equilibrium position

Fig. 2.13 shows an example of a differential capacitance configuration where the central electrode is shared by the two capacitors C_1 and C_2 [17].

In this case, a displacement of the central electrode corresponds to a distance increase between the electrodes of one of the two capacitors (corresponding to a capacitance reduction) and, for the other capacitor, to a distance decrease (of the same quantity) between the electrodes (corresponding to a capacitance enhancement). In terms of linearity, after a Taylor series development, if it is considered the difference between the two capacitances as the output parameter coming from capacitive sensor, we obtain:

$$\Delta C = C_2 - C_1 = \varepsilon \cdot \varepsilon_0 \cdot \frac{A}{d - \Delta} - \varepsilon \cdot \varepsilon_0 \cdot \frac{A}{d + \Delta} \cong \varepsilon \cdot \varepsilon_0 \cdot \frac{A}{d} \cdot \frac{2\Delta}{d}. \tag{2.10}$$

It is important to highlight that, in this case, the sensitivity of the linear approximation expressed in Eq. 2.10 results to be doubled with respect to the case related to only one capacitor. The differential sensor structure is widely utilized in integrated microsensors that employ also Wheatstone bridge configuration circuits. In addition, other capacitive sensors (also with differential structure), having the advantages of low temperature dependence, large dynamic range, simple structure and low power consumption characteristics, have been proposed in the literature [23, 24].

A capacitive sensor probe is, for example, based on a homogeneous parallel-plate capacitor configuration and is suitable for mounting on the flange of a pipe adapter in process automation applications. In particular, it can reveal the quantity, level or kind of liquid substance which flows through the pair of parallel plates. The dielectric properties of the substance influence the relative permittivity between the plates, resulting in a change of the impedance of the same probe [25].

Another kind of capacitive sensor is the chemical sensor which detects the changes in the dielectric properties of the sensing polymeric layer due to absorption of Volatile Organic Compounds (*VOCs*). Such polymer-based chemocapacitive sensors are promising devices in terms of processability, low fabrication cost, reversibility and the wide range of material choice, commercially available, that meets the needs of specific *VOC*-based applications [26].

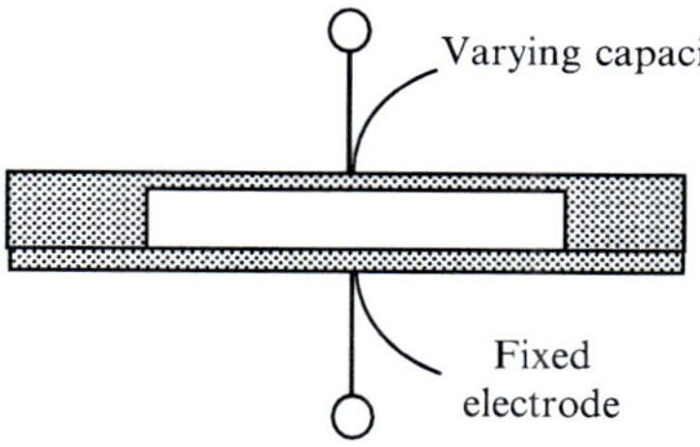

Fig. 2.14 Example of a pressure sensor with silicon-based diaphragm (or membrane): capacitive transductor based on a deflecting membrane (the varying capacitance is due to upper diaphragm motion)

Other capacitive sensors measure the pressure, whose value is one of the most important physical parameters in industry manufacturing, automobile sector, aerospace project, military hardware, consumption electronic, medical application, etc.. One of the most important characteristic, in pressure sensors, is the linearity. These sensors can be used to measure various real-world phenomena like flow, fluid level and acoustic intensities, in addition to pressure. In this area, the most part of researches is focused on piezoresistive or capacitive pressure sensor. The piezoresistive type has a linear sensitivity but the output signal is affected by the temperature and also shows a higher power consumption. On the contrary, the capacitive type is not affected by the temperature and also save the power consumption, but the capacitor variation versus pressure change value typically is not a linear relation [27, 28]. Generally, the pressure is detected through mechanical devices: the sensor mobile element is affected by a displacement due to the force corresponding to the applied pressure which is not compensated by any other force on the opposite surface. Modern pressure capacitive sensors utilize diaphragms based on either silicon or ceramic mounted without any initial strain. The transduction mechanism happens, for an example, through four strain gauges which are orthogonally mounted in the basic structure and can be easily employed in a full-bridge configuration circuit topology. In fact, a possible approach which can be considered for the measurement of the diaphragm deflection is the use of strain gauges. In this case, the maximum stress can be achieved at the diaphragm edge which, therefore, results to be the more appropriate area for the application of strain gauges. In the hypothesis of diaphragm deflection lower than its thickness, the system shows linearity characteristics, so it is possible to easily calculate the electrical signal produced by a capacitive transductor where the sensing diaphragm is an electrode of a plane capacitor, as shown in Fig. 2.14. In this case, for an example, assuming that the diaphragm radius is 1 cm and the membrane distance (distance between the capacitor electrodes) is 50 μm, the initial capacitance value, without any deflection, is about 50 pF [17].

In order to minimize pressure sensor dimensions, the Micro-Electronic-Mechanical-System (*MEMS*) and *CMOS* technology can be combined together to fabricate a novel microsensor that shows also the following advantages: low cost, small area, higher circuit density, lower parasitic effects and fewer I/O pads

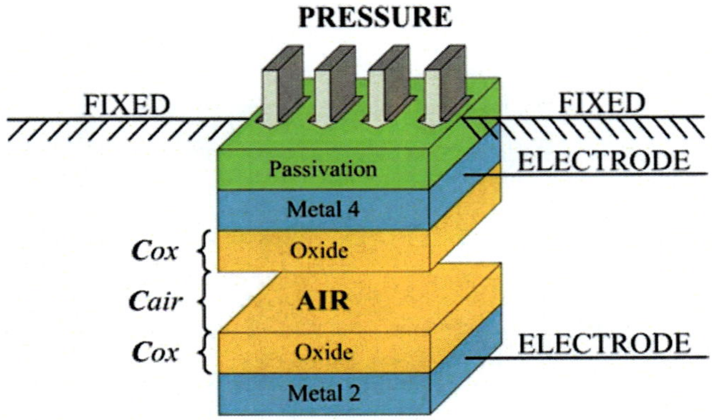

Fig. 2.15 Simplified block scheme of a capacitive pressure sensor based on *CMOS-MEMS* sensing cell

[29, 31, 32]. In this sense, the pressure sensor may utilize different sensing thin films, each of them having the same area. Fig. 2.15 shows the simplified scheme of a *CMOS-MEMS* sensing cell, developed as a capacitive pressure sensor, where, considering the fabrication process for this kind of device, the top and down electrodes of the same sensing cell are covered with oxide layer, air gap and oxide layer again [31, 32]. When top electrode is without any pressure, an initial capacitance is achieved, typically in the order of hundreds of fF. The electrode deformation provides a capacitance variation which can reach also some pF, corresponding to an applied pressure of about a few MPa.

Capacitive sensors can be also used to detect and measure material strains. This is required in many industrial, aerospace and civil applications where the monitoring of an engineering structure health is crucial in maintaining its integrity and avoiding catastrophic structural failure. Measuring strain on selected places of a structure can give information about overall deformation and lead to early detection of potential damage. For these reasons, recently a number of different capacitive strain sensors has been designed [33–35]. They are characterized by temperature independence, low power consumption resulting suitable for wireless and battery-less sensing units. Capacitive strain sensors operate by measuring the capacitance change between two or more electrodes placed on an insulating substrate. In the literature, different interdigital capacitive strain gauges have been developed. These are planar devices based on a collection of interdigitated conductors (fingers) of alternate polarity. As the surface where the gauge is mounted deforms, the distance between the electrodes and their respective capacitance changes [36, 37].

Another important kind of capacitive sensor, utilized in several domestic, industrial and automotive applications, is represented by the accelerometer. This device detects an acceleration proportional to relative displacement. Fig. 2.16 shows a block scheme of an accelerometer with a linear transductor (*LVDT*) [17].

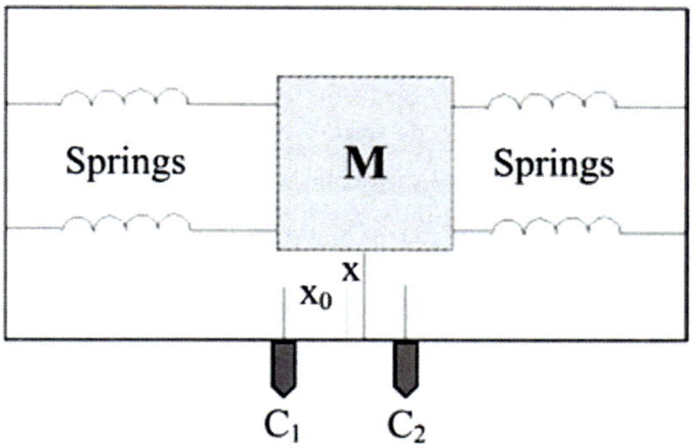

Fig. 2.16 Accelerometer block scheme with differential capacitive *LVDT* transductor

When an acceleration a_x on x axis occurs, the seismic mass M causes a displacement which gives a capacitance variation of C_1 and C_2, as expressed by the following relationship:

$$C_1 = \varepsilon \frac{S}{x_0 + x} = \frac{C_0}{1 + \delta}, \tag{2.11}$$

$$C_2 = \varepsilon \frac{S}{x_0 - x} = \frac{C_0}{1 - \delta}, \tag{2.12}$$

being $C_0 = \varepsilon \cdot S/x_0$ the capacitance value for null acceleration (initial value), $\delta = x/x_0$ the relative displacement of the central electrode connected to the seismic mass and S the electrodes area.

In recent years, Analog Devices has introduced on the commercial world an integrated accelerometer, named *ADXL50*, fabricated through the integrated silicon micromachining technology, which is widely employed as an acceleration sensor in car air-bags [38].

Other capacitive sensors are the so-called tilt sensors which are important in motion detection systems, especially in medical science and health care applications, such as surgical tools, scan and restoration, gait studies and functional electrical stimulation [39, 40]. In self-powered wearable sensor networks, ultra low power tilt sensors could be integrated with other motion detectors and chemical sensors, e.g. glucose or pH sensors, in a highly compact package, to measure physical and biochemical changes simultaneously. Tilt sensors utilize various sensing principles, such as piezoresistive and capacitive. These tilt capacitive sensors can be fabricated also on a silicon-on-isolator (*SOI*) wafer with double sided processing, having an output which periodically changes with respect to tilt angle. Generally, the sensor capacitance value is quite low, ranging from hundreds of fF up to a few pF and the resolution of a sensor could reach about $\pm 1°$. Moreover, these sensors show capability for systems driven by a limited power supply, so a suitable application would be that of wearable body sensor nodes. In fact, when other sensors, like heart pace detectors, are working, the information about the patient movement, such as sleeping, walking or suddenly falling over, are extremely important.

With the ever-increasing demand for miniaturization of electronic devices, the low dielectric constant of polymers may be a critical technological issue in terms of reliable capacitance measurements. A useful method considers the incorporation of high dielectric constant ferroelectric materials (e.g., the perovskite oxide $BaTiO_3$) in the polymer matrix. In addition, the humidity sensing properties of porous ceramic or nanocrystallized polymer, due to water-induced enhancement of its surface electrical conductivity or its dielectric constant, are well-known. Therefore, innovative chemocapacitive sensors, based on polymer layers filled with various amounts of ferroelectric material nanoparticles, have been proposed in the literature [26]. The changes in capacitance response under the presence of different vapour analytes and their mixtures has been studied so to evaluate the effect of incorporated nanoparticles on the sensitivity and selectivity of the pure polymer-based capacitive sensors. Typically, the incorporation of these nanoparticles in the sensing polymeric layer of chemocapacitive sensors results in an increased baseline capacitance value as well as an increased capacitance response ΔC upon vapour analytes exposure.

Other kinds of capacitive sensors are used to evaluate the relative humidity *RH*. They are largely used in industrial, commercial and weather telemetry applications and produced in a wide range of specifications, sizes and shapes including integrated monolithic electronics. These sensors consist of a substrate on which a thin film of polymer or *MOX* is deposited between two conductive electrodes. The sensing surface is coated with a porous metal electrode to protect it from contamination and exposure to condensation. The substrate is typically glass, ceramic or silicon. The incremental change in the dielectric constant of a capacitive humidity sensor is nearly directly proportional to the *RH* of the surrounding environment. The change in capacitance is typically 0.2–0.5 pF for a 1% *RH* change, while the sensor baseline capacitance (even if typically referred to the capacitance base value revealed at 0% *RH* and at room temperature) is between 100 and 500 pF at 50% *RH* at 25°C. Capacitive sensors are characterized by low temperature coefficient, good capability to work at high temperatures (up to 200°C), full recovery from condensation and reasonable endurance to chemical vapours. Generally, the response time of these sensors ranges from 30 to 60 s for about 60% *RH* step change. State-of-the-art techniques for producing capacitive sensors take advantage of many principles used in semiconductor manufacturing to yield sensors with minimal long-term drift and hysteresis. Thin film capacitive sensors may include also monolithic signal conditioning circuitry integrated onto the substrate, which incorporates a *CMOS* timer to pulse the sensor and to produce a near-linear voltage output, as shown in Fig. 2.17. The typical uncertainty of capacitive sensors is about few percents from 5% to 95% *RH* with a two-point calibration [22].

Furthermore, the sensor structure based on *CMOS* interdigitated electrodes (*IDEs*) in combination with a suitable sensing material (e.g., a polymer film on top of the electrodes) is already a well-known design for biochemical sensors [41, 42]

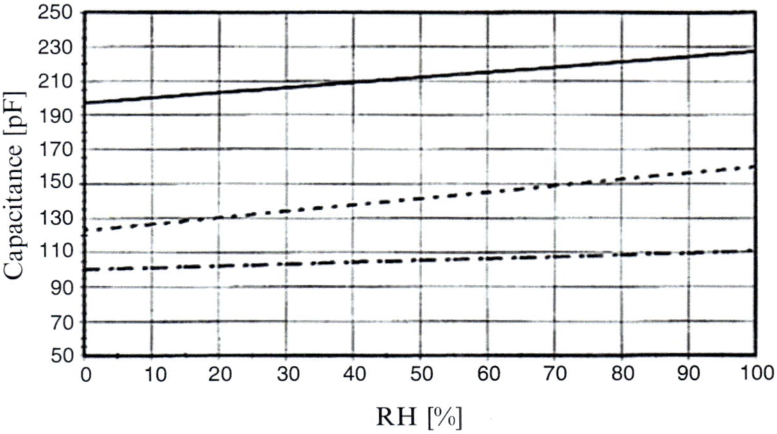

Fig. 2.17 A typical near-linear response of capacitance changes vs. applied RH, at 25°C

and can be easily micromanufactured as well as used for capacitive detection. Also in this case, capacitive sensing approach is leading towards the reduction of power consumption and the microfabrication for simple batch processing, miniaturization and low cost. An *IDE* sensor configuration with polyimide film [42,43] is predominantly used in commercial applications [44,45], and, in particular, the design, fabrication and characterization of a capacitive humidity sensor for very low power applications has been largely proposed in literature [46, 47]. This kind of capacitive sensor is based on *IDEs* covered with a humidity-sensitive polymer (polyimide) that absorbs moisture leading to changes of its dielectric properties. Electrical field lines between the electrodes pass through the polyimide layer so changes in polyimide permittivity lead to changes in sensor capacitance. Polyimide makes a suitable sensing layer due to a high water uptake and a high diffusion rate resulting in high sensitivity and short response time. These sensors show different working capacitances, depending on the size of the designed active areas, which can be about tens of pF and their variation can reach a few pF, around the fixed baseline, for an *RH* variation between 20% and 90%.

Finally, gyroscopes based on *MEMS* structures represent another kind of capacitive sensors that have been introduced into strategic application markets, such as automotive, defence, aviation and space industries as well as, recently, in electronic games. Most of them operate on the principle of detecting an induced Coriolis acceleration to the axis about which the input rotation is applied. Optical, tunnelling, piezoresistive and capacitive sensing mechanisms have been demonstrated to be able to estimate the Coriolis force and, hence, the rotation rate. Among them, capacitive sensing is widely employed because of the relatively easier fabrication, lower power consumption, higher stability and feasibility to realize mechanical feedback. More in detail, *MEMS* gyroscope consists of bar structure proof masses, which can work at atmospheric pressure. Usually, it has a resonance frequency of about 3–4 kHz and

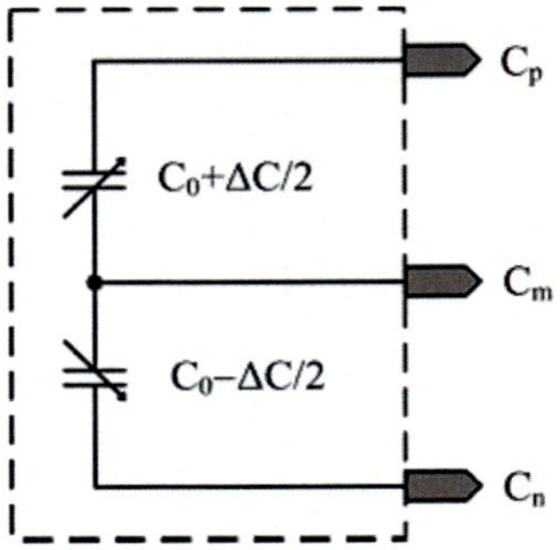

Fig. 2.18 Equivalent schematic circuit of a MEMS vibratory gyroscope

signal band of less than 100 Hz. In this case, and more in general, in the gyroscope design the capacitive sensing is used so to easily convert the input rotation rate to the output capacitance variance, by measuring the displacement of the proof mass in a direction orthogonal to both the driven motion and the axis about which rotational motion has to be sensed [48]. Fig. 2.18 shows the equivalent simple schematic circuit of a *MEMS* gyroscope: it can be seen as a passive capacitive three-terminal device. Typically, these differential capacitive sensors show a relatively low variation of about a few pF around their initial baseline value.

2.3 Temperature and Thermal Sensors

Temperature is an important parameter in many systems, in particular in environmental control systems [49–53]. Several distinct transduction mechanisms have been employed. The mercury thermometer, for an example, is a simple non-electrical temperature sensor. The most commonly used electrical temperature sensors are thermocouples, thermistors and resistance thermometers. Therefore, temperature sensors or thermal sensors can be divided in two main classes: sensors based on resistance variation (more utilized), including both the metallic types (resistance thermometers or thermoresistors, also named Resistance Temperature Detectors (*RTDs*)) and the semiconductors ones (thermistors), and sensors based on thermocouple (thermoelectrical sensors).

Thermoresistors typically show an increase in the resistance of a metal wire with increasing temperature, so exploiting the feature of metallic materials to vary their conductivity with the temperature. As the electrons in the metal gain thermal energy, they move about more rapidly and undergo more frequent collisions one each other and with the atomic nuclei. These scattering events reduce the mobility of the electrons so increasing the resistance. More in detail, thermoresistors consist of a coil of fine metal wire and, generally, are fabricated with platinum because of their main characteristics of long life-time, stability and repeatability. Moreover, platinum wire gives the largest linear range of operation. In order to simply determine the resistance indirectly, a constant current is supplied and the node voltage is measured. On the other hand, a direct measurement can be made by placing the resistor in the sensing arm of the well-known Wheatstone bridge: by adjusting the opposing resistor, it is possible to "balance" the bridge, so to produce a null output voltage. The *RTD* sensitivity is related to its temperature coefficient T_C expressed in units of % resistance per degree of temperature variation as follows:

$$T_C = \frac{\Delta R}{R} \frac{1}{\Delta T}. \tag{2.13}$$

Generally, the resistance of a metal is a complex function of the temperature and in the case of the platinum, the characteristic equation is the Callendar-Van Dusen, which is valid for low temperatures, in particular for those under the water freezing point and down to −200°C:

$$R = R_0 \lfloor 1 + A \cdot \vartheta + B \cdot \vartheta^2 + C(\vartheta - 100) \cdot \vartheta^3 \rfloor, \tag{2.14}$$

being A, B and C constant parameters dependent on the properties of the utilized platinum for sensor fabrication. It is very important to consider that, for a specific temperature range, for example from about 0°C to 650°C, Eq. 2.14 becomes the so-called Callendar equation, constituted by a linear term and a quadratic one, the latter providing its contribute only over a certain temperature range:

$$R = R_0 \lfloor 1 + A \cdot \vartheta + B \cdot \vartheta^2 \rfloor. \tag{2.15}$$

RTD sensors are particularly suitable for absolute temperature measurements. They show good sensitivity and stability and can be interfaced with very simple electronic circuits. Unfortunately, they have non-linear characteristics and show low resistance values. In order to reduce non-linearities, appropriate compensation techniques can be implemented, while to overcome the problem of revealing low resistive values, a great attention in measurement procedures has to be paid (i.e., bridge methods). The platinum *RTD* is the most accurate and stable device in the temperature range 0–500°C, even if it is able to reveal also temperatures up to 800°C (generally, for temperature values higher than 600°C, tungsten-based *RTDs* are used).

Thermistors (the name comes from the contraction of Thermal Resistors) are electrical transducers which exploit the semiconductor electrical properties to vary their conductivity with the temperature. In particular, a thermistor is a resistive element made of semiconductor materials which can have both negative (*NTC* thermistors) and positive temperature coefficients (*PTC* thermistors). The mechanism governing the resistance change of a thermistor is related to a temperature increase which provides an enhancement of the number of conducting electrons through the thermal generation. Thermistors can be measured in the same manner as resistance thermometers, but they have up to 100 times higher T_C values, so they represent the better devices in terms of sensitivity and resolution. In general, the transfer function of a *NTC* thermistor can be approximated with a simplified exponential expression as follows:

$$R = R_0 \exp\left[\beta\left(\frac{1}{T} - \frac{1}{T_0}\right)\right], \tag{2.16}$$

being R_0 the resistance value to the reference temperature T_0 equal to 25°C and β a suitable coefficient. The *PTC* thermistor, on the contrary, has a complex transfer function, which cannot be easily described with a mathematical equation, therefore it is often determined by the designer through a certain number of well-defined points. Moreover, it is able to operate in a small temperature range so is generally used for the protection by overloads and overheating, while, for the measurement of temperature, *NTC* thermistors are almost always employed. Generally, thermistors are more sensitive than *RTDs* and work (in particular the *NTC* thermistors) in a wide temperature range, starting from −100°C up to about +500°C. They provide a very high impedance and, therefore, do not need any particular measurement procedure (i.e., two-wire connection), but, unfortunately, are strongly non-linear.

Then, among the thermoelectric sensors, thermocouples are transducers which employ the Seebeck effect (the thermoelectric property due to the combination of two different conductors placed at different temperatures), which occurs at the junction of two dissimilar metal wires. A voltage difference is generated at the hot junction due to the difference in the energy distribution of thermally energized electrons in each metal. This voltage is measured across the cool terminals of the two wires and changes linearly with temperature over a given range, depending on the choice of metals. In order to minimize measurement errors, the cool terminal of the couple must be kept at a constant temperature, while the voltmeter must show a high input impedance [1, 2].

The traditional approach on integrated temperature sensors makes use of semiconductors, in particular made of bipolar technology; these sensors normally reveal the difference of two base-emitter voltages, biased by different currents, to detect the temperature variation [54]. Recently, bipolar technology has became very costly, when compared to other actual cheaper technologies, so also standard *CMOS* integrated technology has been employed in temperature sensors [51] and for the temperature control of resistive gas sensors, where gas sensing elements are developed on a silicon substrate together with platinum resistors [52].

More in detail, concerning the semiconductor-based electronic devices, since the charge carrier concentrations (n and p), the charge mobility (μ) and the diffusion processes (D) depend on the operating temperature of the same device, the constitutive relationships are related to the temperature. In particular, the current densities (J) for both the electrons and the holes, which highlight the temperature dependence, can be expressed as follows [55]:

$$J_n = q \cdot n(T) \cdot \mu_n(T) \cdot E + q \cdot D_n(T) \cdot \frac{dn(T)}{dx}, \tag{2.17}$$

$$J_p = q \cdot p(T) \cdot \mu_p(T) \cdot E - q \cdot D_p(T) \cdot \frac{dp(T)}{dx}, \tag{2.18}$$

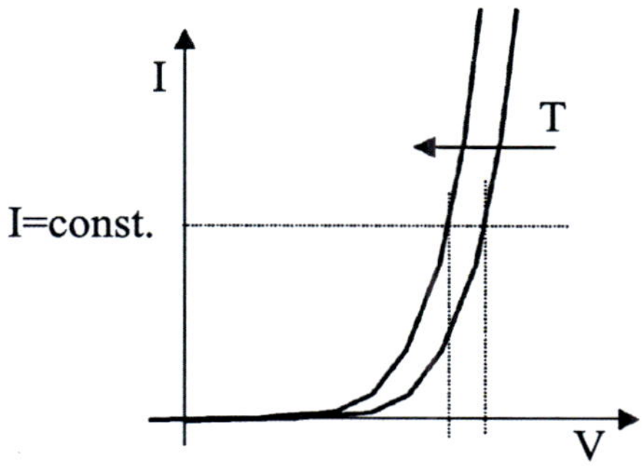

Fig. 2.19 The diode characteristic variation as a function of the temperature

where q is the charge of the electron, n and p indexes refer to electrons and holes relative quantities, respectively, and E is the electric field. As a consequence, the junction diode, which represents the main basic element for the junction-based devices, can be utilized as a temperature sensor, exploiting its temperature-dependent characteristics. In particular, the effect of a temperature variation can be described as a translation of the diode characteristic curve, as shown in Fig. 2.19. It is possible to observe that, if the diode is supplied with a constant current level, when the temperature increases, we observe a reduction of the voltage at the diode terminals. Referring to a semiconductor material, this behaviour can be seen as a resistance decrease. Typically, the diode sensitivity to the temperature is about few mV/K (i.e., considering silicon-based device), which is of the same order of magnitude of a platinum-based *RTD*. Therefore, even if the sensitivity of this junction-based device is smaller than a simple homogenous material (i.e., the thermistor), the diode has the advantage of its simple integrability on chip and, thus, is widely utilized in the integrated circuits as temperature sensor [55].

In addition, it is possible to exploit the diode sensibility to the temperature variation so to implement circuit configurations which provide the so-called Proportional To Absolute Temperature (*PTAT*) signals. The *PTAT* current principle is employed in some commercial integrated temperature sensors (as discrete active components), for example the *AD590* produced by Analog Devices [38], that can be considered a temperature-dependent current generator powered by a constant supply voltage, and the *LM35* produced by National Semiconductor [56], which provides directly a voltage proportional to the temperature to be revealed.

Recently, it has been demonstrated that the so-called thermal $\Sigma\Delta$ modulation (originally conceived for integrated flow sensors) is an attractive technique for temperature control, for example in quartz microbalances (*QMBs*), used as resonating sensors (see Fig. 2.20); in this case, a $\Sigma\Delta$ front-end may be used so that the *QMB* serves as temperature-flow sensor, heater and resonator [57].

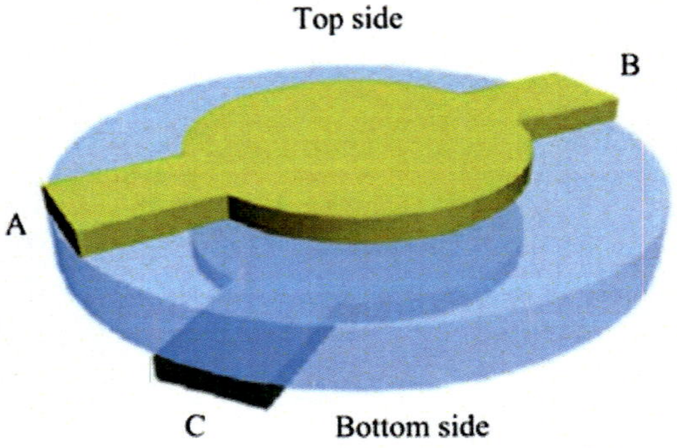

Fig. 2.20 Quartz crystal microbalance scheme

Moreover, nowadays, temperature sensors are attractive because of their low-costs and the possibility of their interfacing in a digital manner (smart sensors). However, the accuracy of current commercial temperature sensors over the industrial temperature range (−55°C to 125°C) is relatively poor. A higher accuracy is feasible, but often requires a costly calibration procedure at multiple temperatures. However, in the literature, a CMOS temperature sensor that achieves a resolution of about ±0.1°C in the range of −55°C to 125°C has been proposed [49]. It has been achieved by using suitable offset cancellation and Dynamic Element Matching (*DEM*) techniques (see Appendix 2) throughout the design, so to make errors contributed by the sensor interface circuitry negligible. As a result, only a single calibration at room temperature is needed and this is still a time-consuming temperature calibration. As a consequence, a much faster alternative calibration technique has also been proposed [50], based on the observation that if the interface circuitry has been designed accurately, the dominant source of error in a temperature sensor is its voltage reference. Therefore, it should only be necessary to calibrate this voltage reference, rather than the complete sensor. Moreover, the voltage measurement associated with this calibration can be performed much faster than an accurate temperature measurement and does not require a temperature-stabilized environment.

Finally, we want to mention thermal conductivity humidity sensors, often used at high temperatures, suitable to measure the absolute humidity by quantifying the difference between the thermal conductivity of dry air and that of air containing water vapour. Thermal conductivity humidity sensors (or absolute humidity sensors) typically consist of two matched *NTC* thermistors: one device is hermetically encapsulated in dry nitrogen and the other is exposed to the environment. They require a calibration process and are typically biased through a constant voltage which provide a suitable operating temperature higher than 200°C. The heat dissipated from the sealed thermistor is greater than the exposed thermistor due to the difference in the thermal conductivity of the water vapour as compared to dry nitrogen. Since the heat dissipated yields different operating temperatures, the thermistor resistance difference results to be proportional to the absolute

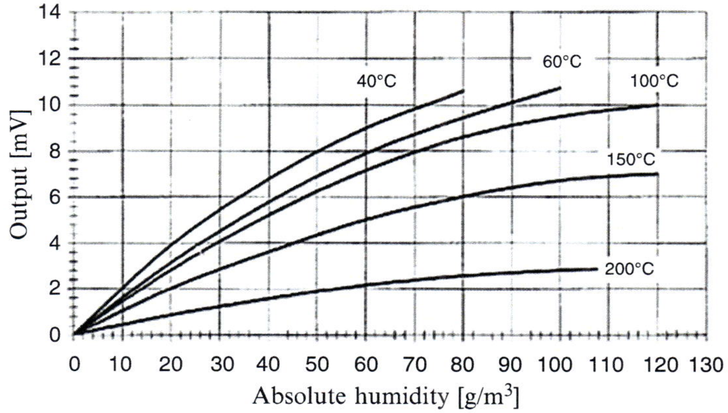

Fig. 2.21 The output signal of the thermal conductivity sensor vs. the absolute humidity as a function of the operating temperature

humidity, as reported in Fig. 2.21 which shows a typical output voltage signal of the thermal conductivity sensor, employed in a resistive bridge circuit configuration, highlighting the fact that this device is affected by the sensing elements operating temperature [22].

2.4 Smart Sensor Systems

The sensor response (i.e., the output signal of the sensor) is typically analog and this is why it is said that "the real world is analog". However, sometimes it can be also convenient to process the information in the digital electrical domain. In this case, a digital electronic system is required for converting the analog sensor response into a suitable digital electrical signal. This is what electronic interfaces perform: they are circuits that convert the sensor responses into proper electric signals easy to be processed. If these interfaces are particularly "intelligent", including special functions such as auto-calibration, sensor biasing, working temperature control, etc., they can be considered "smart".

A smart sensor system is constituted by a sensor with a suitable inherent intelligence given by the related electronic interfaces [58]. More generally, as shown in Fig. 2.22, a smart sensor system may comprise a direct chain (from the measurand M, to the A/D conversion block) and other blocks including power management (energy block), the transducer/receiver block (T/R), a memory, a microcontroller and the actuators, etc. [16, 17].

A smart system (if miniaturized, named microsystem) requires, all together, sensors (if miniaturized, named microsensors), actuators and suitable electronic interfaces. For example, a gas-sensing microsystem typically consists of an array

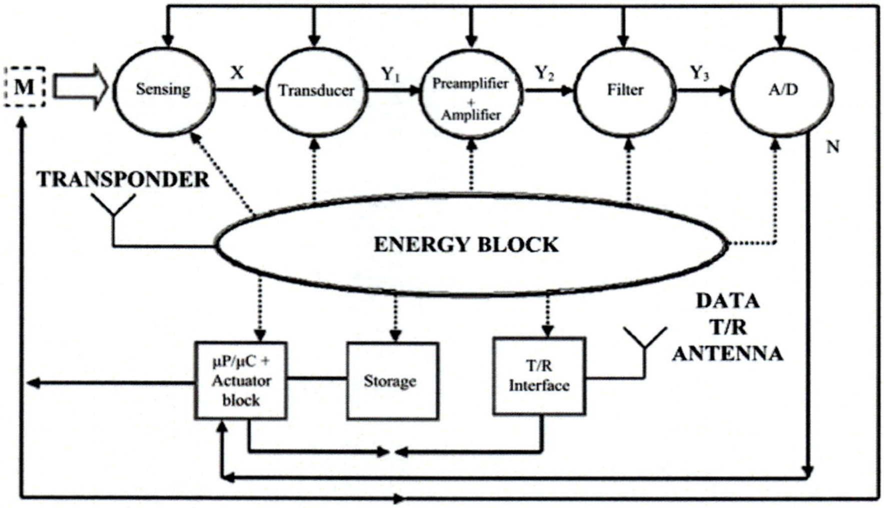

Fig. 2.22 Block scheme of a smart sensor system

of gas-sensors, a temperature control circuit, an electronic readout block and a data processor. In order to develop a really portable device, the system has to be stand-alone, i.e. has to be able to operate without the aid of any laboratory instrument, while sensors, implemented with silicon based technologies, can detect different physical and chemical quantities with acceptable selectivity, sensitivity and resolution. Smart systems can be implemented through two possible ways: the microsystem approach and the micromodule approach [1, 2, 17].

In the microsystem approach, the sensor and the electronic interface are integrated on the same chip. In this case, the complete system is obtained using a standard *IC* process with, eventually, few compatible post-processing steps (typically etching or deposition of materials). Therefore, the microsensor has to be designed taking into account the material features (layer thickness, doping concentrations and design rules) imposed by the standard *IC* process used (*CMOS*, bipolar or *BiCMOS*); any additional processing step required for implementing the sensing devices has to be performed after the completion of the standard *IC* fabrication flow. Obviously, this situation reduces the degrees of freedom available for sensor design, thus introducing additional challenges. Moreover, especially when using sub-micron technologies, this approach can give cost and yield problems. Indeed, the silicon area occupied by the electronic interface circuit typically shrinks with the feature size of the technology, while the sensor area in most cases remains constant, since it is determined by "physical" considerations, such as the mass of the structures or the angle of etched cavities, which are not changed by improvements in the technology. Therefore, while for integrated circuits the increasing cost per unit area is compensated by the reduction in its size, leading to an overall reduction of chip cost with the technology feature dimension, this might not be true for integrated microsystems. In addition, a defect in the sensors may result in the failure of the complete microsystem even if the circuitry is working properly, hence lowering the yield and again increasing the cost (the yield for sensors is typically lower than for electronic circuits). The microsystem approach, however, also has considerable advantages. First of all, the parasitic components due to the interconnections between sensors and electronic interfaces are minimized and, more important, well defined and reproducible, which is very beneficial for the system performances. Moreover, the system assembly is simple, inexpensive and independent from the number of connections needed because all the interconnections are implemented during the *IC* fabrication process. Finally, when required, the use of the same technology allows us to achieve a good matching between the elements of the sensor and those of the interface circuitry, thus allowing an accurate compensation of many parasitic effects [1, 2, 17].

On the contrary, in the micromodule approach, sensors and electronic interface circuits are fabricated on separated chips. However, they are then included in the same package or mounted on the same substrate. The interconnections between the sensor chip and the electronic interface chip can be performed with bonding wires or other techniques, such as flip-chip or wafer bonding. With this approach the two parts can be implemented also with different technologies, optimized for the sensors and the circuitry, respectively. Typically, expensive submicron technologies are used to fabricate the electronic interface circuits, while low cost technologies with large feature size and few masks are used for implementing the sensors. In this case, the material properties of the technology can be adjusted to optimize the performance of the devices. However, the micromodule approach has also drawbacks. First of all, the assembling of the system can be quite expensive and unreliable, allowing only a limited number of interconnections between the sensor and the interfaces. Moreover, sometimes the parasitic components due to the interconnections are orders of magnitude larger, more unpredictable and less repeatable, than in the microsystem approach, thus eventually reducing the sensor performance improvements obtained with technology optimization. Finally, no matching between elements of the sensor and those of electronic interfaces can be guaranteed [1, 2, 17].

In conclusion, the choice of one of the two approaches substantially depends on the application, the quantity to be measured, the kind of sensors, the specifications of the electronic interface circuits and the available fabrication technologies, thus producing a number of trade-offs, which have to be analyzed before taking the best decision.

2.5 Circuits for Sensor Applications: Sensor Interfaces

More specifically, the sensor interface is an electronic circuit which allows to read-out the information coming from the signal generated by a sensor, providing a suitable output signal simple to display or to elaborate [16].

Sensors and electronic interfaces, clearly, are a sub-set of measurement systems and, therefore, their performance should be expressed through parameters as accuracy, precision, sensitivity, resolution, offset, etc.. In this sense, the design or the use of an electronic interface is strictly related to the problem of the detection and measurement of the measurand. More in general, "measuring" means comparing the measurand with a reference quantity (which, ideally, is a constant value). Clearly, the measurand has to be (or to be kept) also constant during all the measurement process. In other words, the measurement process must be much faster than all the possible variations of the measurand (in many, but not all, electronic interfaces, this is not a problem, because the input signals coming from the sensor are typically much slower than electrical systems). As an important consequence, interface designers may conveniently find a suitable trade-off between accuracy and speed [16, 17].

An ideal measurement system converts input signals into output signals according to a desired transformation, while a non-ideal system does this not instantaneously and, unfortunately, introduces an error. In the case of instantaneous systems, the error may be defined as the difference between the measured output and the theoretical ideal output. Therefore, as also mentioned before, the accuracy of a system may be qualitatively defined as the capability of the system to produce small errors. More in detail, as an example, if the interface is implemented by a voltage amplifier showing a negligible input offset voltage and a very small relative gain error, the system has a high accuracy. However, if the amplifier has a significant input equivalent noise (with zero mean value), its precision could be poor; then, if the amplifier is inserted in the measurement chain, the precision of the electronic interface (and, hence, of the measurement system) could be poor as well. If the error must be small for every measurement, we need a both accurate and precise electronic interface; if only the mean value of the error (with reference to a high number of repeated measurements) is important, an accurate system is sufficient. These specifications may be translated into accuracy and precision requirements. Furthermore, it is helpful to consider some sources of errors in a measurement. Accuracy and precision of a measurement may not be better than those of the reference quantity; this is why "high-quality" references are very important. In some cases, they are available; in other cases, the reference signal must be generated by the interface itself (e.g., since voltage references are essential building blocks for many electronic interfaces, sometimes the design of accurate and precise integrated band-gap references is a main issue). Beside the errors of the reference, errors also occur in the comparison process; the errors of *ADCs*, for instance, fall in this class of errors. Additionally, the perturbation introduced by the measurement action should be negligible for the desired level of accuracy; in this sense, impedance loading effects must always be taken into account and properly evaluated. Therefore, in most practical cases, some preliminary simulations are necessary for the accurate analysis of these interfaces [16, 17].

Another fundamental parameter of a system is the sensitivity and, as for the sensor, can be defined as the ratio between the generated output variation and the input signal variation (system transfer function), coming from the sensor. In general, the sensitivity depends on the operating point of the system and, clearly, is a pure number if and only if the input and the output signals are homogeneous. In this case, the sensitivity is also called gain (e.g., the sensitivity of a voltage amplifier is a gain, since both the input and the output signals are voltages). If the sensitivity is not a pure number (as typically happens), it may not be considered a gain and must be expressed with proper dimensional units (e.g., for a current to voltage converter, which has a current input and a voltage output, the sensitivity must be expressed in Ω). However, in general, it must be possible to regulate it by choosing, for example, suitable values for the employed passive components. In a given operating point, there is a minimum variation of the output signal which the system is able to detect (this quantity is generally not zero because of noise and interferences). This minimum variation of the measurand which may be revealed is defined the resolution of the system (also this quantity is related to the fixed operating point). Moreover, the transfer function of a linear time-invariant system is generally a constant. On the other hand, since that instantaneous systems, strictly, may not exist because of the finite speed of real systems, transfer functions of non-ideal systems always depend on frequency; in practical cases, transfer functions may be only approximately constant (e.g., within 3 dB of variation) within a certain range of frequencies, called as bandwidth (e.g., 3 dB bandwidth). All the non-ideal systems have a limited speed and, therefore, have a finite bandwidth. Since non-ideal systems are slowly time-variant, in many practical cases the time invariance hypothesis is possible and useful. As an example, a temperature resistive sensor is already a time-variant system because its resistance changes with time (due to temperature variations). Depending on the application, this may or may not be an issue: for instance, if the variations of the temperature dependent resistance are very slow when compared with all the other variations in the system, we may consider a constant resistance and make sure that the complete system properly works with all the possible resistance values. In order to get high accuracy, low interferences, high reliability and low cost characteristics, it is often convenient to integrate sensors and electronic interfaces in the same chip; generally, this can only be done in standard *CMOS* processes especially for the low cost constraints [16, 17].

Finally, there is an additional consideration to be done: generally it is necessary to develop an accurate model of the considered sensors, independently from its complexity. In some cases, transducers are just electronic devices; even in these cases, models which are satisfactory for most electronic designs may be not enough accurate for the design of high performance electronic interfaces and sensors. In other cases, transducers are non-electrical devices and it may be not obvious how to simulate these transducers together with the rest of the electronic interface. Almost always, the best practical solution is to model non-electrical signals and systems by means of equivalent signals and systems in the electrical energy domain, so that the complete system may be analyzed by means of standard simulators for electronic circuits such as ORCAD PSpice or CADENCE [59].

2.5.1 *Low-Voltage Low-Power Voltage-Mode and Current-Mode Analog Sensor Interfaces*

Recently, the development of *VLSI* technology, together with the request of a larger number of elements on a single chip, has led to an improved interest in analog circuit design, especially for what concerns *ICs*. The main aim of analog *IC* is to satisfy particular specifications through circuit architectures showing the required performances. Moreover, *IC* designers have been putting an increasing effort into the reduction of supply voltage and power dissipation of analog, digital and mixed signals integrated circuits and systems. The *LV LP* analog integrated circuit design, widely utilized in portable single-cell battery operated applications (e.g., biomedicals, cellular phones, etc.), has led to implement new design strategies in low cost *CMOS* integrated technology [60–65].

LV analog design techniques differ considerably from traditional supply design and the basic analog blocks have to be reconsidered in a *LV* environment. Especially for portable applications, *LV* circuits need to be compatible with common battery voltage values. In this sense, traditional architectures available for working at low supply rails are generally inadequate as well as typical models for transistors which have to be implemented with a new particular attention in the boundary region between weak and strong inversion, where transistors are often biased. For example, in all the basic blocks, as the *OA*, the new constraints concern both the full input swing (performed by two complementary pairs in parallel) and the complete output range (so to have the rail-to-rail operation, e.g., by a class-AB stage with low output quiescent current and output current control). Amplifier input stages have also to show a transconductance independent from the input common mode voltage, so to present the same circuit characteristics in any biasing condition. As a sum of all these factors, we can say that in *LV* design it is fundamental an efficient use of the supply voltage range. In the literature, a *CMOS* circuit can be included in the *LV* category according to the number of stacked gate-source (threshold) and drain-source (saturation) voltages, (V_{TH} and V_{DSAT}, respectively). There is not a predefined value which exactly determines the boundary between a non-*LV* and a *LV* topology. In particular, the term *LV*, considering a standard *CMOS* technology, can be typically used for circuits that are able to operate at a supply voltage of $2V_{TH} + 2V_{DSAT}$, while Very Low Voltage (*VLV*) circuits have also to work at only $V_{TH} + V_{DSAT}$. Of course, this is only a possible definition but, in this sense, numerical supply values are strictly related to the technology used and tend to decrease during the years with the scaling of circuit sizes.

In analog circuits, the reduction of the supply voltage does not necessarily correspond to a decrease of related power consumption. In this case, the "folding" technique can replace the traditional "stacking" of transistors. In order to keep the power low, analog circuits have to be designed as much simple as possible. Moreover, it is important to consider that a trivial decrease of biasing currents, which can reduce circuit dissipation, degrades the circuit performance, first of all bandwidth and dynamic range. As a consequence, chip area cannot be drastically reduced with the lowered feature dimensions. Nevertheless, power limitations are mainly related to: parasitic capacitances; traditional current-inefficient amplifiers, not optimised for a low quiescent dissipation; peak-to-peak limitations. As a result, *LP* design is characterised by an efficient use of the supply current, through the utilisation of class-AB output stages and an efficient frequency compensation strategy.

The combination of these constraints and requirements gives the basic rules to be followed to design *LV LP* circuits, even if in a large number of analog applications designers focus their studies only on the development of topologies able to work at reduced supply. In this case, using typical values of biasing current in the μA range, circuits show generally a reduced power consumption (e.g., not higher than 1 mW). In this sense, for *LV LP* applications, a special care has to be used in the design of suitable current sources: the design of the biasing currents independent from the supply voltage variations, so as to avoid performance reduction (or degradation) when the supply battery discharges, is one of main aspects to consider.

Concerning the integrated technology, the continuous reduction of the threshold voltage in standard *CMOS* has definitively directed *LV* design towards *CMOS* itself, which is also typically characterized by a very low quiescent power consumption. Reducing the supply voltage, *CMOS* transistor is often biased to work in weak inversion region: in this sense, the use of good transistor models is of a fundamental importance [66]. In addition, the interfacing of the sensitive element with a suitable integrated circuit is a fundamental characteristic. In this sense, *CMOS* technology is widely used, because it allows to match the reduction of costs of the silicon with the possibility of designing new *LV LP* interface circuits to be easily dedicated to the portable sensor applications market. Nevertheless, since *CMOS* transistors show high input offset voltages and high input low frequency noise voltages, accurate *CMOS* amplifiers, in integrated sensors interface applications, are possible only if the effects of these non-idealities are well compensated.

Starting from these considerations, it is important to highlight that, for *LV LP* applications, the *CM* approach can be considered an alternative to traditional *VM* circuit to obtain high performance architectures, because the designer deals with current levels for circuit operation instead of node voltages. In this manner, as well known, *CM* circuits, which are able to overcome the limitation of the constant Gain-Bandwidth (*GBW*) product and the trade-off between speed and bandwidth typical of *OA*, give good alternative solutions. In particular, *CM* topologies improve integrated circuit performances in terms of *LV LP* characteristics, such as slew-rate and bandwidth, through the development and the use of suitable Second Generation Current Conveyors (*CCIIs*, see Appendix 1), which represent the main basic building block in the *CM* approach [67–71].

All the *CCII*-based topologies, designed with *LV LP* techniques, have a low operating supply voltage, related to the drain-source (saturation) voltage required by the biasing transistors, which has to be minimised so to reduce the circuit total supply voltage. Several *CCIIs* topologies presented in the literature are based on a differential pair followed by a class-*AB* output stage. Theoretical analyses have confirmed that this solution ensures good performance also in terms of *LV LP* characteristics [71].

2.6 Basic Sensor Interfacing Techniques: Introduction to Signal Conditioning

A signal conditioning system (or, in other words, electronic interface, read-out circuit, front-end, etc.) takes the output from a sensing element and converts it into a more suitable form for further processing (e.g., amplification, analog-digital conversion, frequency-voltage conversion, etc.), as described in Fig. 2.23 at block scheme level. Therefore, a signal conditioning circuit provides a functional transformation needed for accurate and consistent measurement of electrical quantities that, typically, have very small changes.

The simpler interface circuits, often utilized, for example, as basic signal conditioning stages in resistive sensors, are the voltage divider, shown in Fig. 2.24, and its differential version, the Wheatstone bridge, depicted in Fig. 2.25, where V_{IN} is the supply voltage and one (or more) of the bridge elements (impedances) are the sensors. These simple basic solutions are able to perform, more in general, a conversion from an impedance (e.g., a resistance) variation into a voltage one [6,74]. In particular, in Fig. 2.26, some examples of impedance-based passive bridges, together with the related balance conditions, have been reported.

Usually, bridge circuits can be accompanied by an additional conditioning circuitry (e.g., a voltage amplifier connected to the bridge output terminals) which amplifies the bridge output always giving a signal proportional to the sensor parameter variation, with an increased sensitivity. Alternatively, especially for large variations of the sensing element, a conversion towards a periodic output waveform is generally performed. Typically, in this case, the output period is proportional to the measurand or to its variations.

In the following Sects. 2.6.1–2.6.3, we will describe synthetically the basic concepts related to the sensor interface circuits design and the main sensor signal condition techniques concerning more specifically the three main sensor typologies: resistive, capacitive and temperature.

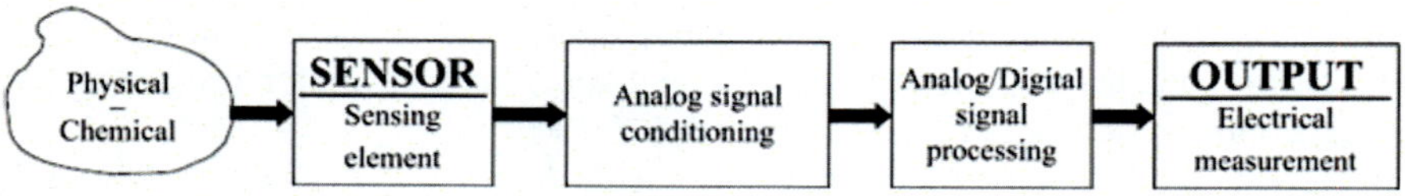

Fig. 2.23 Block scheme of a complete signal conditioning system

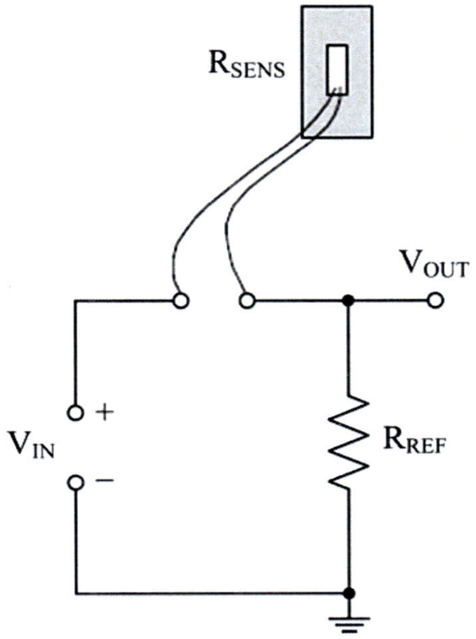

Fig. 2.24 Block scheme of a voltage divider as a signal conditioning circuit for resistive sensors (V_{OUT} represents the circuit output signal)

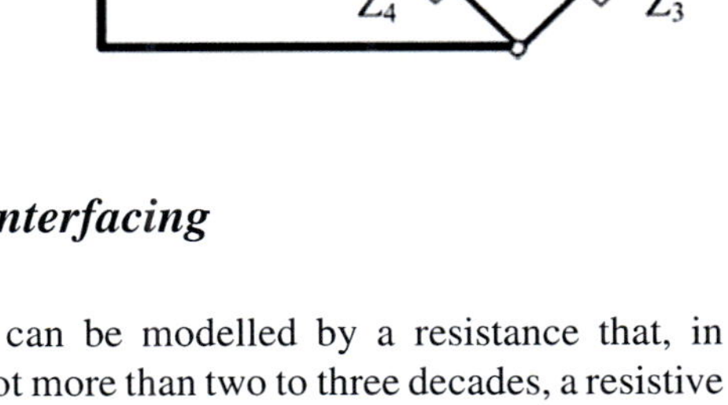

Fig. 2.25 The Wheatstone bridge schematic circuit for sensor interfacing

2.6.1 Resistive Sensors Basic Interfacing

When the sensor electrical parameter can be modelled by a resistance that, in particular, varies into a reduced range, not more than two to three decades, a resistive voltage divider circuit, operating a Resistance-to-Voltage (R-V) conversion (as yet shown in Fig. 2.24), can be utilized as simple resistive sensor interface circuit. Typically, it applies a constant voltage so to measure the change of conductivity of the resistive sensing element.

Another very simple interfacing circuit for resistive sensors, varying into a reduced range, can be implemented by the well-known Wheatstone bridge which

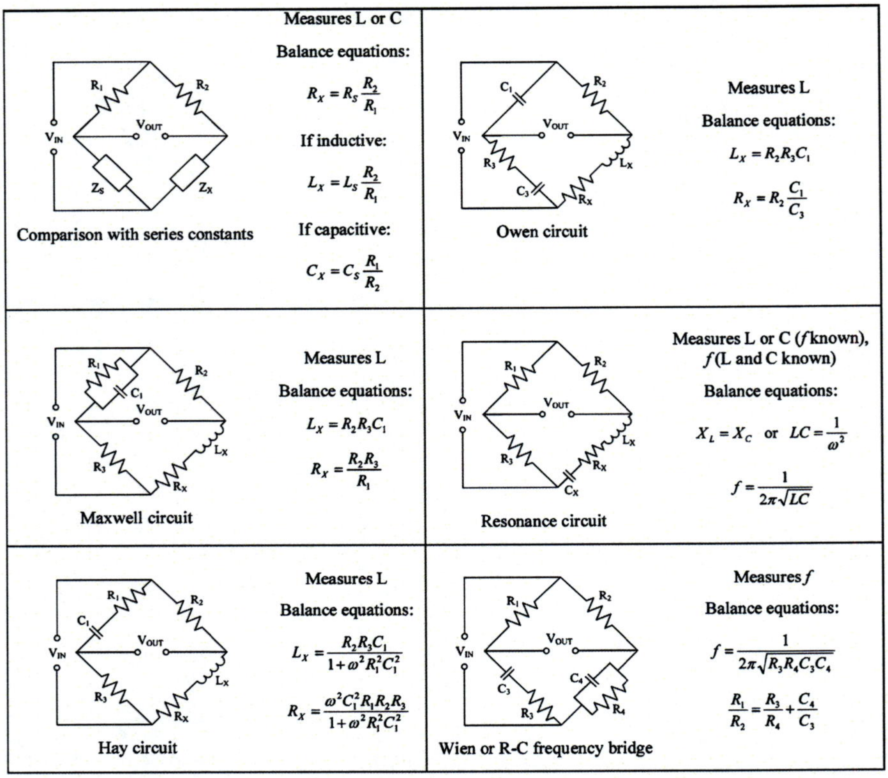

Fig. 2.26 Different examples of impedance bridge configurations

operates also an *R-V* conversion (see Fig. 2.25 where all the impedances are pure resistances). This circuit configuration represents the "fully-differential" version of the basic voltage divider and shows its same sensitivity. In this case, one of the four resistances is the resistive sensor whose sensing element varies when an external physical or chemical phenomenon occurs. The main drawback of this kind of resistive sensor interface is in its unsettable and low sensitivity, only dependent on the total supply voltage (in this case, as in the simple voltage divider, the sensitivity is constant and equal to a quarter of the total supply voltage, when a low variation of only one resistance of the bridge occurs).

Beside the use of expensive pico-ammeters, alternative solutions for resistive sensor interfaces are available in the literature; they are based on the resistance estimation utilizing high-resolution *ADCs*. In order to guarantee the best resolution for each resistance value in the considered range, a variable gain stage (scaling factor system) is adopted. However, such systems need difficult and expensive calibration procedures, especially when very high resistance values are considered. In fact, the scaling factors need the use of either multistage amplifiers or resistors whose value is on the order of the resistances to be estimated: in the first case, noise is the main limit, whereas in the second it is hard to manage resistors on the order of GΩ, in terms of accuracy, stability, practical circuit implementation, integration in a possible single-chip solution, etc..

Therefore, if larger variations of sensor resistive values happen, we can employ a Resistance-to-Time (R-T) conversion, which can be also considered as a Resistance-to-frequency (R-f) conversion when the "time" (period) is related to a periodic waveform. The R-T based interfaces exploit the easiness of measuring time intervals over a wide range of variation. As a consequence, no more scaling factor systems are needed. Typically, an R-T basic scheme is based on an oscillator architecture which exploits the sensor to be excited by a switched voltage (the AC excitation voltage). In this case, the simpler electronic interface which operates an R-T (or R-f) conversion can be implemented by an *OA* (or a *CCII*) in an astable multivibrator configuration. In fact, this circuit solution implements a square wave generator, whose output voltage period T (or frequency f) is dependent on the sensor resistance value. More in general, a wide range integrated circuit interface for resistive sensors is an oscillating circuit which generates an *AC* periodic signal whose oscillation period T is dependent on sensor resistance value so to operate a suitable R-T conversion. Usually, in these kinds of front-ends, a constant current, whose value only depends on sensor resistance, is generated and utilized to charge and discharge a capacitor, alternatively, providing a periodic signal at the interface output. In order to have a reduced error in oscillation period measurements, so in sensor resistance estimations, the interface must be designed with good performances in terms of time responses (high Slew Rate (*SR*) values of the active components) and both voltage and current very low offset values. In this case, also high-valued resistive sensors and their variations (starting from tens of kΩ it can reach tens, hundreds of GΩ) can be accurately revealed. In particular, for about six to seven decades of resistance variations, the interface circuit generally has to show also good linearity and sensitivity. Typical applications of these wide range electronic interfaces are in environmental gas monitoring systems, where *MOX*-based resistive gas sensors are often utilized.

The main interface circuits for resistive sensors will be described in a deep detail in the next Chapters, considering both the *OA* (*VM* approach) and the *CCII* (*CM* approach) as active elements, and exciting the sensors both with a DC and with an AC supply.

2.6.2 Capacitive Sensors Basic Interfacing

The typical simplest way to measure a capacitance is to convert it (or its variation) into a suitable voltage level, performing the so-called Capacitance-to-Voltage (C-V) conversion. This can be simply done by one of the bridge configurations shown in

Fig. 2.26, once that all the other passive components are known or can be accurately measured. Otherwise, a charge-pump configuration (or charge pre-amplifier), typically based on *OA* in an inverting topology, can convert proportionally the sensor capacitance variation into an output voltage. Generally, together with the basic capacitive sensor module (e.g., containing the basic signal conditioning circuit), these sensing system include also an instrumentation amplifier, an *ADC* and a suitable digital signal processing block.

Alternatively, some *RC*-based oscillators as frequency output sensing circuits, which needs resistors and capacitors, can be employed. They typically implement a ring oscillator, whose output frequency shifts, in this case, because of capacitance change. The output signal can be automatically quantified by a digital counter, therefore the entire system become simpler and smaller. Therefore, actually, the capacitive sensors are often interfaced with read-out electronic circuits that perform a Capacitance-to-frequency (C-f) conversion (e.g., oscillators and phase shifters for oscillating circuits, etc.). Moreover, the sensor capacitance can be charged and discharged by a constant current and the frequency of the signal revealed at the output of the designed system is inversely proportional to the sensor capacitance value (e.g., the simple basic interface circuit for capacitive sensors, operating a C-f conversion, can be implemented by an *OA* in the well-known astable multivibrator configuration). Then, an automatic storage of the oscillation frequency can be also performed, using a digital frequency counter. These kind of solutions are very flexible for any research field and, in particular, suitable, for example, for capacitive pressure microsensors which show variable frequency output signal and also for other portable applications such as implantable bio-medical and industrial systems.

Other capacitance read-out circuits could be based on switched-capacitor (*SC*), continuous-time current generator (*CTCG*) and continuous-time voltage generator (*CTVG*). Usually, the *CTVG* sensing has superior noise performances compared to the other two, therefore is more suitable for high precision capacitive sensor interfacing.

Nevertheless, the main problem related to all these interface solutions concerns the detection of either very low capacitance values or its small variations. In this sense, the proper design of a suitable read-out circuit, which has to be able to provide the smallest parasitic capacitances at its terminals, is another important task, while a special consideration for shielding to still reduce parasitic capacitances of the electronic front-end, which is essential to have suitable performances, has to be also done avoiding the need for large connectors. Therefore, the key aspect of the problem is related to the sensing system, where the sensitivity to parasitic elements, interconnection wires and noise has to be the lowest possible. For these reasons, differential capacitive sensors have often to be taken into account, developed and utilized.

Also for capacitive sensors, the utilized interfacing techniques will be proposed and described in detail in the next Chapters, with both *VM* and *CM* approaches.

2.6.3 Temperature Sensors: Basic Interfacing and Control Systems

The simpler readout circuit for temperature sensors (considering that they are often resistive sensors) can be implemented, also in this case, by the Wheatstone bridge, which operates, as well known, an *R-V* conversion. It has to be composed by four resistances, whose temperature coefficients are positive for two of them (those diagonally opposed) and negative for the other ones (diagonally opposed too).

Alternatively, a Temperature-to-Time conversion can be also adopted. This solution shows often digital components, so it is able to measure the temperature with over 10 bit accuracy. The time-to-digital converter replaces, in this case, the conventional *ADC* and its output is a sequence of pulses whose number is proportional to temperature (e.g., a difference of delay times can be built through a logical *EX-OR* of two outputs) [73, 74].

Read-out electronic circuits often need a suitable temperature control system, formed by a temperature sensor and a heater resistance, so to achieve an optimal sensor operating temperature. In this sense, the sensor interfacing can improve and optimize sensor sensitivity and selectivity. More in detail, a higher selectivity with respect to different physical or chemical measurands can be obtained by using an array of different sensors, while a higher sensitivity can be achieved through a specific pattern to be applied to properly regulate the operating temperature of sensors (i.e., the application of the so-called thermal modulation technique).

The accurate control of the sensor operating temperature is particularly important in gas monitoring. Usually, sensor gas responses have to be carried out between about 20°C and 400°C operating temperatures and with different target toxic gas concentrations, ranging in about 1–100 ppm. Therefore, an electronic interface can be completed with a suitable electronic system performing the accurate control of the sensor working temperature, generally implemented through a proper control sub-system in a feedback configuration. In order to exploit this technique with enough accuracy in the chemical measurement, an embedded temperature control loop is necessary, because the temperature of the sensor should be accurately controlled, so to operate a suitable sensor sensitivity improvement. In addition, a data elaboration system can be required, implementing a pattern recognition algorithm for the post-processing of the data acquired from the designed front-end, so to have a more complete electronic sensor system able to perform target gas concentration measurements providing directly numerical values.

References

1. T. Grandke, W.H. KO, Fundamentals and General Aspects, in *Sensors: A Comprehensive Survey*, ed. by W. Gopel, J. Hesse, J.H. Zemel (Wiley VCH, Weinheim, 1996). ISBN 3527293299
2. R.C. Dorf, *The Electrical Engineering Handbook* (CRC Press LLC, Boca Raton, 2000). ISBN 0849385741

3. G. Sberveglieri, C. Baratto, E. Comini, G. Faglia, M. Ferroni, A. Vomiero, Single crystalline metal oxide nano-wires/tubes: controlled growth for sensitive gas sensor devices, in *2nd IEEE International Conference on Nano/Micro Engineered and Molecular Systems*, Bangkok, Thailand, Jan 2007, pp. 227–229
4. A. Hierlemann, O. Brand, C. Hagleitner, H. Baltes, Microfabrication techniques for chemical biosensors. Proc. IEEE **91**(6), 839–863 (2003)
5. E. Martinelli, C. Falconi, A. DAmico, C. Di Natale, Feature extraction chemical sensors in phase space. Sensors Actuator B **95**, 132–139 (2003)
6. R. G. Longoria, Resistive sensors, lectures at the University of Texas at Austin
7. M. Knite, V. Teteris, A. Kiploka, J. Kaupuzs, Polyisoprene-carbon black nanocomposites as strain and pressure sensor materials. Sensors Actuator A **110**(1–3), 143–150 (2004)
8. A. Varfolomeev, V. Filippov, S. Lazarev, E. Meylichov, V. Pokalyakin, A. Volkov, A. Volynskiy, S. Yakimov, A. Zharkovsky, Piezoresistive sensor based on nanostructured metal layer on polymer film, in *Proceedings of Eurosensors*, Dresden, Sept 2008, pp. 152–153
9. W. Thomson, The electro-dynamic qualities of metals. Phil. Trans. Royal. Soc Lond. **146**, 733 (1856)
10. M. Messina, F. Franze, N. Speciale, E. Cozzani, A. Roncaglia, Thermofluid analysis of ultra low power hotplates for a MOX gas sensing device. IEEE Sens. J. **9**(5), 504–511 (2009)
11. A. Hackner, A. Habauzit, G. Muller, E. Comini, G. Faglia, G. Sberveglieri, Surface ionization gas detection on platinum and metal oxide surfaces. IEEE Sens. J. **9**(12), 1727–1733 (2009)
12. J. Courbat, D. Briand, L. Yue, S. Raible, N. F. De Rooij, Ultra-low power metal-oxide gas sensor on plastic foil, in *International Conference on Solid-State Sensors, Actuators and Microsystems*, Transducers, June 2009, pp. 584–587
13. S. Bicelli, A. Depari, G. Faglia, A. Flammini, A. Fort, M. Mugnaini, A. Ponzoni, V. Vignoli, Model and experimental characterization of dynamic behaviour of low power Carbon monoxide MOX sensors with pulsed temperature profile, in *Proceedings of IEEE Instrumentation and Measurement Technology Conference* – IMTC, Victoria, May 2008, pp. 1413–1418
14. I. Elmi, S. Zampolli, E. Cozzani, M. Passini, G. Pizzochero, G. C. Cardinali, M. Severi, Ultra low power MOX sensors with ppb-Level VOC detection capabilities, in *Proceedings of IEEE Sensors*, Oct 2007, pp. 170–173
15. S. Bicelli, A. Flammini, A. Depari, D. Marioli, A. Ponzoni, G. Sberveglieri, A. Taroni, low-power carbon monoxide MOX sensors for wireless distributed sensor networks, in *Proceedings of IEEE Instrumentation and Measurement Technology Conference*, May 2007, pp. 1–5
16. C. Falconi, E. Martinelli, C. Di Natale, A. DAmico, F. Maloberti, P. Malcovati, A. Baschirotto, V. Stornelli, G. Ferri, Electronic interfaces. Sensors Actuator B **121**, 295–329 (2007)
17. A. D'Amico, C. Di Natale, *Introduzione ai sensori* (Aracne, Roma, 2008). ISBN 9788854816633
18. Internet resource: http://www.figarosensor.com. Datasheet TGS826
19. A. Ponzoni, C. Baratto, S. Bianchi, E. Comini, M. Ferroni, M. Pardo, M. Vezzoli, A. Vomiero, G. Faglia, G. Sberveglieri, Metal oxide nanowire and thin-film-based gas sensors for chemical warfare simulants detection. IEEE Sens. J. **8**(6), 735–742 (2008)
20. J. Lozano, J. P. Santos, M. Aleixandre, M. C. Horrillo, Electronic nose applied to off flavours detection in wine, in *Proceedings of Eurosensors*, Dresden, Sept 2008, pp. 479–482
21. C. Di Natale, F. Davide, A. D'Amico, G. Sberveglieri, Sensors for domestic applications, in *Proceedings of the First European School on Sensors*, (World Scientific Publ., Singapore, 1995)
22. Internet resource: http://www.sensorsmag.com/sensors/humidity-moisture/choosing-a-humidity-sensor-a-review-three-technologies-840
23. S. Pennisi, High-performance and simple CMOS interface circuit for differential capacitive sensors. IEEE Trans. Circ. Syst. II **52**(6), 322–326 (2005)
24. L. Zhao, E. M. Yeatman, Inherently digital micro capacitive tilt sensor for low power motion detection, in *Proceedings of Eurosensors*, Dresden, Sept 2008, pp. 621–624
25. T. Schneider, S. Doerner, P. Hauptmann, Wide-band impedance spectrum analyzer for monitoring of dielectric and resonant sensors, in *Proceedings of Eurosensors*, Dresden, Sept 2008, pp. 1399–1402

26. K. Manoli, P. Oikonomou, D. Goustouridis, E. Karonis, I. Raptis, M. Sanopoulou, Interdigital chemocapacitive sensors, based on polymer/batio3 composites, in *Proceedings of Eurosensors*, Dresden, Sept 2008, pp. 1101–1103
27. E. Pritchard, M. Mahfouz, B. Evans, S. Eliza, M. Haider, Flexible capacitive sensors for high resolution pressure measurement, in *Proceedings of IEEE Sensors*, Lecce, Oct 2008, pp. 1484–1487
28. C. Hierold, B. Clasbrummel, D. Behrend, T. Scheiter, M. Steger, K. Oppermann, H. Kapels, E. Landgraf, D. Wenzel, D. Etzrodt, Low power integrated pressure sensor system for medical applications. Sensors Actuator A **73**(1–2), 58–67 (1999)
29. T.G. Constandinou, J. Georgiou, C. Toumazou, Micropower front-end interface for differential capacitive sensor systems. IET Electron. Lett. **44**(7), 470–472 (2008)
30. Y. H. Hsueh, J. H. Lin, Low power integrated capacitive pressure microsensor design, in *Proceedings of Eurosensors*, Dresden, Sept 2008, pp. 284–287
31. C.T. Ko, S.H. Tseng, M.S.C. Lu, A CMOS micromachined capacitive tactile sensor with high frequency output. J. Microelectromech. S. **15**(6), 1708–1714 (2006)
32. C.L. Dai, Y. Tai, P.H. Kao, Modeling and fabrication of MicroFET pressure sensor with circuits. Sensors **7**, 3386–3398 (2007)
33. L. Kin-tak, C. Chi-chiu, Z. Li-min, J. Wei, Strain monitoring in composite-strengthened concrete structures using optical fibre sensors. Composites B **32**(1), 33–45 (2001)
34. H. Guckel, T. Randazzo, D.W. Burns, A simple technique for the determination of mechanical strain in thin films with applications to polysilicon. J. Appl. Phys. **57**(5), 1671–1675 (1985)
35. S. Kon, K. Oldham, R. Ruzicka, R. Horowitz, Design and fabrication of a piezoelectric instrument suspension for hard disk, in *Proceedings of SPIE 6529*, Part 2, art. no. 65292V, 2007
36. J. Guo, H. Kuo, D. J. Young, W. H. Ko, Buckled beam linear output capacitive strain sensor, in *IEEE Solid-State Sensor and Actuator Workshop*, 2004, pp. 344–347
37. P. Broutas, D. Goustouridis, S. Chatzandroulis, P. Normand, D. Tsoukalas, Capacitive strain sensors using polymer embedded thin Si membranes, in *Proceedings of Eurosensors*, Dresden, Sept 2008, pp. 186–188
38. Internet resource: http://www.analog.com. Datasheet ADXL50
39. R. Dai, R.B. Stein, B.J. Andrews, K.B. James, M. Wieler, Application of tilt sensors in functional electrical stimulation. IEEE Trans. Rehabil. Eng. **4**(2), 63–72 (1996)
40. L. Zhao, E. M. Yeatman, Micro capacitive tilt sensor for human body movement detection, in *Proceedings of International Workshop on Wearable and Implantable Body Sensor Networks*, Aachen, 2007
41. A.V. Mamishev, K. Sundare-Rajan, F. Yang, Y. Du, M. Zahn, Interdigital sensors and transducers. Proc. IEEE **92**, 808–845 (2004)
42. J. Laconte, V. Wilmart, D. Flandre, J. P. Raskin, High-sensitivity capacitive humidity sensors using 3-layer patterned polyimide sensing film, in *Proceedings of IEEE Sensors*, 2003, pp. 372–377
43. Z. Chen, C. Lu, Humidity sensors: a review of materials and mechanisms. IEEE Sensor Lett. **3**, 274–295 (2005)
44. A.D. DeHennis, K.D. Wise, A wireless microsystem for the remote sensing of pressure, temperature and relative humidity. J. Microelectromechanical S. **14**(1), 12–22 (2005)
45. A. Oprea, N. Barsan, U. Weimar, M. L. Bauersfeld, D. Ebling, Capacitive humidity sensors on flexile RFID labels, in *Proceedings of Transducers*, 2007, pp. 2039–2042
46. C.Y. Lee, G.B. Lee, Humidity sensors: a review. Sensor Lett. **3**, 1–15 (2005)
47. L. Löfgren, B. Löfving, T. Pettersson, B. Ottosson, S. Haasl, C. Rusu, K. Persson, O. Vermesan, N. Pesonen, P. Enoksson, Low-power humidity sensor, in *Proceedings of Eurosensors*, Dresden, Sept 2008, pp. 231–234
48. C. Zhang, T. Yin, Q. Wu, H. Yang, A large dynamic range CMOS readout circuit for MEMS vibratory gyroscope, in *Proceedings of IEEE Sensors*, Lecce, Oct 2008, pp. 1123–1126
49. M.A.P. Pertijs, K.A.A. Makinwa, J.H. Huijsing, A CMOS smart temperature sensor with a 3σ Inaccuracy of $\pm 0.1°$C from -55°C to 125°C. IEEE J. Solid State Circ. **40**(12), 2805–2815 (2005)

50. M. A. P. Pertijs, A. L. Aita, K. A. A. Makinwa, J. H. Huijsing, Voltage calibration of smart temperature sensors, in Proceedings of IEEE Sensors, Lecce, Oct 2008, pp. 756–759
51. G.C.M. Meijer, G. Wang, F. Fruett, Temperature sensors and voltage references implemented in CMOS technology. IEEE Sens. J. **1**(3), 225–234 (2001)
52. G. Ferri, N. Guerrrini, V. Stornelli, C. Catalani, A novel CMOS temperature control system for resistive gas sensor array, in *Proceedings of ECCTD*, Cork, 2005, pp. 351–354
53. S.S.W. Chan, P.C.H. Chan, A resistance-variation tolerant constant-power heating circuit for integrated sensor applications. IEEE J. Solid-State Circuits **34**(4), 432–437 (1999)
54. C. Falconi, C. Di Natale, A. D'Amico, J. Huijsing, A model of bipolar transistors for thermal sensors applications, in *Proceedings of IEEE Sensors*, Orlando, 2002
55. A. Sedra, K.C. Smith, *Microelectronic Circuits*, 5th edn. (Oxford University Press, New York, 2007). ISBN 0195142527
56. Internet resource: http://www.national.com. Datasheet LM35
57. E. Zampetti, C. Falconi, S. Pantalei, E. Martinelli, C. Di Natale, A. D'Amico, Thermal sigma delta modulation for quartz crystal microbalances, in Proceedings of *AISEM (Associazione Italiana Sensori e Microsistemi) Conference*, 2005
58. J. Huijsing, Integrated smart sensors. Sensors Actuator A **30**, 167–174 (1992)
59. G. Massobrio, P. Antognetti, *Semiconductor Device Modeling with SPICE* (Mc Graw Hill, New York, 1993). ISBN 0070024693
60. R. Hogervorst, J.H. Huijsing, *Design of Low-Voltage Low-Power Operational Amplifier Cells* (Kluwer Academic Publishers, Boston, 1996). ISBN 1441951652
61. W.A. Serdijin, A.C. van der Voerd, A.H.M. van Roermund, J. Davidse, Design principle for low-voltage low-power analog integrated circuits. Analog Integr. Circ. Signal Process. **8**, 115–120 (1998)
62. W.A. Serdijn, A.C. van der Woerd, J.C. Kuenen, Low-Voltage *Low-Power Analog Integrated Circuits* (Kluwer Academic Publishers, Boston, 1995). ISBN 9780792396086
63. S. Sakurai, M. Ismail, *Low-Voltage CMOS Operational Amplifiers* (Kluwer Academic Publishers, Boston, 1995). ISBN 9780792395072
64. G. Ferri, Low power adaptive biased integrated amplifiers. Analog Integr. Circ. Signal Process. **33**, 251–264 (2002)
65. G. Ferri, P. De Laurentiis, A. D'Amico, G. Stochino, Low Voltage Design, Electronics World, pp. 714–722, 1999
66. C. Enz, F. Krummenacher, E. Vittoz, An analytical MOS transistor model valid in all regions of operation and dedicated to low-voltage and low-current applications. Analog Integr. Circ. Signal Process. **8**, 83–114 (1995)
67. C. Toumazou, A. Payne, D. Haigh, *Analogue IC design: The Current Mode Approach* (Peter Peregrinus, London, 1990)
68. C. Toumazou, J. Lidgey, Universal Current Mode Analogue Amplifiers, in *Analogue IC design: The Current Mode Approach*, ed. by C. Toumazou, F.J. Lidgey, D.G. Haigh (Peter Peregrinus, London, 1990)
69. G. Palumbo, S. Palmisano, S. Pennisi, *CMOS Current Amplifiers* (Kluwer Academic Publishers, Boston, 1999)
70. K. Koli, K. Halonen, *CMOS Current Amplifiers* (Kluwer Academic Publishers, Boston, 2002)
71. G. Ferri, N. Guerrini, *Low voltage Low Power CMOS Current Conveyors* (Kluwer Academic Publishers, Boston, 2003). ISBN 1402074867
72. T.G. Beckwith, N.L. Buck, R.D. Marangoni, *Mechanical Measurements*, 3rd edn. (Addison-Wesley, New York, 1982)
73. P. Chen, C.C. Chen, C.C. Tsai, W.F. Lu, A time-to-digital-converter-based CMOS smart temperature sensor. IEEE J. Solid-State Circuits **40**(8), 1642–1648 (2005)
74. M. Landwehr, H. Grätz, A low-power, low-area, delay-line based CMOS temperature sensor, in *Proceedings of Eurosensors*, Dresden, Sept 2008, pp. 1392–1394

Chapter 3
The Voltage-Mode Approach in Sensor Interfaces Design

Electronic sensor interfaces, developed in *VM* approach, generally use a conversion towards an output DC voltage signal, especially where the variations of the sensing element (resistance or capacitance) are relatively small (one to two decades). On the contrary, if the sensor variations are larger, i.e., three decades or more, a conversion towards an output periodic AC voltage signal is mandatory. In fact, in the latter case, the conversion to an output voltage is not advisable owing to the limitations given by the noise (for low output voltage levels) and by the supply voltage (for high output voltage levels). In this Chapter, different *VM* readout circuit solutions for resistive, capacitive and temperature sensors are described. These circuits have been also implemented as discrete element *PCBs*, using commercial components and sometimes, in the case of integrated circuit design, with *LV LP* characteristics, in a standard *CMOS* technology.

3.1 Introduction to Voltage-Mode Resistive Sensor Interfaces

The choice of the first analog interface circuit for resistive sensors depends on the range of resistance variation that is related to the kind of sensor and to the amount of its variation. For example, a platinum resistive temperature sensor typically exhibits rather low relative resistance variations; on the contrary, *MOX*-based resistive gas sensors may change their resistance by orders of magnitude as a consequence of physisorption, chemisorption and catalytic reactions. In addition, the parasitic (typically capacitive) component of the sensing element can also affect the sensor estimation, in the case of an AC-excitation of the sensor.

As mentioned before, when the resistive sensing element varies into a reduced range (about one to two decades) and its capacitive contribution has not to be detected, a simple resistive voltage divider circuit, operating an *R-V* conversion, can be utilized as first analog interface. More in detail, considering Fig. 3.1 and, as an example, according to typical properties of semiconductor-based resistive gas

A. De Marcellis and G. Ferri, *Analog Circuits and Systems for Voltage-Mode and Current-Mode Sensor Interfacing Applications*, Analog Circuits and Signal Processing,
DOI 10.1007/978-90-481-9828-3_3,

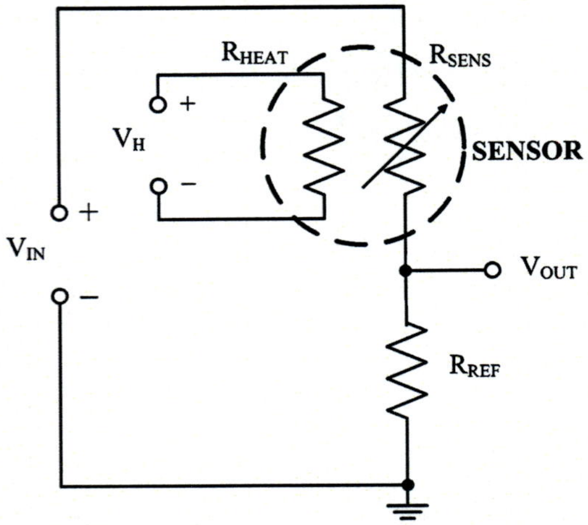

Fig. 3.1 A simple interface circuit suitable for the measurement of the resistance of gas sensor sensing element (V_{IN} = circuit excitation voltage; V_H = heater voltage; R_{REF} = load reference resistance; R_{HEAT} = sensor heater resistance; V_{OUT} = circuit output voltage)

sensors, if a DC supply voltage V_{IN} is applied to drive the sensing element R_{SENS} and utilizing a load reference resistance R_{REF} in the circuit, the output voltage V_{OUT} can be revealed and processed instantaneously, so to determine the sensor resistance. In Fig. 3.1 a heater resistance R_{HEAT} has been evidenced allowing the sensor resistance R_{SENS} to work at a suitable operating temperature (typically, for example in gas sensors, its best value that ensures a higher sensitivity and selectivity of the measurand); in the next figures, this aspect will be neglected for the sake of simplicity [1].

From the voltage divider, changes of sensing element resistance R_{SENS} can be evaluated, once R_{REF} and V_{IN} are known, by measuring the circuit output voltage V_{OUT}, as follows:

$$R_{SENS} = R_{REF}\left(\frac{V_{IN}}{V_{OUT}} - 1\right). \tag{3.1}$$

The fully differential version of the voltage divider (for what concerns the output voltage) is the well-known Wheatstone bridge (resistive bridge), whose schematic circuit is shown in Fig. 3.2, which still operates an *R*-*V* conversion, better rejecting the common-mode. In particular, it can be used for converting low sensor resistance variations into a differential voltage signal V_{OUT} [1]. It is composed by four resistances and, usually, a resistive sensor is one of the four branches of the bridge whose resistive sensing element varies when an external physical or chemical phenomenon occurs.

Referring to Fig. 3.2, the bridge is balanced when the ratio of resistances of any two adjacent arms is equal to that of the remaining two arms (taken in the same sense): $R_1/R_2 = R_3/R_{SENS}$ or $R_1/R_3 = R_2/R_{SENS}$. As a particular case, the bridge is also balanced when all the four resistances are the same value: $R_1 = R_2 = R_3 = R_{SENS}$. In these cases, the generated differential output voltage signal V_{OUT} is equal to zero. On the contrary, starting from equilibrium condition (balanced bridge), when the sensor varies its resistance R_{SENS}, a non-zero differential voltage V_{OUT} can be revealed at the output of the bridge, whose value is proportional to

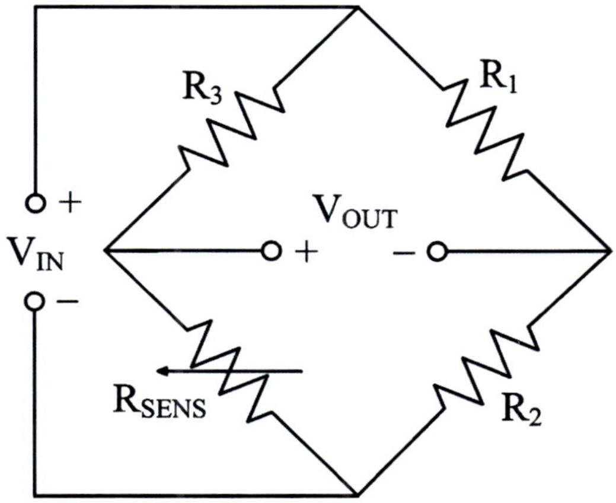

Fig. 3.2 A resistive sensor interface based on Wheatstone bridge: the *R-V* conversion

the sensor resistance variation (but only when these variations are low). More in general, the generated output voltage can be expressed as:

$$V_{OUT} = \left[\frac{R_1 R_{SENS} - R_2 R_3}{(R_1 + R_2)(R_3 + R_{SENS})} \right] V_{IN}. \tag{3.2}$$

As mentioned before, the main drawback of this kind of resistive sensor interface is its unsettable and low sensitivity, dependent only on the total supply voltage for low resistance variations. In fact, if V_{IN} is the total supply voltage, in the basic Wheatstone bridge, the sensitivity, defined as the ratio between the differential output voltage change and the relative variation of the sensor resistance R_{SENS}, is constant and equal to $V_{IN}/4$ for the variation of only one resistance of the bridge (note that the value of the sensitivity is the same of the simple voltage divider). In fact, if the relative variation of the sensor resistance is relatively low (e.g., about less than 5% with respect to the sensor resistance base-line), an almost linear relation between the differential output voltage and the relative variation itself exists as follows:

$$V_{OUT} = V_{IN} \frac{x}{4 + 2x} \cong V_{IN} \frac{x}{4}, \tag{3.3}$$

being x the relative resistance variation, determined with respect to sensor resistance base-line. As shown in Fig. 3.3, through a suitable null detector (e.g., a simple multimeter or voltmeter), which reveals the balanced condition of the bridge (i.e., the output voltage equal to zero), by changing the value of a variable resistor R_{VAR}, it is possible to determine the unknown resistance value provided by the resistive sensor R_{SENS}, that differently changes as a function of an external physical or chemical phenomenon to be detected and measured.

The use of a differential input *OA*-based voltage amplifier allows to enhance the front-end circuit sensitivity. This *VM* circuit, performing also the single-ended conversion, can be placed at the output nodes of the bridge (V_{OUT} terminals). In this

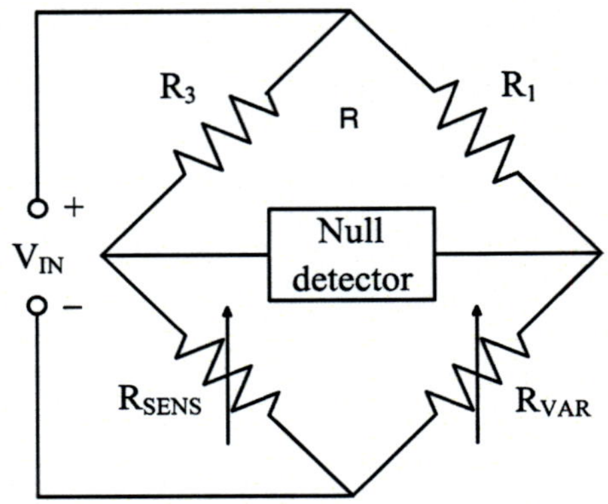

Fig. 3.3 The use of a null detector in the Wheatstone bridge circuit

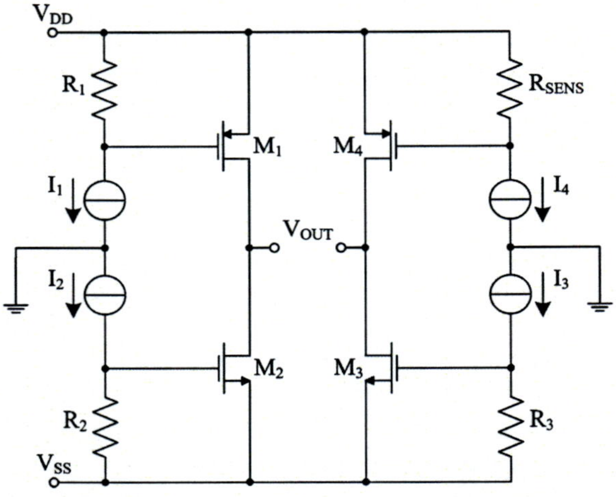

Fig. 3.4 Active Wheatstone bridge as resistive sensor interface

case, an instrumentation amplifier is the best possible topology since it shows a very high input impedance and, through the feedback configuration, a well-defined and controlled amplification factor. An additional important characteristic of this amplifier must be its low input voltage offset (see Appendix 2 for further details).

An improved topology of the bridge is based on the conversion of the passive resistances into active ones, utilizing *CMOS* transistors, with the aim to obtain better sensitivity and resolution values. The modified topology, whose schematic circuit is shown in Fig. 3.4, introduces *CMOS* transistors to implement the four branches of the bridge [2]. This circuit has a symmetrical structure to achieve a high *CMRR* performance so, at the output terminals, a common mode feedback circuit (*CMFB*) must be added to fix the output voltage V_{OUT} at the half of the total supply level and to guarantee the maximum output dynamic range. This circuit can be designed to work with a low supply voltage (e.g., 1.2 V total power supply) and also with a low power consumption.

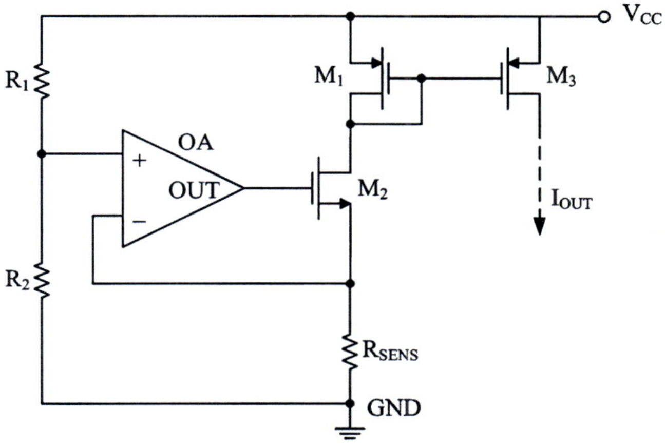

Fig. 3.5 Resistance-to-current converter as resistive sensor interface

Through an accurate circuit design, it is possible to achieve an improvement of both the sensitivity and the resolution by almost two orders of magnitude, in comparison with the passive topology (the resistive Wheatstone bridge). Moreover, an improvement of the bridge sensitivity can be obtained also employing an *ISFET* sensor and three MOSFET devices as bridge components [2, 3]. Also in this case, the circuit sensitivity can be further improved by the use of an *OA*-based voltage amplifier.

Another simple resistive sensor interface is shown in Fig. 3.5. This solution (to be implemented as an integrated circuit because of the presence of *MOSFETs*) is based on a Resistance-to-Current (*R*-*I*) converter which allows to generate an output current I_{OUT} dependent on the sensor resistance value R_{SENS}.

Through a simple analysis it is possible to evaluate the generated current as follows:

$$I_{OUT} \cong V_{CC}\frac{R_2}{R_1 + R_2}\frac{1}{R_{SENS}}, \tag{3.4}$$

assuming that M_1 and M_3 are matched and equal transistors. Obviously, the output current I_{OUT} can be easily converted into a voltage output signal through a further Current-to-Voltage (*I*-*V*) conversion [1].

3.2 The DC Excitation Voltage for Resistive Sensors

Sensors that behave as pure resistors as well as those sensing elements which do not bear an alternating voltage (i.e., an AC excitation signal) since they give bad responses and lower lifetimes [4, 5], can be excited by a constant voltage value (i.e., a DC excitation signal), especially when, for several specific applications, it is

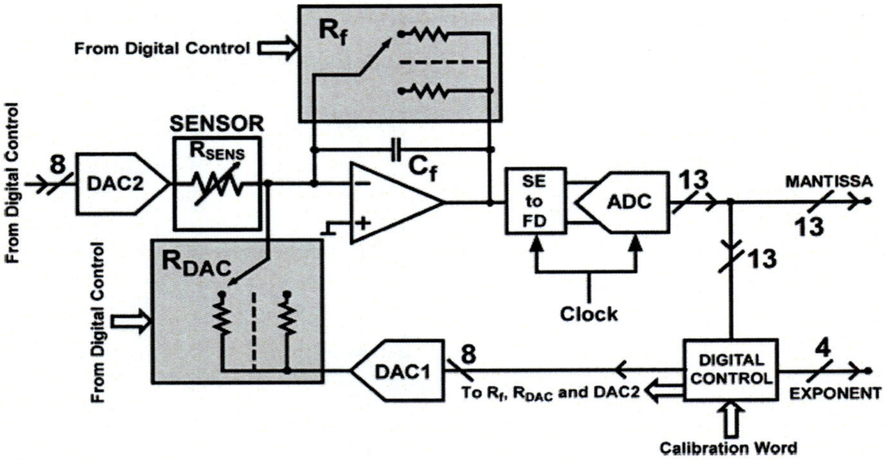

Fig. 3.6 Block scheme of a resistive sensor interface with scaling factor and DC excitation voltage of the sensor

also possible to neglect the effect of the resistive sensor parasitic component (e.g., parasitic capacitance). In addition, sometimes, especially in gas sensors [6–8], the base-line resistances of the sensors may typically vary from a small value (e.g., 200 Ω) up to a very big one (e.g., 10 MΩ); furthermore, the sensor resistance must be measured with a precision near to 0.1% in order to detect the different gases with a sufficient resolution (i.e., 1 ppm). These constraints would require, without any range compression, a particular interface solution which could perform the compression of the sensor resistance value (i.e., R_{SENS}) through a logarithmic-based algorithm. Unfortunately, even if a wide range is guaranteed by this technique, it is difficult to get an accuracy better than 1% [9, 10]. An alternative interface which, after calibration, allows a final worst case measurement with an accuracy better than 0.1% in about 10 ms per sensor query, fast enough for allowing dynamic pattern recognition algorithms, which gather important information from the derivatives of the sensor responses, has been recently proposed [11–15]. This interface, whose block scheme is reported in Fig. 3.6, utilizes a DC excitation voltage for the resistive sensor R_{SENS} [14, 15], so, clearly, it is not able to detect any capacitive component of the resistive sensing element. It operates an *R*-*V* conversion, giving a digital output; the desired resolution all over the required dynamic range has been satisfied by splitting the system scale in ten sub-intervals, each of them having an operative width of about half a decade. The calibration is necessary so to compensate the offset and gain error mismatch by means of two *DACs* which regulate, respectively, a programmable current for the offset error and sensor bias constant voltage for the gain error. Furthermore, since the measured dynamic range of the proposed circuit is more than five decades [100 Ω–20 MΩ], the interface circuit fulfils all the requirements for both static and dynamic pattern recognition algorithms. More in detail, considering the gas sensor a pure resistor and the gas concentration proportional to the resistance variation, the interface circuit channel consists of a single-ended continuous-time trans-resistance stage that converts the current flowing through the sensor into a voltage and by a differential switched capacitor oversampled incremental A/D converter that reads the output of the trans-resistance amplifier.

The desired resolution over the whole required dynamic range has been satisfied by dividing the dynamic range in ten sub-intervals (or scales, each of them with a width of about half a decade), but, because of this split, a calibration technique is needed to compensate offset and gain error mismatch between different scales. Furthermore, a partial overlap of adjacent sub-intervals of about a quarter of decade helps in the calibration phase, which consists in the actual conjunction of consecutive scales in the analog response. The integrated circuit of the interface shown in Fig. 3.6 has a power dissipation of about 6 mW from a single 3.3 V supply voltage, while the nominal system read-out rate is 100 Hz, considering pre-amplifier settling, A/D conversion and scale selection time. The designed integrated circuit (developed and fabricated in a standard 0.35 μm *CMOS* technology), operating at 3.3 V single supply voltage, requires a silicon area of about 3.1 mm^2, not including chip pads [11, 13]. The regulated current source used to compensate inter-scale system offset mismatch is performed with a 8-bit buffered resistive Digital-to-Analog Converter (*DAC1* in the schematic) and a programmable resistor R_{DAC}, that also needs to be selected from an array. In the design, it has to be $R_f = R_{DAC}$, so to keep the operational amplifier with gain and feedback factors of the same order of magnitude over the entire dynamic range of the interface circuit. In this way, a good matching between the integrated resistors R_f and R_{DAC} is also obtained. Furthermore, if a fine regulation of the sensor voltage reference (i.e., V_{REF}) is provided, it is possible to correct separately the gain-error of each of the ten scales available in the circuit. This has been achieved in the design by introducing an additional buffered *DAC* (*DAC2*). The two 8-bit *DACs* (*DAC1*, *DAC2*) and the two selector circuits for R_f and R_{DAC} are all controlled by a common digital unit, whose tasks are the choice of the measurement range and the actual correction of offset and gain inter-scale errors by applying the information provided as "calibration words" during initial setup phase. The sensor query is performed in two steps: in the first step, the scale in which falls the value to be measured is found with successive coarse measurements during which the *ADC* is used at reduced resolution, 6 bits, to decrease the search time, which is performed by decrementing each time the feedback resistance value R_f. The sub-range, which does not saturate the A/D converter with the adequate safety margin of about 150 mV, is used for the fine 13-bit measurement. In fact, the digitized resistance value will consist of a 13-bit mantissa, supplied by the *ADC*, and of a 4-bit exponent, which actually is the identification number of the scale used for the fine measurement.

As just underlined, for these kind of interfaces, which operate an *R-V* conversion for a wide resistive range, the system calibration is mandatory, so when resistive sensor base-line or its variation can change of different decades (also more than 5–6), the *R-V* conversion is not practically suitable and an *R-T* conversion is decidedly better.

In the following Section we will describe some solutions of low-cost uncalibrated fully-integrable front-ends, in *VM*, being based on *OA* as active block, for high valued resistive sensor interfacing, performing an *R-T* conversion and always utilizing a DC excitation voltage for the sensor.

3.2.1 Uncalibrated DC-Excited Sensor Based Solutions

In Fig. 3.7 an integrable *OA*-based resistive sensor interface, performing an *R-T* conversion, is presented. In order to evaluate only the resistive behavior of the utilized sensor, this front-end excites the sensor with a DC voltage (V_{IN}) [16]. The proposed circuit, based on an oscillator topology, is able to reveal more than four decades of high resistance variations (from about 1 MΩ to more than 10 GΩ), typical of some resistive gas sensors (i.e., *MOX*-based gas sensors). The proposed front-end has been designed, as integrated circuit, in a standard *CMOS* technology (AMS 0.35 μm), so to be suitable in low-cost portable applications. It is formed by five main parts: a resistance to voltage converter (OA_1, R_{SENS}, R_1), two buffers (one of which non-inverting, B_1, and the other inverting, B_2), an inverting integrator (OA_2, R_2, C_1) and a Schmitt Trigger (OA_3, R_3, R_4). Furthermore, a couple of switches, S_1 and S_2, operates in opposite phase and provides two different DC voltages (depending on R_{SENS}), with opposite values, to the inverting integrator input. Fig. 3.8 shows the main voltage signals generated at output (V_{OUT}) and internal (V_{TH} and V_A) nodes of the circuit.

Referring to Figs. 3.7 and 3.8 and considering an ideal behaviour for all the components, through a straightforward analysis, it is possible to observe a linear relation between the period T of generated output signal and the sensor resistance R_{SENS}, according to the following expression:

$$T = 2\frac{R_4}{R_3 + R_4}\left(V_{SAT+} - V_{SAT-}\right)\frac{R_2}{R_1}\frac{C_1}{V_{IN}}R_{SENS}. \tag{3.5}$$

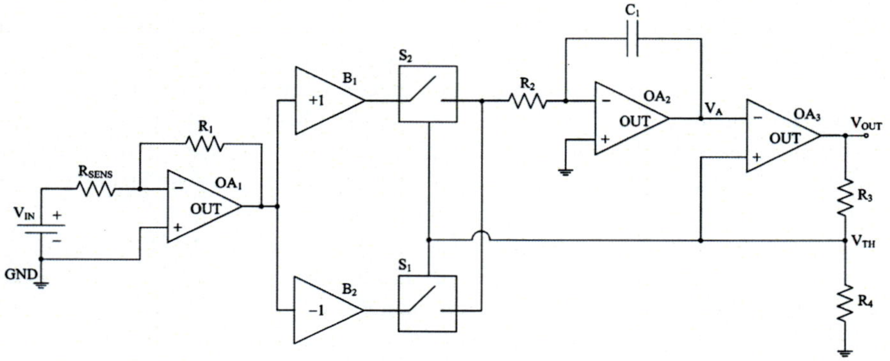

Fig. 3.7 The proposed *OA*-based interface with a DC resistive sensor excitation voltage

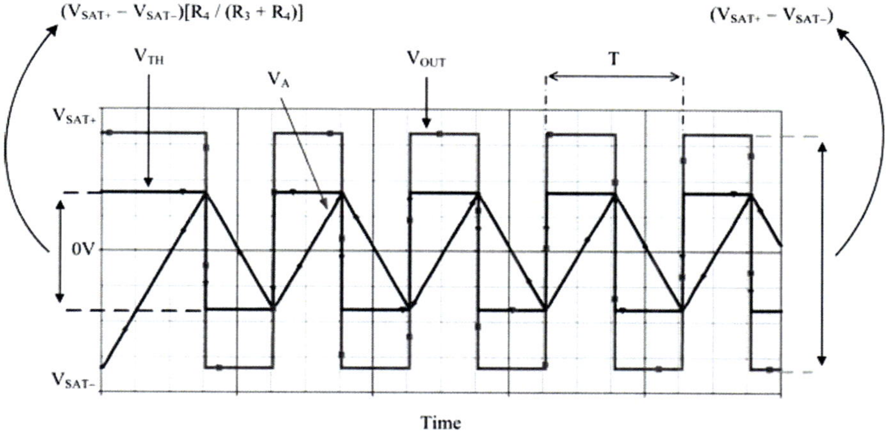

Fig. 3.8 The main voltage signal behaviours (V_A, V_{TH} and V_{OUT})

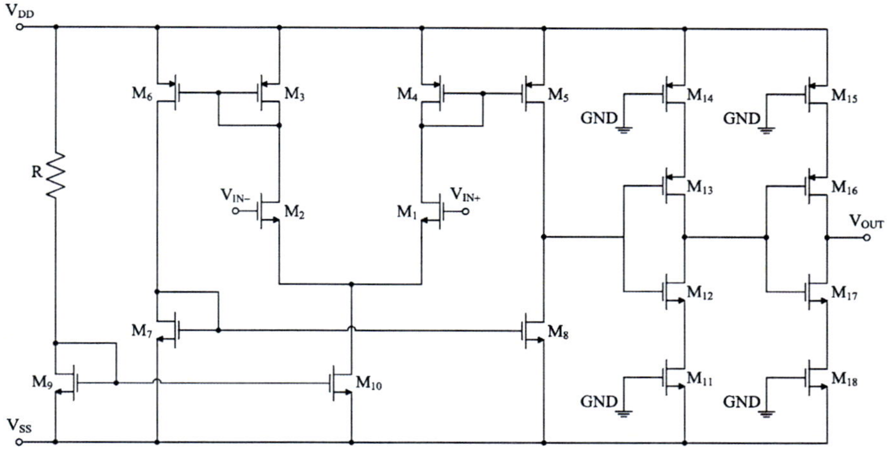

Fig. 3.9 Designed *OTA* schematic at transistor level

Simulations on the complete integrated solution, designed with dual supply voltage (±1.65 V, so that $V_{SAT+} = -V_{SAT-} \approx 1.65$ V) and using the *OTA* shown in Fig. 3.9, whose characteristics are summarized in Table 3.1, have confirmed the possibility to estimate high resistive values for more than four decades of resistance variations (see Table 3.2).

Experimental measurements have been performed using sample components on a discrete-element board, in particular utilizing LF411 as amplifier, supplied at ±15 V. The period of the generated square-wave signal, evaluated at the output node of Schmitt Trigger (see Fig. 3.7), has shown a good linearity with a reduced

Table 3.1 *OTA* main characteristics: post-schematic simulation results

OTA parameter	Value
Voltage supply	±1.65 V
Power dissipation	1.2 mW
GBW	34 MHz
Output dynamic range	Full
Open loop voltage gain	97 dB
Slew-Rate	18 V/μs
Input voltage offset	1.5 μV
Input equivalent noise	205 nV/√Hz@1kHz

Table 3.2 Simulated and theoretical period vs. R_{SENS} (integrated solution)

R_{SENS} [Ω]	Simulated period [s]	Theoretical period [s]	Relative error [%]
1 M	225.839 μ	200 μ	+12.91
5 M	962.991 μ	1 m	−3.70
10 M	1.923 m	2 m	−3.84
50 M	9.605 m	10 m	−3.95
100 M	19.210 m	20 m	−3.95
500 M	96.165 m	100 m	−3.83
1 G	192.401 m	200 m	−3.79
5 G	971.895 m	1	−2.81
10 G	1.968	2	−1.58
50 G	10.587	10	+5.87

Table 3.3 Experimental results: measured and theoretical period vs. R_{SENS} (using prototype board)

R_{SENS} [Ω]	Measured period [s]	Theoretical period [s]	Relative error [%]
5 M	856.10μ	842.50 μ	+1.59
10 M	1.72 m	1.69 m	+1.74
50 M	8.54 m	8.43 m	+1.29
100 M	16.81 m	16.85 m	−0.24
500 M	86.19 m	84.25 m	+2.30
1 G	176.20 m	168.50 m	+4.57
5 G	898.90 m	842.50 m	+6.69

relative error, as reported in Table 3.3, also for high resistive values (for these measures, sample commercial resistors have been utilized). These experimental results, performed considering the following values: $V_{IN} = 1.65\,\text{V}$, $V_{SAT+} = -V_{SAT-} \approx 13.9\,\text{V}$, $C_1 = 100\,\text{pF}$, $R_1 = 10\,\text{M}\Omega$, $R_2 = 1\,\text{M}\Omega$, $R_3 = R_4 = 100\,\text{k}\Omega$, have confirmed the theoretical expectations for about three decades of resistance variations (in this case, front-end sensitivity has been set to about 168 μs/MΩ).

A simplified version of the circuit described above is depicted in Fig. 3.10. This new version employs only three *OAs* (reducing the utilized active blocks) and four switches in order to properly control the voltage signal V_1 generated by the first stage, dependent on the sensor resistance value R_{SENS}. In this way, some problems due to the implementation of the two buffers in the previous solution (as voltage

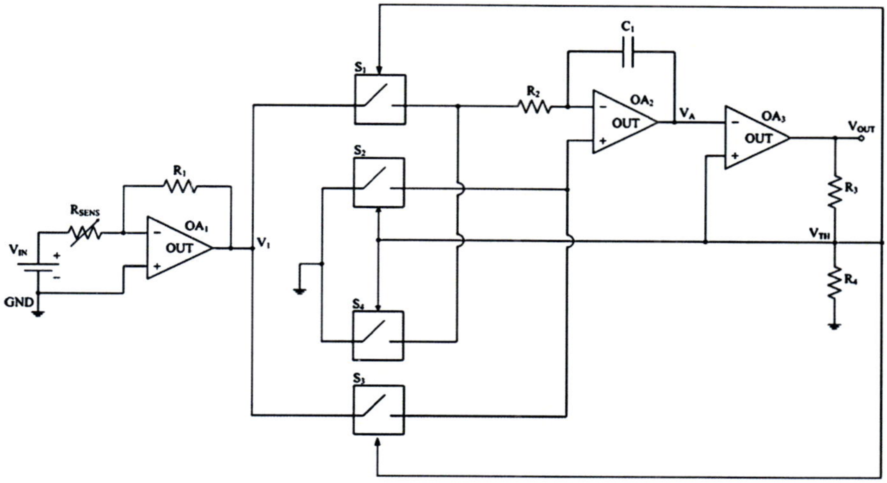

Fig. 3.10 The modified *OA*-based interface with sensor DC excitation voltage

offsets, also of different values for the two buffers) have been overcome. Moreover, it is important to highlight that also the first stage can be easily replaced by a simple voltage divider (the sensor resistance R_{SENS} and a reference one R_1). More in detail, in this interface topology, OA_2 operates both as an inverting integrator and as a non-inverting one, through the suitable use of the four switches, while the same generated voltage signal V_1 represents always its input signal, which has to be integrated (S_1–S_2 closed, S_3–S_4 opened and vice-versa). Through a straightforward analysis, it is possible to evaluate the relationship between the sensor resistance R_{SENS} and the period T of the output square wave signal, as follows:

$$T = 2R_2C_1\left[\left(\frac{R_4}{R_3+R_4}\frac{V_{SAT+}-V_{SAT-}}{R_1V_{IN}}\right)R_{SENS}-1\right]. \tag{3.6}$$

Since the voltage integrator has a double operating function, the presence of the capacitance C_1 involves a charge effect, which influences instantaneously the ramp signal when there is the operating function commutation (from inverting to non-inverting and vice versa), through a vertical edge on V_A, evidenced in Fig. 3.11, whose value depends on the V_1 level. PSpice simulations have confirmed the validity of this solution, for about three decades (see Table 3.4).

3.2.2 *Fast DC-Excited Resistive Sensor Interfaces*

As described in the previous Paragraph, *R-T* converters, which exploit the easiness of measuring times and intervals over a wide range of variation, are widely

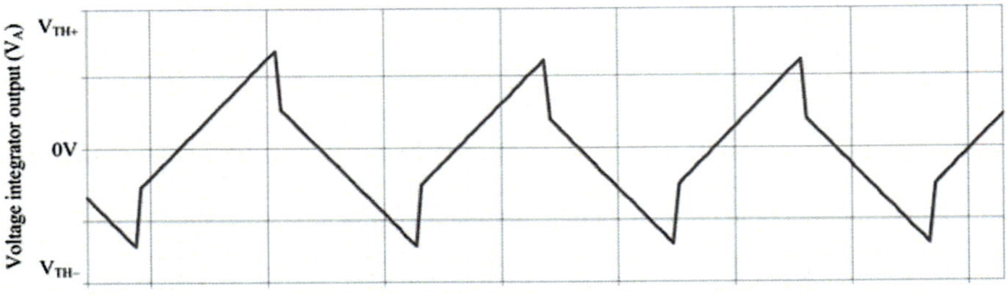

Fig. 3.11 A particular of the ramp signal V_A generated by integrator: the voltage gap is due to charge effect dependent on R_{SENS} (low value sensor resistance provides a high voltage gap)

Table 3.4 Simulated and theoretical period vs. R_{SENS} (PSpice simulations)

R_{SENS} [Ω]	Simulated period [s]	Theoretical period [s]	Relative error [%]
10 M	1.746 m	1.8 m	−3
100 M	19.405 m	19.8 m	−2.02
500 M	97.917 m	100 m	−2.08
1 G	195.681 m	200 m	−2.15
5 G	976.651 m	1	−2.33
10 G	1.948	2	−2.58

used in electronic interfaces thanks to their low-cost, low-noise and high-range characteristics. However, *R-T* main limit is in the variable and, in some cases, long measuring time, typically ranging from microseconds (corresponding to tens of kilohms) to several seconds (related to tens of gigohms), thus preventing an accurate analysis of fast transients. Moreover, recent studies about some gas sensors (e.g., *CO*) have demonstrated the opportunity of a more detailed analysis of the fast transients, for example during the issue of heating pulses [17].

For all these reasons, an interface system for resistive sensors has been recently implemented so to obtain a fast read-out feature [18]. Particularly, a low-cost electronic circuit has been developed to allow a regular sampling frequency on the order of 100 Hz, still keeping the measuring range over six decades or more. This solution introduces a different approach based on a combination of the *R-T* method with a technique based on the Least Mean Square (*LMS*) algorithm, covering a range of about 10 kΩ ÷ 10 GΩ and allowing reduced measurement times (maximum T_{meas} = 10 ms). Therefore, the circuit is suitable for the fast thermal transients analysis of resistive gas sensors as, for an example, the SnO_2 nanowire *MOX* sensor.

The main block of the proposed interface system, based on an inverting voltage integrator, is reported in Fig. 3.12. The sensor is considered to be in a very stable environment, so a DC excitation voltage V_{EXC} has been adopted. The current I_S flowing through the sensor resistance R_{SENS} is converted, through the capacitor C, in a voltage V_{OUT}, varying in a linear way, with a fixed slope α, depending on the sensor itself (i.e., the R_{SENS} value), as shown in Fig. 3.13.

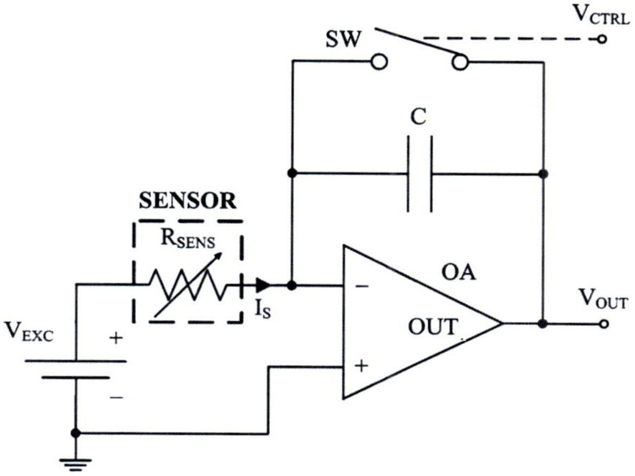

Fig. 3.12 Scheme of the integrator circuit

Fig. 3.13 Output signal behavior

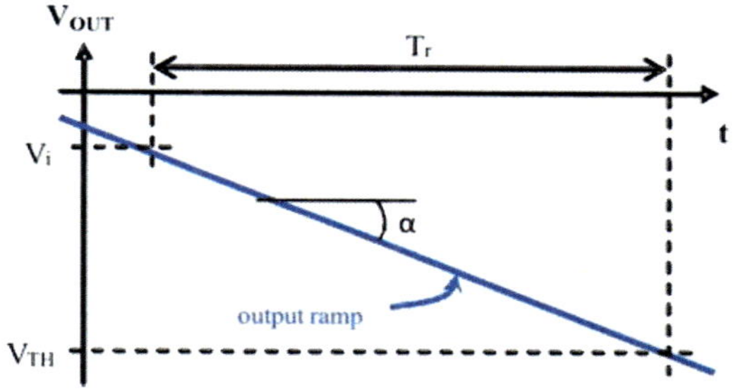

The relation between the sensor resistive value R_{SENS} and the slope α of the output voltage ramp V_{OUT} is the following:

$$|\alpha| = \frac{V_{EXC}}{R_{SENS}C}. \tag{3.7}$$

The estimation of the ramp slope α can be performed in several ways. In the classical *R-T* converter circuits, the time T_r required by the ramp V_{OUT} to reach a fixed and well-known voltage value V_{th} (threshold) is measured and the sensor resistive value R_{SENS} can be estimated using the following inverse relation:

$$R_{SENS} = \frac{V_{EXC}}{|V_{th} - V_i|}\frac{T_r}{C} \tag{3.8}$$

being V_i the value of V_{OUT} voltage at the beginning of the measurement. The switch *SW* needs to be suitably driven, by a control voltage V_{CTRL}, so to reset the output voltage to the initial value (in this case, it is $V_i = 0\,\mathrm{V}$) and to allow

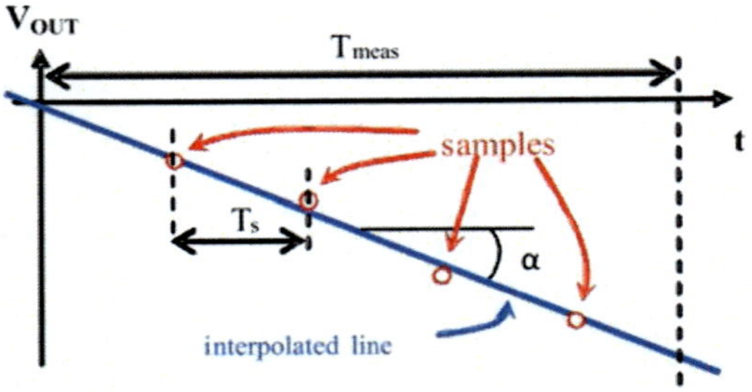

Fig. 3.14 Least mean square interpolation algorithm applied to the integrator output ramp

continuous measurements of the time T_r. It should be noticed that to perfectly reset the integrator output V_{OUT} is not a trivial job, because very high insulation switches have an on-resistance in the order of kΩs. Otherwise, a two-thresholds (V_i, V_{th}) circuit can overcome uncertainty due to this aspect. A comparator can be used to detect when the output signal V_{OUT} reaches the threshold V_{th}. Using the comparator output signal, a digital electronic system can be used to easily estimate the T_r interval and to drive the switch *SW*. The choice of the V_{th} value is a tradeoff between the desired time resolution in the measurement and the time required to perform the estimation. In fact, the less the time T_r is (small V_{th}), the worse the resolution related to the time estimation is. On the contrary, if a high V_{th} value is chosen, the time T_r becomes bigger than the desired measuring time T_{meas}, when high sensor resistance values are considered.

For example, if $C = 100\,\text{pF}$, $V_{EXC} = 1\,\text{V}$, $T_{meas} = 10\,\text{ms}$, and $V_{th} = 10\,\text{V}$, the maximum R_{SENS} value which can be estimated is 10 MΩ. If the threshold value is lowered to $V_{th} = 1\,\text{V}$, then the measurement range is extended up to 100 MΩ. However, in the first case, the T_r value with $R_{SENS} = 10\,\text{k}\Omega$ is 10 μs, while in the second case it is only 1 μs, requiring a high-resolution timing measurement system (better than 10 ns). Nevertheless, even in case of using both thresholds according to the R_{SENS} value, the problem in measuring resistances greater than 100 MΩ still exists.

Therefore, the proposed approach intends to keep the *R-T* conversion technique for small sensor resistance values adding new estimation methods if the threshold is not reached in the desired measurement time (high sensor resistance values). In fact, if the slope α of the ramp is too slow, the proposed solution is based on the estimation of α value by using an interpolation method starting from few samples of the ramp acquired in a limited time, less than the desired T_{meas}. More in detail, the *LMS* interpolation method allows the determination of the slope of a line which minimizes the squared error with respect to the acquired experimental points, as shown in Fig. 3.14.

Depending on the measuring time T_{meas} and on the number N or the samples needed for the application of the *LMS* method, the sample frequency $F_s = 1/T_s$ can be determined considering that $N \cdot T_s < T_{meas}$. Theoretically, such a method can be used for any resistive value, but actually there are limitations for its applicability both for high resistive values and for small ones. In fact, when high resistive values

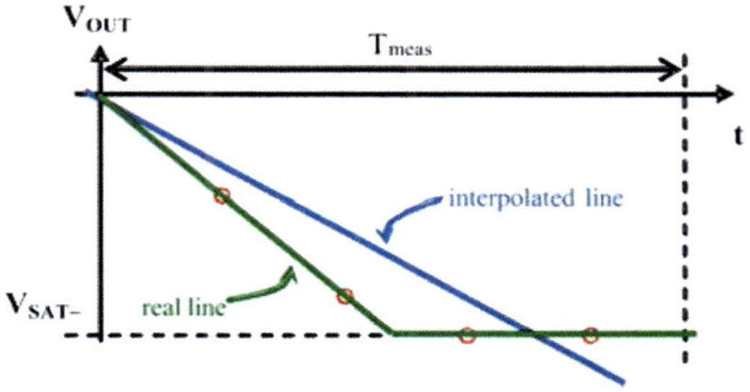

Fig. 3.15 Least mean square interpolation fails when the output voltage reaches the saturation value

are considered, the slope of the ramp is very small and the variation of the V_{OUT} voltage within the measuring time T_{meas} can be on the order of the A/D converter resolution or of the noise present in the circuit. On the other hand, when small resistive values are considered, the ramp slope is very high and the limitation of the output range of the operational amplifier or the input range of the *A/D* converter can occur, as visible in Fig. 3.15. In this situation, the integrator output V_{OUT} reaches the negative saturation voltage V_{sat-} before the last sample is taken (e.g., with the previous value for V_{EXC}, C and T_{meas}, if we select $V_{sat-} = -10\,\text{V}, N = 5$, then $Fs = 500$ Sample/s, the lower limit is about $10\,\text{M}\Omega$ that is the minimum R_{SENS} value which can be estimated).

For this reason, the circuit shown in Fig. 3.16 has been developed as a prototype *PCB*. It is based on an integrator whose output voltage V_{OUT} is used by two comparators tuned to different threshold values $V_{TH,L}$ and $V_{TH,H}$. In such a way, the *R-T* method can be applied with improved accuracy and/or range. In addition, V_{OUT} is also sampled by an *ADC* to allow the use of the *LMS* interpolation method when the *R-T* technique fails. The time estimation is performed by simple counters implemented in a programmable logic device (i.e., a Cyclone *FPGA* from Altera) which is also devoted to the control of the reset switch *SW* through a suitable control voltage V_{CTRL}. In addition, the *FPGA* sends the measured data to a *PC* by means of an RS232 link and generates the correct trigger signal to control the A/D conversion within the measuring cycle. The A/D conversion is performed by a *PCI* acquisition board from National Instruments (i.e., NI-6110), with a 12 bit resolution.

The value of R_{SENS} can be easily computed starting from α value by inverting Eq. 3.7. If, in that cycle, both time T_i (related to the first threshold V_i interception) and time T_{th} (related to the second threshold V_{th} interception) are available, then R_{SENS} can be computed applying Eq. 3.8, where $T_r = T_{th} - T_i$. It should be noticed that the R_{SENS} estimation by means of Eq. 3.8 (*R-T* method), if available within T_{meas}, should be preferred. On the other hand, the estimation by means of α value, computed according to the *LMS* method, works properly only if the ramp does not saturate within T_{meas} and allows to complete the estimation before the ramp reaches the thresholds. Experimental results have been conducted using: $V_{EXC} = 1\,\text{V}$, $V_{TH,L} = 1\,\text{V}$, $V_{TH,H} = 10\,\text{V}$, $F_s = 10\,\text{k}$ Sample/s, $N = 100$, $T_{meas} = 10\,\text{ms}$ and power supply equal to $\pm 12\,\text{V}$. The voltage limitation for the *LMS* method (V_{sat-}, see Fig. 3.15) is not determined by the integrator output range, but by the input range

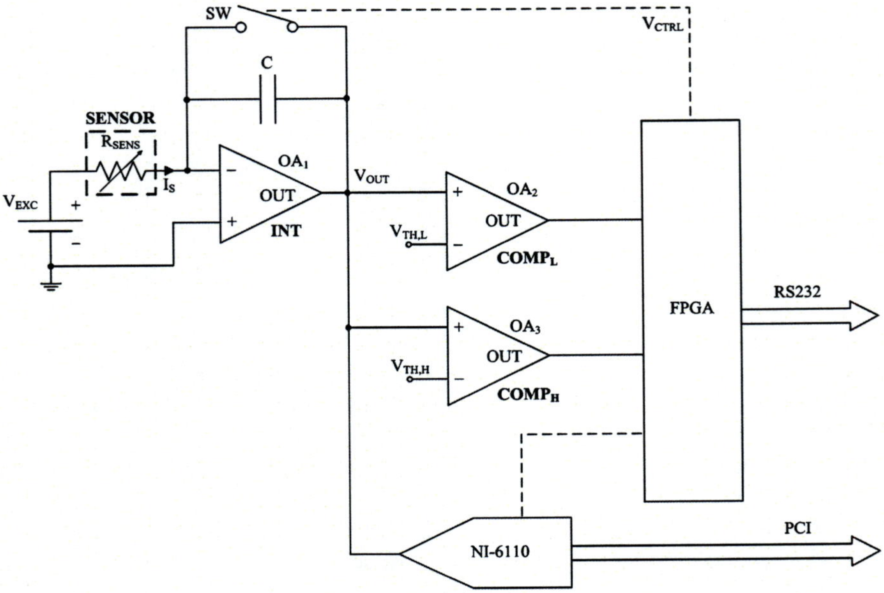

Fig. 3.16 The proposed system, combining the *R-T* approach with *LMS* interpolation method

Table 3.5 Experimental results with commercial resistors

R_{SENS} "true" $[M\Omega]$	*R-T* mean $[M\Omega]$	*R-T* std [%]	*R-T* error [%]	LMS_{100} mean $[M\Omega]$	LMS_{100} std [%]	LMS_{100} error [%]	LMS_8 mean $[M\Omega]$	LMS_8 std [%]	LMS_8 error [%]
0.0102	0.0110	0.00	8.20						
0.0996	0.1005	0.02	0.92						
0.9998	1.0022	0.01	0.24						
9.9690	9.9931	0.01	0.24	10.075	0.01	1.06	10.072	0.02	1.04
100				100.85	0.06	0.85	100.89	0.24	0.89
1,000				1009.3	0.51	0.93	1017.7	1.97	1.77
10,000				9338.7	4.84	−6.61	10, 597	39.66	5.97
20,000				16561	8.56	−17.19			
50,000				34030	22.40	−31.94			

of the NI-6110 acquisition board, which is $\pm 10\,V$ (we consider $V_{sat-} = -10\,V$). A set of commercial resistors in the range from $10\,k\Omega$ up to $100\,G\Omega$ has been used to characterize the measuring performances of the complete system (resistors have been also measured using the Fluke 8840 A multimeter). Table 3.5 shows the estimation results obtained using three different approaches: the *R-T* technique, the *LMS* algorithm using all the 100 samples for every cycle (LMS_{100}) and the *LMS* algorithm using only 8 samples for every cycle (LMS_8). The reported "error" is the relative error computed as the difference between the estimated and the "true" value (measured). The *R-T* approach estimation is computed considering the ramp time between the two thresholds, that is applying Eq. 3.8 with $V_i = V_{TH,L}$ and

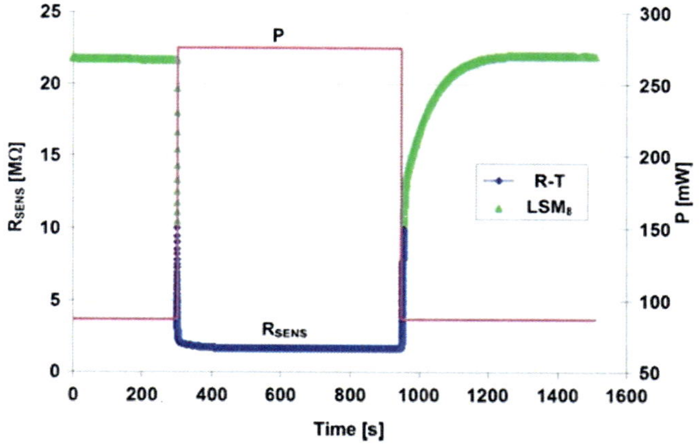

Fig. 3.17 Sensor response to a fast variation of the power issued to the heater

$V_{th} = V_{TH,H}$. In this situation, the upper operative limit of the technique is about $10\,\mathrm{M}\Omega$. As expected, the estimation error looks significant with resistance values on the order of $10\,\mathrm{k}\Omega$, due to the poor time resolution (50 ns) for the measure of the threshold interceptions. However, when available, the *R-T* method should be preferred thanks to the very low value of the relative standard deviation ("*std*", see Table 3.5). On the contrary, the *LMS* technique lower limit is about $10\,\mathrm{M}\Omega$ if the slope estimation is performed using 100 samples, whereas it is about $8\,\mathrm{M}\Omega$ if only 8 samples are considered (in this case, samples are taken after 1, 2,... 8 ms from the beginning of the ramp, therefore no matter if the saturation limit is reached after the last sample has been taken).

From these results, the LMS_8 method leads to worse performances than the LMS_{100} one (both in terms of estimation error and measuring range) since with very high resistance values (>$10\,\mathrm{G}\Omega$) 8 samples seem not to be enough to estimate with sufficient reliability the ramp slope. Moreover, even if 100 samples are used, the 12-bit resolution (corresponding to about 5 mV) of the A/D converter leads to a significant error in the ramp slope estimation if high resistance values are considered (with a $10\,\mathrm{G}\Omega$ resistance, the ramp decreases of only 10 mV in 10 ms).

Furthermore, the system has been tested using a commercial sensor and examining its behavior when changing the heater power, so its operating temperature. The sensor used in this test is a SnO_2–based nanowire sensor. Fig. 3.17 shows the sensor response during such a test, where the heater voltage has been quickly changed from 1 to 2 V and then again to 1 V, causing a change of the sensor working temperature. The R_{SENS} values are obtained from the *R-T* estimation, when available (for resistance values up to $10\,\mathrm{M}\Omega$), otherwise using the LMS_8 approach. Details of the sensor response in Fig. 3.17, during the falling transients, are reported in Fig. 3.18. In this figure the point where the method estimation changes (from LMS_8 to *R-T*) has been highlighted with an alteration in the line color (from light grey to dark grey). The proposed method allows to track with regular sampling the sensor

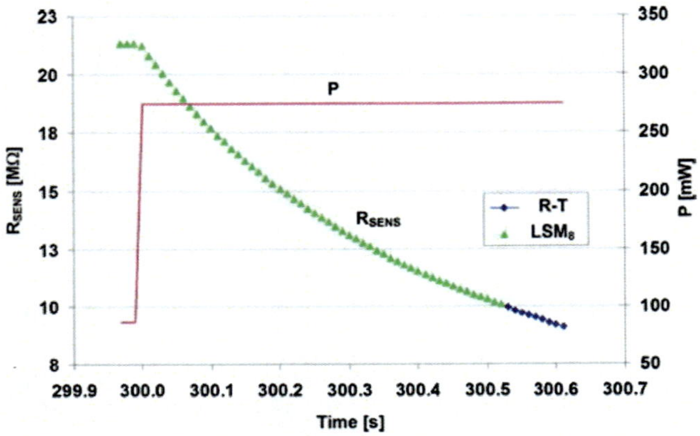

Fig. 3.18 Particular of the falling transient of Fig. 3.17

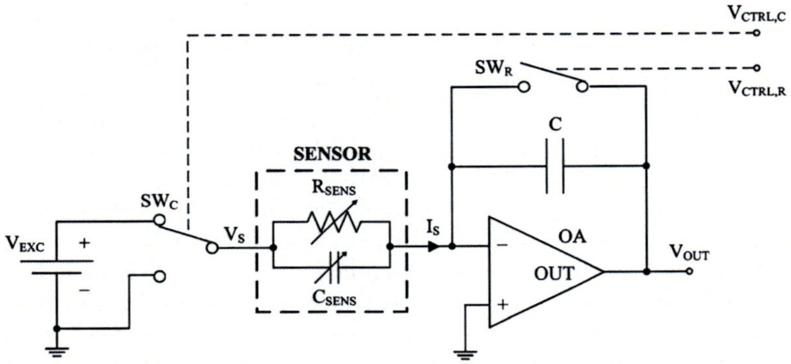

Fig. 3.19 Scheme of the *R-T* circuit based on an integrator

response in very fast transients, allowing a detailed analysis of the sensor behavior, through a resistance estimation with a measuring time T_{meas} =10 ms in the range from 10 kΩ to 10 GΩ (the relative estimation error is below 10%).

In order to estimate also the sensor parasitic capacitance, a modified version of the previously described topology has been developed [19]. It is based on the *R-T* approach, as shown in Fig. 3.19, where, in the first analysis, the switch SW_C is kept to the higher position. Thus, the sensor supply V_S is a DC constant voltage V_{EXC} and the current I_S, flowing through the sensor, is transformed in the voltage V_{OUT} by means of the voltage integrator composed by the capacitor C and the *OA*.

As in the previous circuit solution, the switch SW_R is used to reset the integrator output voltage (when closed) at the beginning of each measuring cycle. If the R_{SENS} value can be supposed to be constant within the whole measuring cycle, then the output voltage V_{OUT} is a falling ramp, starting from the initial value V_i ($V_i \cong 0$ V). The slope α of V_{OUT} depends on the R_{SENS} as is expressed again by Eq. 3.7.

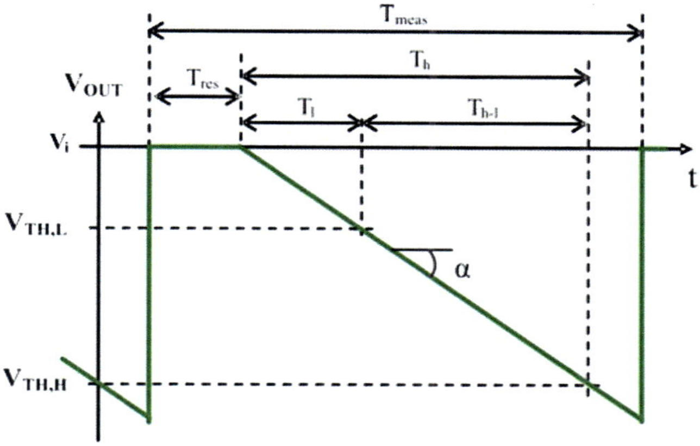

Fig. 3.20 Time diagram for the *R-T* circuit in Fig. 3.19

Therefore, α can be easily evaluated by measuring the time interval necessary to have a well-known V_{OUT} voltage drop. The timing diagram of the previous circuit is shown in Fig. 3.20, where two different threshold voltages ($V_{TH,L}$ and $V_{TH,H}$), have been defined. The R_{SENS} estimation can be done using one of the equivalent formulas in following expression:

$$R_{SENS} = \frac{V_{EXC}}{|V_{TH,L} - V_i|}\frac{T_l}{C} = \frac{V_{EXC}}{|V_{TH,H} - V_i|}\frac{T_h}{C} = \frac{V_{EXC}}{|V_{TH,H} - V_{TH,L}|}\frac{T_{h-l}}{C}. \quad (3.9)$$

The duration of the considered time intervals directly depends on the R_{SENS} value. This means that the higher the R_{SENS}, the longer the time intervals. Also in this case, the choice of the circuit parameters (C, $V_{TH,L}$, $V_{TH,H}$, V_{EXC}) is a tradeoff between the desired time resolution in the measurement and the time required to perform the estimation. In fact, the smaller $V_{TH,H}$, the less the time interval T_h, the worse the resolution related to the T_h estimation. On the contrary, if a high $V_{TH,H}$ value is chosen, the time T_h can become longer than the desired measuring time T_{meas}, when high R_{SENS} values are considered (see Fig. 3.21). Considering the same example, if $C = 100\,\text{pF}$, $V_{EXC} = 1\,\text{V}$, $T_{meas} = 10\,\text{ms}$ and $V_{TH,H} = 10\,\text{V}$, the maximum R_{SENS} value which can be estimated is about $10\,\text{M}\Omega$. If the threshold value is lowered to $V_{TH,H} = 1\,\text{V}$, then the measurement range is extended up to $100\,\text{M}\Omega$. However, in the first case, the T_h value with $R_{SENS} = 10\,\text{k}\Omega$ is 10 μs, whereas in the second case it is only 1 μs, requiring a high-resolution timing measurement system (e.g., better than 10 ns if a 1% resolution is desired). The estimation of the ramp slope α, and so R_{SENS} by inverting Eq. 3.7, can be performed using a linear fitting by means of the *LMS* algorithm. Obviously, it must be ensured that all the desired samples can be collected inside the measuring time T_{meas}, before the ramp V_{OUT} of the voltage integrator implemented by the *OA* reaches the

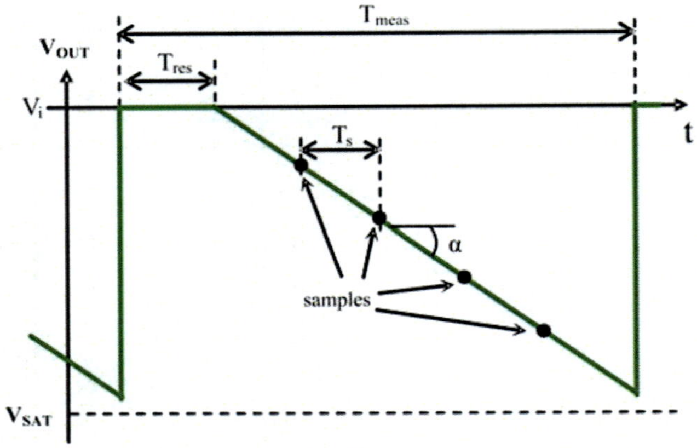

Fig. 3.21 Sample acquisition inside the measuring time

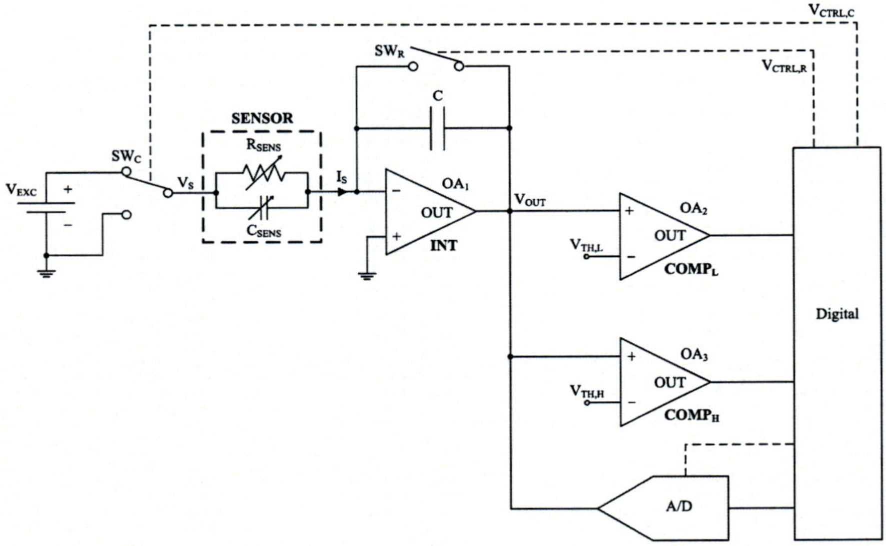

Fig. 3.22 Block scheme of the proposed system

saturation voltage V_{SAT}, as visible in Fig. 3.21. The number of samples N and the sampling time T_s influence the *LMS* interpolation performances and they need to be chosen so to have $N \cdot T_s < T_{meas}$, as before.

Therefore, considering the complete scheme of the system, at block level, reported in Fig. 3.22, by suitably choosing the parameters of the interface circuit, it is possible to divide the R_{SENS} measuring range in two intervals. For the lower part of the range, the *R-T* estimation method is preferred, because the threshold

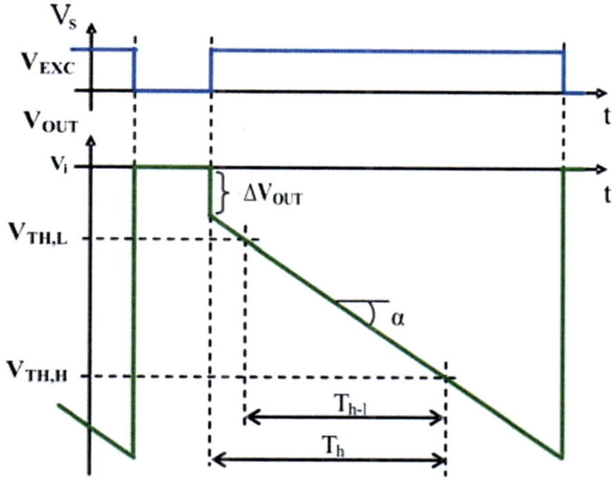

Fig. 3.23 Charge transfer effect due to the parasitic capacitance

voltages can be reached within the measuring time T_{meas}. In the upper part of the range, the *LMS* method is suitable, because the ramp V_{OUT} of the voltage integrator *INT*, implemented by the *OA*, is slow enough to avoid the *OA* saturation. A partial overlap of the two methods, in the middle part of the measuring range is advisable, to help with the calibration procedures, which, in this case, is required.

The digital block in the system, as depicted in Fig. 3.22, accomplishes, once again, many tasks: the estimation of the time intervals T_l and T_h by means of high-resolution counters; the generation of the conversion trigger to the *A/D* converter and the acquisition of the digitalized samples; the suitable control voltage $V_{CTRL,R}$ of the reset switch SW_R. Moreover, it manages the switch SW_C, through the control voltage $V_{CTRL,C}$, needed for the sensor parasitic capacitance estimation. In particular, if the switch SW_C is kept in the upper position, the sensor supply V_S is a constant voltage V_{EXC}. Thus, the parasitic capacitance C_{SENS} has no effect on the estimation of the resistance value R_{SENS} and this is one of the best advantages of the proposed system. However, if the parasitic capacitance needs to be estimated as well (e.g., for diagnostic purposes or to extract more information from the sensor behavior), the excitation voltage V_S of the sensor needs to be somehow changed. Thus, in the proposed system, this is obtained by commutating the switch SW_C from the upper to the lower position during the reset phase; in this way, the parasitic capacitance induces a voltage step ΔV_{OUT} of the output ramp, as visible in Fig. 3.23, in the correspondence of the initial next measuring time.

In fact, a sudden commutation of the sensor voltage V_S causes a charge transfer effect between C_{SENS} and C, leading to the vertical edge of the integrator output V_{OUT} whose magnitude is related to the parasitic capacitance as follows:

$$\Delta V_{OUT} = \frac{C_{SENS}}{C} V_{EXC}. \tag{3.10}$$

When the output voltage V_{OUT} crosses both the threshold voltages $V_{TH,L}$ and $V_{TH,H}$, a simple equation correlating ΔV_{OUT} to circuit parameters and measured time intervals T_h and T_{h-l} can be found, as reported in the following expression:

$$\frac{T_h}{T_{h-l}} = \frac{V_{TH,H} - \Delta V_{OUT}}{V_{TH,H} - V_{TH,L}}. \tag{3.11}$$

Thus, the combination of Eqs. 3.10 and 3.11 leads to the parasitic capacitance estimation, whereas Eq. 3.9 can again be used to perform the R_{SENS} measurement without being affected by C_{SENS}.

On the other hand, if the output V_{OUT} is too slow to intercept the threshold voltages, the *LMS* algorithm can be used both for the resistance and the parasitic capacitance estimation. In fact, the slope α is not influenced by the parasitic capacitance, therefore the R_{SENS} estimation can be performed as in the previous case (constant sensor excitation voltage). In addition, the *LMS* method furnishes the offset of the ramp with respect to the reference axis as well and such offset is exactly the quantity ΔV_{OUT} needed to estimate the C_{SENS} by means of Eq. 3.10. Obviously, these relations are true only if the initial value V_i of the ramp V_{OUT} is assumed to be zero. If this condition cannot be verified (e.g., because of a too high on-state resistance of the switch SW_R), a significant error can influence the parasitic capacitance estimation. Therefore, in order to limits this trouble, the proposed system estimates the initial value V_i and compensates the non-perfect integrator reset, by sampling the output voltage V_{OUT} also during the reset phase.

Also for this solution, a prototype *PCB* has been fabricated so to verify the feasibility of the proposed method. A Texas Instrument device (i.e., ADS8422, 16 bits of resolution) has been used as A/D converter, acquiring 16 samples per cycle ($T_s = 0.5$ ms) plus one sample during the reset phase. The overall cycle time T_{meas} has been set to 10 ms. The chosen threshold voltages are $V_{TH,L} = 1$ V and $V_{TH,H} = 10$ V, whereas the excitation voltage is $V_{EXC} = 1$ V. With such values, the upper limit for the *R-T* method, which corresponds to the lower limit for the *LMS* one, is around 10 MΩ. For the digital system, an Altera *FPGA* (Cyclone) has been adopted, implementing all the control functions and a 50 ns-resolution counter for the time interval estimations. Once again, the *LMS* interpolation algorithm and the resistance estimation are performed off-line by means of a *PC* (data are sent by the *FPGA* via an RS232 serial link), but they could be implemented directly by the digital block, leading to a stand-alone system.

In order to characterize the method, the sensor has been emulated by means of commercial resistors (10 kΩ ÷ 100 GΩ) and capacitors (1 ÷ 47 pF). Table 3.6 reports the results related to the sensor resistive component R_{SENS}. For both the methods, the relative standard deviation and the linearity error (computed by means of the *WLMS* linearization) are shown. In addition, experimental results on the C_{SENS} estimation feature have shown a good linearity of the system, with a reduced linearity error (about 0.3% full scale) for both the *R-T* and *LMS* algorithms, in the range 0 ÷ 47 pF (with $R_{SENS} = 10$ MΩ). Table 3.7 shows the results when a 10 MΩ resistor has been used. Then, the C_{SENS} estimation feature has been furthermore investigated using different R_{SENS} values and results concerning the linearity error are shown in Table 3.8.

Table 3.6 Sensor resistance estimation

	R-T		*LMS*	
R_{SENS} $[M\Omega]$	Rel std %	Lin. Err %	Rel std%	Lin. Err %
0.01	<0.01	−9.83	NA	NA
0.1	0.03	−0.82	NA	NA
1	0.01	0.08	NA	NA
10	0.01	0.17	<0.01	−0.58
100	NA	NA	0.03	−1.08
1,000	NA	NA	0.29	−1.13
10,000	NA	NA	2.48	−1.14
10,0000	NA	NA	11.65	−1.14

Table 3.7 Parasitic capacitance estimation with $R_{SENS} = 10\,\mathrm{M}\Omega$

	R-T ($R_{SENS} = 10\,\mathrm{M}\Omega$)		*LMS* ($R_{SENS} = 10\mathrm{M}\Omega$)	
C_{SENS} [pF]	Rel std %	Lin. Err %	Rel std %	Lin. Err %
0	0.14	0.02	0.07	0.02
1	0.16	0.12	0.07	0.13
2.2	0.13	−0.20	0.07	−0.20
4.7	0.17	0.03	0.07	0.02
15	0.15	0.07	0.07	0.09
10	0.16	−0.15	0.06	−0.16
22	0.14	0.31	0.06	0.29
33	0.15	−0.32	0.07	−0.33
47	0.15	0.12	0.06	0.12

Table 3.8 Parasitic capacitance estimation with different R_{SENS} values

	Lin Err [%] (*R-T*)		Lin Err [%] (*LMS*)			
C_{SENS} [pF]	$R_{SENS} =$ 100 kΩ	$R_{SENS} =$ 1 MΩ	$R_{SENS} =$ 100 MΩ	$R_{SENS} =$ 1 GΩ	$R_{SENS} =$ 10 GΩ	$R_{SENS} =$ 100 GΩ
0	0.01	0.01	−0.01	0.01	0.02	−0.01
1	0.30	0.12	0.11	0.13	0.14	0.14
2.2	−0.12	−0.25	−0.21	−0.22	−0.27	−0.20
4.7	−0.02	0.04	−0.03	0.01	0.00	0.02
15	0.10	0.17	0.15	0.10	0.12	0.09
10	−0.12	0.04	0.01	−0.02	0.00	−0.02
22	−0.18	0.07	0.19	0.19	0.18	0.16
33	−0.24	−0.43	−0.34	−0.33	−0.33	−0.33
47	0.26	0.22	0.13	0.13	0.13	0.14

3.3 The AC Excitation Voltage for Resistive Sensors

Generally, when large variations of sensor resistive values occur, the most used strategy is related to an AC-excitation voltage for the gas sensor, operating an *R-T* (or *R-f*) conversion. This allows both to avoid the use of high-resolution pico-ammeters, scaling factors, switches, etc., and to employ the same output periodic waveform to provide the AC-excitation to the sensor [11,20–28]. Moreover, sensor signal conditioners with frequency output offer a number of benefits compared to voltage output circuits, such as improved noise immunity, easiness in multiplexing, insulation and signal processing.

This approach is particularly useful in interfacing *MOX* resistive gas sensors, that, interacting with oxidizing gases such as NO_2 and O_3, increase the resistance of n-type metal oxides like SnO_2 and WO_3 [29], while decrease that of p-type metal oxides like *NiO* and *CoO* [30, 31]. These sensors show base-line values varying in a very high range as well as the sensitive elements heavily change their resistance according to their preparation or structure. Moreover, the resistance value of the sensitive element could change substantially also for small reagent concentrations (e.g., CO, CO_2, NO, H, *GPL*, etc.). This wide range can include very high values, also in the order of tens of GΩ. This is also related to the use of new materials, together with new fabrication processes, which often show very high resistance values (in the order of the GΩ) [32–35]. Furthermore, in order to reduce the power consumption, the measurement of high resistive values is necessary: in fact, many sensors are able to operate also for low temperature values, but in this way they show very high resistances and a different level of selectivity and sensitivity [36]. Microsensor applications and, especially, electronic noses, use several different gas sensors and electronic circuits operating on a very wide range, suitable to avoid the setting of scaling [9,37,38]. In the literature, different oscillators as sensor interfaces have been proposed, but, in general, there is a lack of integrated circuits representing a compact version of the interface itself, able to give an output frequency which can easily be processed in a digital manner [39–42].

The simpler electronic interface which converts a pure resistive variation into a frequency can be implemented by an *OA* in astable multivibrator configuration, as shown in Fig. 3.24 [43]. This circuit solution implements a square wave generator, whose output voltage signal period is linearly dependent on the sensor resistance value. Due to the limited frequency characteristics of non-ideal *OAs*, the circuit is only suitable for relatively low frequencies (in the kHz range or lower); the capacitance value determines both the frequency range and the circuit sensitivity.

Through a straightforward circuit analysis, it is possible to evaluate the output period T of the generated square waveform V_{OUT}, dependent on sensor resistance value R_{SENS}, according to the following equation:

$$T = 2R_{SENS}C \ln\left(\frac{1+\beta}{1-\beta}\right), \tag{3.12}$$

where

$$\beta = \frac{R_1}{R_1 + R_2}. \tag{3.13}$$

Obviously, Eq. 3.12 is valid only for an ideal *OA*, but this interface is suitable for *R-T* conversion for square wave signal periods higher than about tens of μs (corresponding to an oscillation frequency of about hundreds of kHz, typical *BW*

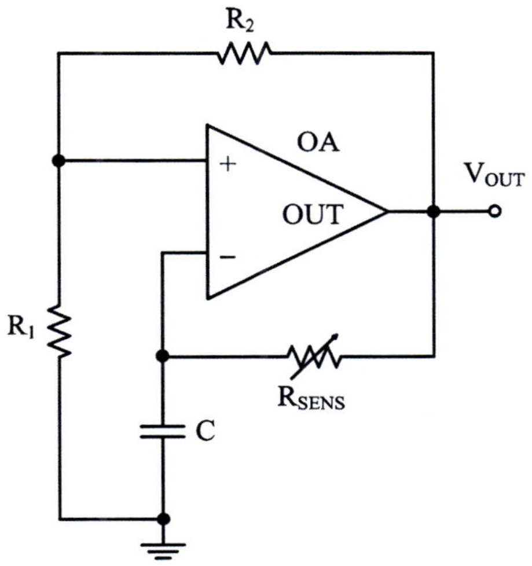

Fig. 3.24 Astable multivibrator circuit as resistive sensor interface

of non-ideal *OA*). More in detail, if we consider $R_1 = R_2$, the previous Eq. 3.12 becomes the following:

$$T \cong 2.2 R_{SENS} C, \tag{3.14}$$

from which R_{SENS} value can be easily evaluated.

Considering ideal conditions, the circuit has no limitations for high period values (except for the fact that a long measurement time occurs), so it is able to operate, for an example, at least for six decades of resistance variations, which correspond to a period span of the same number of decades. The sensitivity, for this kind of resistive sensor interface, is very low and, consequently, the main problem related to this front-end concerns the detection of small resistance values or variations. Moreover, it is also important to employ precise values of R_1 and R_2 resistances and non-linear effects (among which the temperature) have to be taken into account and verified so to be considered negligible in the period measurement and, consequently, in the sensor resistance estimation.

Starting from the astable multivibrator circuit, Fig. 3.25 shows the block scheme of an improved wide range resistive sensor integrable interface, also based on an oscillator topology [39].

Because of its simple structure, it can be easily designed at transistor level so to achieve a complete integrated solution. This circuit works as an *R-T* converter, where the oscillation period T of the generated output square wave signal V_{OUT} is directly proportional to sensor resistance value R_{SENS}. The current which flows into R_{SENS} alternatively charges and discharges the capacitor C. Through a straightforward analysis it is possible to determine the following expression of the oscillation period T:

$$T = 4 (A - 1) R_{SENS} C. \tag{3.15}$$

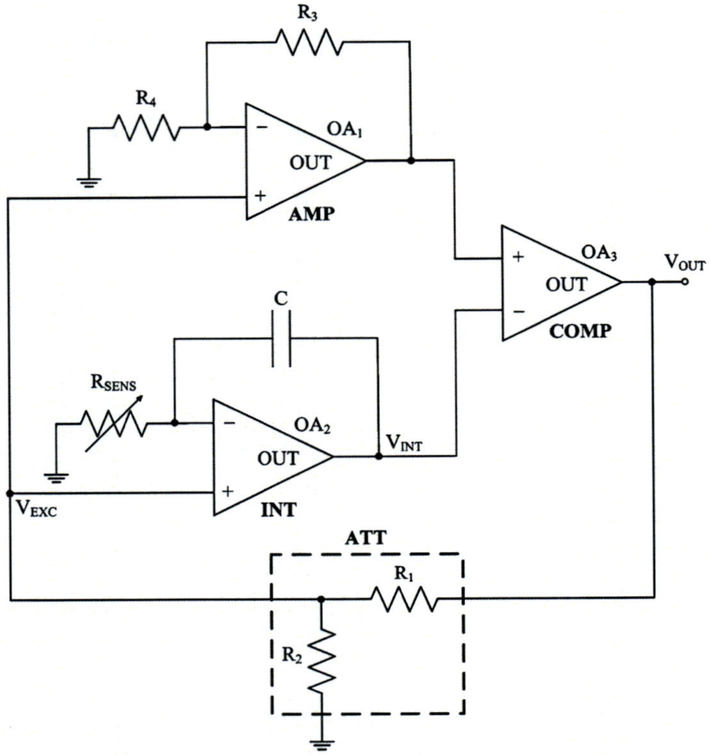

Fig. 3.25 Block scheme of a resistive sensor interface based on an *R-T* conversion (*AMP* = non-inverting amplifier; *COMP* = voltage comparator; *INT* = non-inverting integrator; *ATT* = attenuator constituted by a resistive voltage divider)

Sensitivity can now be regulated by choosing suitable values of two parameters: A and C, being $A = 1 + (R_3/R_4)$ the voltage gain of the non-inverting amplifier *AMP* implemented through a suitable feedback configuration by OA_1. A possible internal structure of each active block, which can be easily implemented with a microelectronic approach through a *CMOS* Operational Transconductance Amplifier (*OTA*), at transistor level and in a standard *CMOS* technology, is shown in Fig. 3.26. It can be designed with both low supply voltage and low power consumption, therefore is suitable for integrated portable sensor applications. The *OTA* has to show a very high *SR* and a very low input voltage offset. In this way, the error between ideal and measured oscillation periods becomes negligible. The interface shown in Fig. 3.25 is able to reveal about over six decades of sensor resistance variations, e.g., between tens of kΩ up to tens of GΩ, showing a very good linearity and a low percentage relative error. A suitable improved modification of this interface, using inverting blocks and able to evaluate also the sensor parasitic capacitance, is shown in the next paragraph.

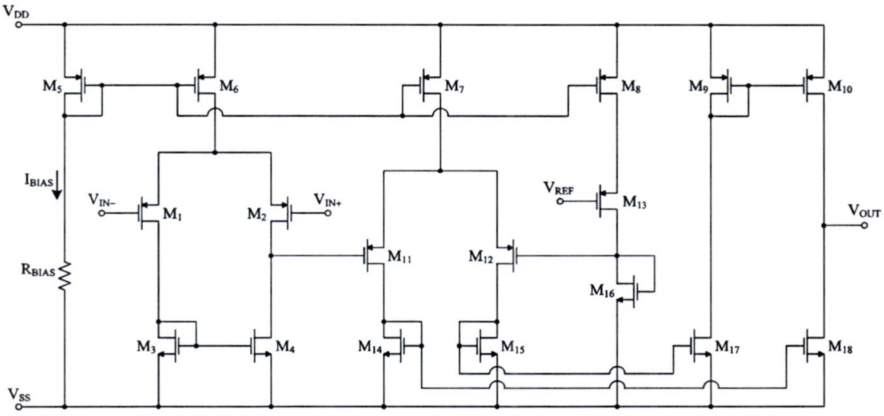

Fig. 3.26 Schematic circuit at transistor level of the implemented *OTA*

3.3.1 Uncalibrated AC-Excited Sensor Based Solutions

As mentioned before, oscillating circuits, operating an *R-T* conversion, represent the best solution when large variations of sensor resistive values occur. Unfortunately, sensor behaviour is not exactly equal to a pure resistance; in fact, physical connections, miniaturization processes employed in realization of sensor and of its heating element create parasitic capacitances. The equivalent circuit of the resistive gas sensor, for example *MOX*-based sensors, appears therefore as the parallel between a resistance and a capacitance, whose value is quite low (few pF) [4]. In this model, the equivalent sensor resistance presents values ranging from hundreds of kΩ up to hundreds of GΩ and the parasitic capacitances must be determined to reduce errors in AC-excited sensor resistance estimations [41, 42]. The capacitance evaluation can be useful also to extract more information from the sensor either for diagnostic purposes or to better characterize new experimental sensors, based, for example, on nanowire structures.

Concerning the AC-excitation approach, an evolution of the circuit reported in Fig. 3.25, always performing the *R-T* conversion, is shown in Fig. 3.27 [23, 41, 44, 45]. This oscillating circuit allows to measure both high resistive sensor variations and parasitic capacitive behaviours without any initial calibration of the complete measurement system. It is a compact, *LV LP*, low-cost and simple interface, completely integrated on silicon with a standard *CMOS* technology, suitable for portable applications. The circuit is able to reveal over six orders of magnitude of gas sensor resistance variation and, moreover, can estimate, with high accuracy, sensor parasitic capacitances up to 50 pF, showing high linearity and reduced percentage error between measured and theoretical results. The integrated solution has been designed in order to be both independent from supply voltage variations and temperature drifts.

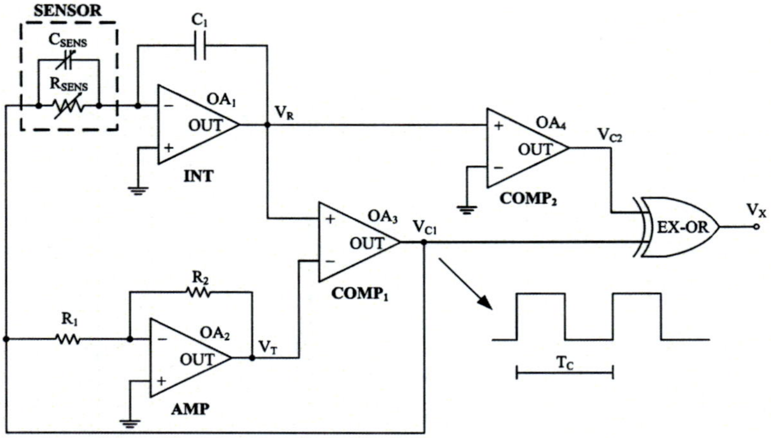

Fig. 3.27 Block scheme of the proposed interface

The interface is composed by five main blocks: two comparators, an inverting integrator, an inverting amplifier and an exclusive *OR* (*EX-OR*) logic block. The first comparator ($COMP_1$) generates a square wave voltage signal (V_{C1}), whose amplitude is equal to the total supply voltage $2V_{CC}$ and its period is proportional to the sensor resistance R_{SENS} and depends also on parasitic capacitance C_{SENS} (under the hypothesis of R_{SENS} and C_{SENS} being constant during the measuring time). The second comparator ($COMP_2$) and the *EX-OR* block have been added with the aim to also estimate the value of the sensor parasitic capacitance C_{SENS}. In fact, the *EX-OR* gate generates also a square-wave signal, whose duty-cycle depends on C_{SENS}. In order to better explain the system behaviour, a timing diagram is reported in Fig. 3.28, showing the voltage signals at each node of the interface under the hypothesis of constant R_{SENS} and C_{SENS} values during the analyzed time.

More in detail, the output voltage of the first comparator ($V_{C1} = \pm V_{CC}$) represents both the periodic signal, from which it is possible measure the period T_C, and the input signal for three other blocks (inverting integrator, inverting amplifier and *EX-OR*). Then, this voltage signal is attenuated by G factor ($G < 1$) and successively applied to the inverting node of the same $COMP_1$, so, the voltage reference V_T can assume only two values: $\pm GV_{CC}$. The *R-T* conversion is performed by an integrator stage, composed by the operational amplifier OA_1 and the capacitor C_1, which integrates the current, dependent on voltages $\pm V_{CC}$, flowing through the sensor, modelled using a resistance R_{SENS} in parallel with a capacitance C_{SENS}. Since sensor excitation voltage V_{C1}, for a time period, is constant, the integrator output V_R is, in the same period, a falling or a rising ramp, depending on the sensor current direction (a rising ramp, when $V_{C1} = -V_{CC}$ and a falling ramp if $V_{C1} = +V_{CC}$). Therefore, the constant current which flows in R_{SENS} alternatively

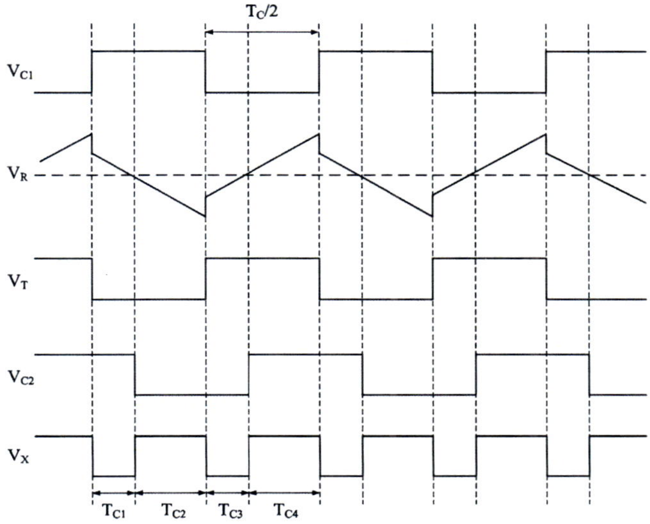

Fig. 3.28 Voltage signal levels generated by active blocks

charges and discharges the capacitor C_1 and, since V_{C1} is also the integrator input signal, we can write:

$$V_R(t) = V_R(0) - \int \frac{V_{C1}}{R_{SENS}C_1}dt. \tag{3.16}$$

If we call T_H the time between two consecutive commutations, the value of V_R, determined at $t = T_H$ through Eq. 3.16, must be equalized to the threshold level $V_T = GV_{C1}$ as follows:

$$V_R(T_H) = -V_{C1}\left(G - \frac{T_H}{R_{SENS}C_1}\right) = GV_{C1} = V_T, \tag{3.17}$$

being G the ratio between R_2 and R_1, typically lower than 1. Therefore, since V_{C1} is constant for a half-period, the integrator output V_R is, in the same time interval, a falling or a rising ramp, depending on the sensor current direction. The ramp signal V_R is compared by the comparator $COMP_1$ with a reference threshold voltage V_T that follows, with the opposite sign, the excitation voltage V_{C1}. In this manner, the comparator $COMP_1$ generates a square wave voltage signal V_{C1}, with an amplitude equal to the total supply voltage (i.e., $2V_{CC}$), from which it is possible to determine the ideal oscillation period T_C, whose value is proportional to the sensor resistance R_{SENS} value, according to the following equation:

$$T_C = 2T_H = 4GC_1R_{SENS}. \tag{3.18}$$

The presence of the sensor parasitic capacitance, C_{SENS}, involves a fast charge transfer between the same C_{SENS} and the integrator capacitor C_1 when the sensor voltage V_{C1} commutates. Therefore, the ramp signal V_R is instantaneously affected through a vertical edge, during the instantaneous commutation of V_{C1} from $-V_{CC}$ and $+V_{CC}$, and vice-versa, as in the following equation:

$$|\Delta V_R| = \frac{C_{SENS}}{C_1}|\Delta V_{C1}|. \tag{3.19}$$

Thus, this parasitic capacitance effect acts only at the beginning of both the rising and falling ramp of V_R by means of a step of the ramp itself, giving rise to the following relation for the period:

$$T_C = 4GC_1 R_{SENS}\left[1 - \left(\frac{C_{SENS}}{GC_1}\right)\right]. \tag{3.20}$$

From Eq. 3.20, it comes that the proposed interface shows two degrees of freedom, in particular C_1 and G (then R_1 and R_2), that helps to choose its sensitivity. Moreover, if GC_1 is designed to be much higher than parasitic sensor capacitance C_{SENS} value, the output period T_c can be considered as independent from C_{SENS}, still achieving the previous Eq. 3.18. The second comparator, $COMP_2$, is a "zero-comparator" and allows to separate the ramp signals in two different parts: the first, immediately after the commutation of V_{C1}, presents the charge transfer effect due to the presence of the C_{SENS}, while the second part depends only on R_{SENS}. More precisely, the *EX-OR* logic block generates a square wave signal V_X which allows to estimate both C_{SENS} and R_{SENS} values, according to the next equations, where increasing and decreasing ramp contributions have been averaged. In particular, referring to timing diagram in Fig. 3.28, it should be clear that time intervals T_{C2} and T_{C4} depend on the resistive value R_{SENS} only, while T_{C1} and T_{C3} depend on both R_{SENS} and C_{SENS}. Concluding, R_{SENS} value can be simply estimated measuring the time intervals T_{C2} and T_{C4}, whereas the capacitive effect due to C_{SENS} can be evaluated comparing times intervals T_{C1} with T_{C2} and T_{C3} with T_{C4}, through the following expressions:

$$R_{SENS} = \frac{T_{C2} + T_{C4}}{2GC_1}, \tag{3.21}$$

$$C_{SENS} = GC_1\frac{(T_{C2} + T_{C4} - T_{C1} - T_{C3})}{2T_{C2} + 2T_{C4}}. \tag{3.22}$$

Resuming, the proposed front-end can be used in two ways. If the sensor capacitance can be neglected, the output "semi-period" T_H of signal V_X directly provides the R_{SENS} value by a scaling factor according to Eq. 3.18 (in this case, it should be noticed that T_C period is referred instead of T_H, being $T_C = 2T_H$). Otherwise, times $T_{C1}, T_{C2}, T_{C3}, T_{C4}$ of the output signal V_X should be measured and the R_{SENS} and C_{SENS} values can be computed according to Eqs. 3.21 and 3.22.

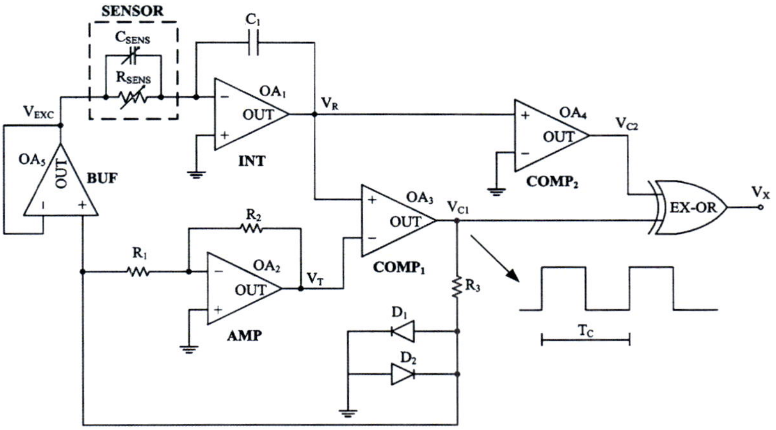

Fig. 3.29 Modified scheme of the proposed interface implemented as *PCB*

The circuit in Fig. 3.27 has been modified so to fabricate a suitable prototype board, by inserting a diode voltage limiter at the comparator $COMP_1$ output and a voltage follower just before the sensor, as shown in Fig. 3.29. In fact, in order to achieve a good symmetry between the two values of the comparator $COMP_1$ output, a voltage limiting circuit, simply based on two diodes (D_1 and D_2), has been introduced. This is an important issue, because this signal becomes the sensor excitation voltage and generally sensors could show different resistance values if they are supplied with different voltages. With this solution, the sensor supply voltage switches between a positive and a negative voltage whose value is the diode forward voltage (i.e., about 0.65 V). Another practical problem comes from sensor location during experimental tests. The sensor is usually positioned in a measuring chamber and is connected to the circuit by shielded cables; the cables could have capacitive effects which may slow the commutation of the comparator $COMP_1$. In order to reduce this effect, a voltage follower, named *BUF* in Fig. 3.29, has been inserted.

The fabricated discrete-elements prototype *PCB*, used for experimental tests, has been developed using commercial components; in particular, OA_1 and OA_2 have been implemented by low input bias current operational amplifier OPA350 ($I_{b-typ} < 1\text{pA}$), while $COMP_1$ and $COMP_2$ are fast response comparators TLC3702 (commutation time less than 5 μs), all from Texas Instruments, with ±3.3 V power supply. 1N4148 diodes and HCF4070BE *EX-OR* gate have been also utilized. The R_1/R_2 ratio has been chosen so to obtain a gain factor G of about 2, while the integrator capacitance C_1 has been set to 100 pF. Due to the very low values of the circuit currents, especially if high-value resistances are under consideration, during the measurements the prototype has been located in a shielded box to prevent electromagnetic interferences to affect the measure. The output signal V_X has been managed (automatically measured) through an 8-bit microcontroller (PIC

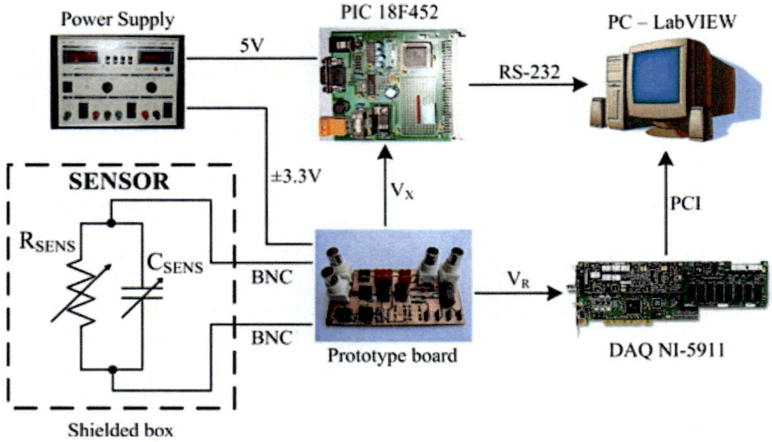

Fig. 3.30 Block scheme of the experimental setup

18F452) with time-measuring resolution of 200 ns. Exploiting the input capture capability of this microcontroller, a simple program allows to estimate all four intervals (T_{C1}, T_{C2}, T_{C3} and T_{C4}) simply using the V_X commutations as interrupt events. Each event determines the value of a free-running counter to be read, so that the length of a time interval could be estimated as the difference between two consecutive counter readouts. Time interval measures have been subsequently sent by means of a RS-232 link to a *PC*, equipped with LabView environment, which visualizes data and store them for further offline analysis. The block scheme of the whole experimental setup is shown in Fig. 3.30.

Experimental tests have been performed using commercial resistors (1% from 100 kΩ to 1 GΩ, 5% for other values till 100 GΩ) and capacitors (5% from 1 to 22 pF) to emulate the sensor behaviour. The measure time is about 1 s and a repetition of 100 measurements has been taken for each resistor-capacitor combination. Test resistors and capacitors are also lodged in a metal box and are connected to the prototype, by mean of shielded *BNC* cables, so that effects due to electromagnetic interferences are strongly reduced. As it is not simple to retrieve high-accuracy commercial instrumentation for these resistance values, a reference estimation method has been designed to avoid limits due to low-precision high-value resistors (5%). As shown in Fig. 3.31 (considering $R_{SENS} = 100\,\text{M}\Omega$, $C_{SENS} = 22\,\text{pF}$), if the ramp signal V_R is acquired by a fast high-resolution acquisition board and the Least Mean Square (*LMS*) line is computed starting from acquired samples, then R_{SENS} can be simply estimated dividing V_{C1}/C_1 by the line slope α, while C_{SENS} is determined dividing the voltage gap β by V_{C1}/C_1 ("ramp method"). In order to sample the ramp signal V_R, a National Instruments PCI-5911 data acquisition board (12.5 M sample/s, 16 bit) has been used.

The ramp method, which avoids non-idealities due to component delays and comparator offsets, has been validated with low-value resistors (<15 MΩ), measured

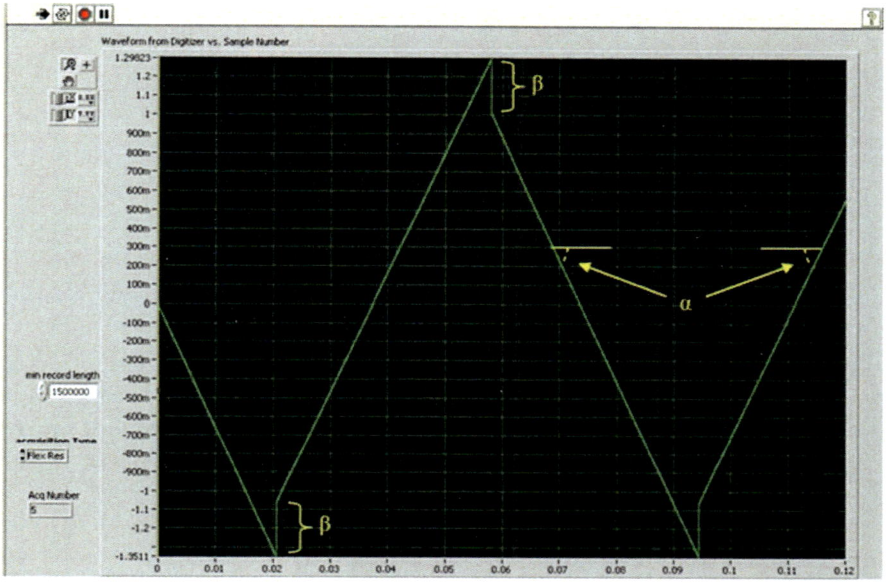

Fig. 3.31 Acquired samples ($R_{SENS} = 100\,\text{M}\Omega, C_{SENS} = 22\,\text{pF}$)

Table 3.9 Experimental results concerning R_{SENS} estimation (expected $C_{SENS} = 0\text{pF}$): mean value (*AVERAGE*) and standard deviation (*STD-DEV*) computed over 100 read-out $\langle R_{SENS} \rangle$

Nominal value (expected) [MΩ]	"Ramp" value *AVERAGE* [MΩ]	"Ramp" value *STD-DEV* [MΩ]	Estimated value *AVERAGE* [MΩ]
0.22	0.223130	0.0000002	0.218
1	1.001029	0.0000012	0.988
10	10.02033	0.0000106	9.883
100	100.1038	0.0000439	99.235
1,000	1,004.068	0.0001421	996.901
10,000	10,032.09	0.0068210	9927.146
100,000	100,100.6	0.2330870	102173.600

by a precision multimeter (Fluke 8840A) with overall performance better than 0.1%. Concerning the linearity evaluation, the *LMS* line is not the best linear approximation of the input-output characteristic, because of the wide input range (more than four orders of magnitude) and the logarithmic distribution of samples. In fact, low resistances always present a small absolute error if compared with high ones, therefore they do not contribute to *LMS* minimization process. In order to better fit experimental results over a wide range (resistance values are spread along six decades), the so-called *WLMS* line, which minimizes the relative error, has been considered as the linear approximation of the input-output characteristic, in particular for R_{SENS} and C_{SENS} estimations. The system has been tested over six decades (from $100\,\text{k}\Omega$ to $100\,\text{G}\Omega$), yielding to a relative standard deviation of about 0.02% with respect to the mean value (100 continuous readouts). Table 3.9

Table 3.10 Experimental results concerning C_{SENS} estimations

Nominal value (expected) [pF]	$R_{SENS} = 100\,\mathrm{M\Omega}$ *AVERAGE* [pF]	$R_{SENS} = 1\mathrm{G\Omega}$ *AVERAGE* [pF]	$R_{SENS} = 10\,\mathrm{G\Omega}$ *AVERAGE* [pF]
0	−1.525	−0.179	0.244
1	−0.391	0.872	1.376
2.2	0.700	2.015	3.054
4.7	3.298	4.710	5.640
10	8.729	10.066	10.655
22	21.094	22.559	23.309
47	46.298	47.678	48.425

shows results about R_{SENS} estimation, showing, in particular, its mean value and standard deviation. The $220\,\mathrm{k\Omega}$ results replace the $100\,\mathrm{k\Omega}$ ones, because, in this case, component delays are in the same order of the ramp period and the used microcontroller is unsuitable to measure very short times. Relative displacement between estimation method (see Eq. 3.21) and "ramp", evaluated by a *WLMS* line, is about 1% over about six decades. The *WLMS* line has been computed in order to minimize the mean square value of relative range.

The C_{SENS} estimation capability has been proved using different values of resistors in parallel with the test capacitors. Results concerning C_{SENS} evaluation are shown in Table 3.10; calculation, through Eq. 3.22, works well with high-value resistances. Deviation of estimated value (for $R_{SENS} = 10\,\mathrm{G\Omega}$) with respect to the value calculated by the ramp method is in the order of 100 fF. Comparing the capacitance estimated value referring to $R_{SENS} = 1\,\mathrm{G\Omega}$ with the one relative to $R_{SENS} = 10\,\mathrm{G\Omega}$, a sort of negative offset appears. This negative offset increases if sensor resistance decreases and if ramp slope is too fast ($R_{SENS} < 10\,\mathrm{M\Omega}$), the absolute error is in the order of few pF. In this case, capacitance estimation can be used only for diagnostic purposes. Offset is due to the component delays and their non-linear effects; for instance, the non-instantaneous commutations, which have not been taken into account by complete equations. Negative offset increases if component delays and non-linearities are in the same order of times T_{C1}, T_{C2}, T_{C3}, T_{C4} (see Fig. 3.28) and their differences. Standard deviation of capacitance estimation is always lower than 100 fF.

The circuit behaviour has been furthermore evaluated using a MoW-based *MOX* resistive gas sensor which needs to be heated at a suitable operating temperature so to properly work; therefore, an external power supply has been used to drive the sensor embedded heater element. In fact, sensor resistance strongly depends on the sensor temperature and, as a consequence, on the power furnished to the heater element.

In this experiment the sensor heater has been driven with different voltages in order to achieve a set of working temperatures; the sensor resistance R_{SENS} and the sensor capacitance C_{SENS} have been monitored with the fabricated *PCB* prototype. Fig. 3.32 shows the measurement results about the estimation of R_{SENS} and C_{SENS} during the same experiment and how they vary when heater voltage, therefore

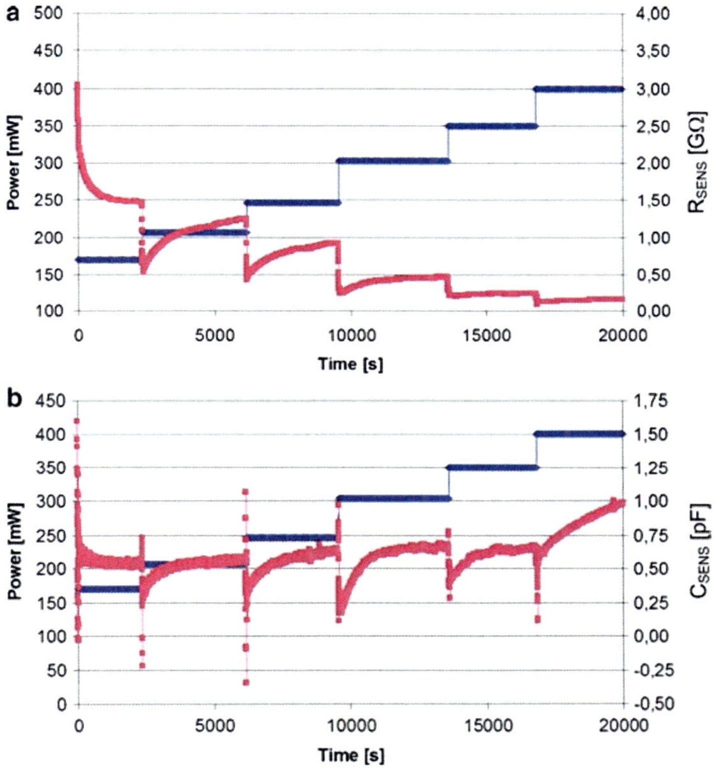

Fig. 3.32 Sensor component estimations vs. time when heater power is varied: (**a**) R_{SENS} behaviour; (**b**) C_{SENS} behaviour

heater power, is suddenly changed. C_{SENS} variation is very noisy in correspondence of sudden heater power variation (in particular, during a sudden R_{SENS} variation) yielding to negative values. It should be also noticed how the proposed system is able to evaluate the resistance value of the sensor even when a large excursion of resistance occurs. Fig. 3.33 shows a zoom of R_{SENS} and C_{SENS} behaviour when heater power is varied from 150 to 200 mW. Also these measurements are very noisy in correspondence of sudden heater power variation, because C_{SENS} estimation works properly only if resistance is stable during the whole ramp (that is the ramp is perfectly linear), therefore C_{SENS} evaluation is accurate only when sensor is in a steady state (if sudden variations of R_{SENS} occurs, C_{SENS} values should be suitably filtered).

Usually, when the experimental measurements performed through the prototype *PCB* have given satisfactory results, the interface is then designed, at transistor level, in a standard *CMOS* technology, so to implement a completely on-silicon integrated circuit suitable for very compact portable applications. In this case, the integrated version of the interface does not require any external component (e.g., resistors,

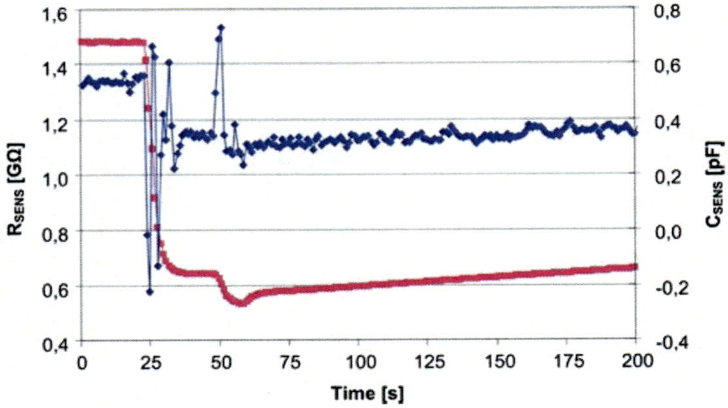

Fig. 3.33 A detail of R_{SENS} and C_{SENS} behaviours during a sudden heater power variation (from 150 mW to 200 mW)

capacitors, etc.). The main non-idealities of the implemented active components have been taken into account, in particular the finite *SR* value and the non-zero input voltage offset of the operational amplifiers. Internal circuit topologies have been developed to obtain better performances in this sense and to operate at reduced supply voltage with low power consumption. The active block used as amplifier has been implemented by a suitable *OTA*, whose internal topology, designed at transistor level in AMS 0.35 μm standard *CMOS* technology, is reported in Fig. 3.34. Through simple calculations, it has been possible to evaluate both the relative error, due to input voltage offset V_{OFF} of the inverting integrator, and the absolute error, owed by *OTA SR* finite value of the amplifiers used as comparators. Both these errors affect the output voltage measurement on the first comparator, according to the following expressions:

$$e_{OFF} = \frac{V_{OFF}}{V_{SAT} + V_{OFF}}, \tag{3.23}$$

$$e_{SR} = \frac{4V_{SAT}}{SR}, \tag{3.24}$$

being V_{SAT} the output saturation voltage of the comparator which excites the sensor (e.g., $V_{SAT} = V_{SAT+} = -V_{SAT-} \cong V_{DD} = -V_{SS}$) and the last error e_{SR} expressed in seconds. As it shown in Eq. 3.24, the absolute error due to the finite *SR* introduces a higher error in the case of lower R_{SENS} values.

The *OTA* shown in Fig. 3.34 is composed by two stages: the input stage (formed by transistors $M_1 - M_9$), which is a symmetrical *OTA*, and the output stage (formed by transistors $M_{10} - M_{13}$), that is an AB-class inverter amplifier, based on a push–pull configuration ($M_{11}M_{12}$), that allows to obtain a full dynamic output range, with a source degeneration ($M_{10}M_{13}$). In particular, transistors M_{10} and M_{13} allow a

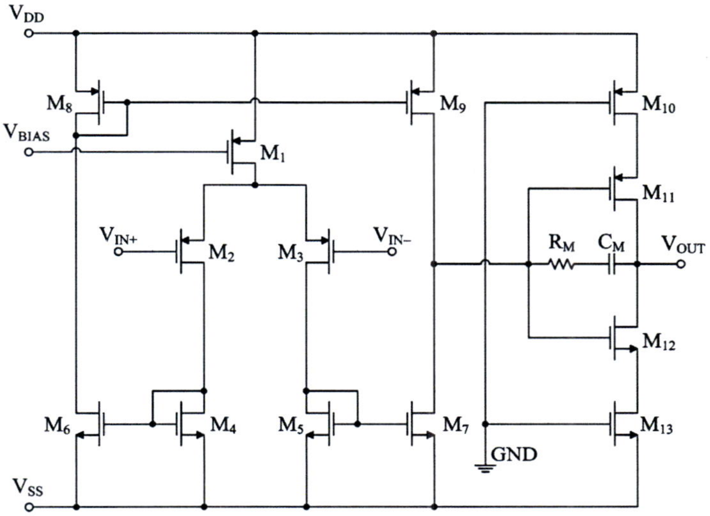

Fig. 3.34 *OTA* schematic at transistor level

better control of the current flowing in the output branch, even if, through the source degeneration, this same current has been done slightly dependent on supply voltage variations. Moreover, this second stage allows to get a high open loop voltage gain, so to make the amplifier more ideal. Frequency stability has been obtained by an *RC* series Miller compensation (i.e., R_M and C_M components, see Fig. 3.34) [46]. The choice of a symmetrical configuration as *OTA* input stage and a careful layout implementation have reduced both the systematic and the random input offset voltages. Furthermore, two *pMOS* matched transistors, M_2 and M_3, showing in the chosen technology, a lower K_F value than *nMOS* ones, have been utilized as input differential pairs so to have a low input equivalent noise. In order to bias the circuit with current enough to have a high *SR* value, a current reference generator, whose value depends on V_{TH} (threshold voltage of the *MOS* transistor), has been implemented. In this manner, the generated current is independent both from supply voltage variation (e.g., a battery discharge), also for relatively high variations respect to the nominal value (higher than $\pm 10\%$), and from temperature drifts of the whole *OTA*. This current reference needs a "start-up" circuit [47], which allows to operate in the correct non-zero operating point. In Fig. 3.35, the complete current generator for *OTAs* biasing, formed by transistors M_1–M_4 and resistor R_{BIAS} (which fixes the desired current), with its relative start-up circuit (M_5–M_7), is shown.

Fig. 3.36 shows the transistor level implementation of the *EX-OR* logic function. It is the simple traditional internal topology of this digital block which shows a very low power dissipation.

The integrated solution has been designed in order to cover a reduced silicon area (lower than 1 mm^2), making it suitable to be replied on silicon substrate and

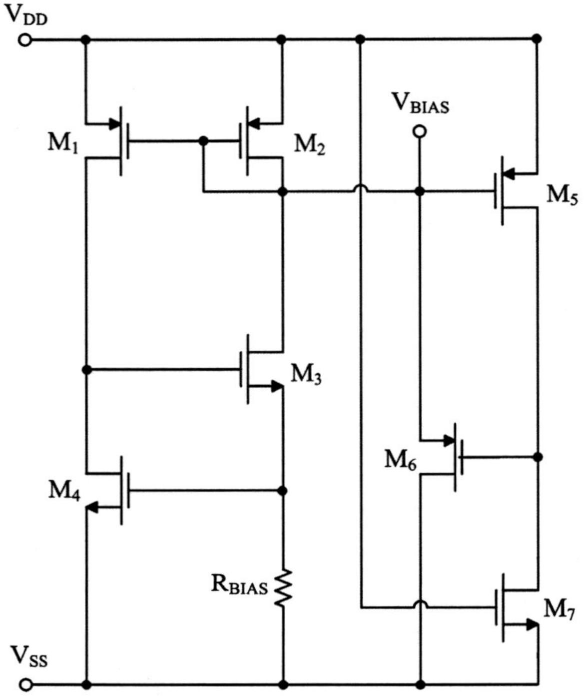

Fig. 3.35 Complete current generator schematic at transistor level: start-up circuit and V_{TH}– based current reference

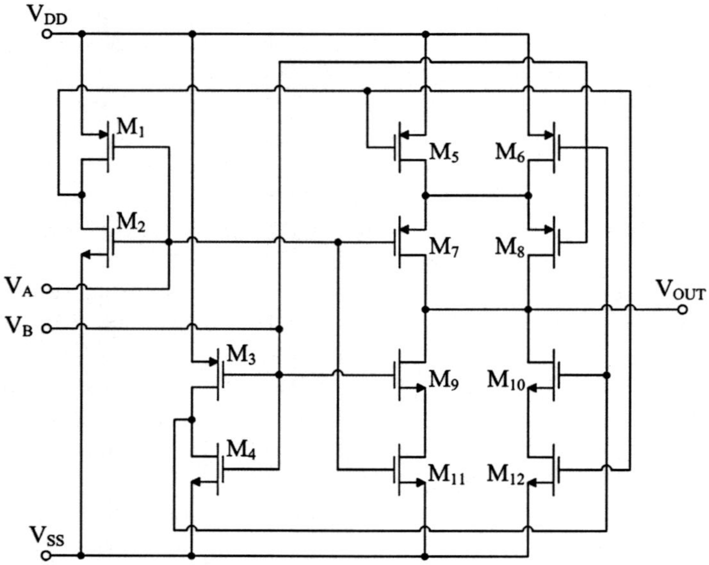

Fig. 3.36 *EX-OR* schematic at transistor level

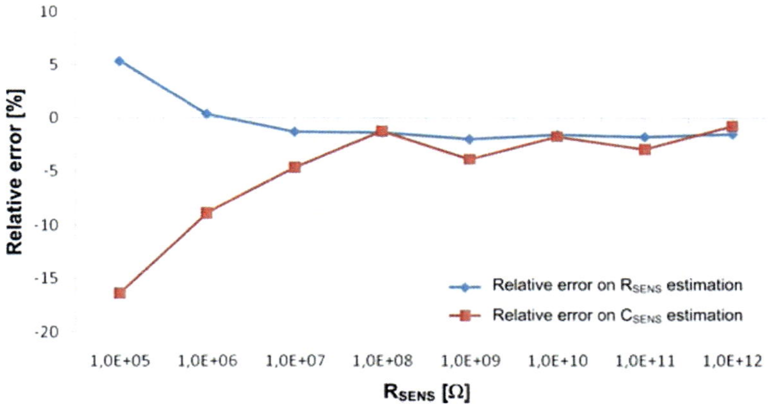

Fig. 3.37 Post-layout simulation results: relative error on R_{SENS} and C_{SENS} estimations vs. R_{SENS} (C_{SENS} fixed to 10 pF)

having the possibility to acquire and manage sensor (e.g., gas chemical sensor) arrays information through a complete *CMOS* device (a low-cost single chip). Moreover, each of the four *OTAs* has been designed with the same transistors sizes and same internal topology, depicted in Fig. 3.34, so to simplify the layout design and integration operations.

Post-layout simulations on the designed integrated solution, which have demonstrate also stability and independence of the circuit from temperature drifts, and experimental results, on MoW-based *MOX* sensors, have been performed showing high linearity and reduced percentage error with respect to the theoretical expectations, for more than six decades of resistance values (from 100 kΩ to more than 100 GΩ), as well as for the sensor in-parallel capacitive component estimation (in the order of few pF), without any initial calibration of the system. In particular, R_{SENS} and C_{SENS} estimations have been operated by the evaluation of the four time intervals (T_{C1}, T_{C2}, T_{C3} and T_{C4}). As an example, having chosen $R_1 = 50\,\text{k}\Omega$, $R_2 = 40\,\text{k}\Omega$, $C_1 = 100\,\text{pF}$, Fig. 3.37 shows the relative errors (in %) on sensor resistance and parasitic capacitance estimations, extracted from the time domain post-layout simulations. In both cases, relative error is high only for low resistance values, while there are no limits for very high sensor resistance values, except for the fact that a long measure time occurs.

Further post-layout simulations, in CADENCE environment, have been also performed in terms of parametric analyses both at different operating temperatures and considering supply voltage variations (e.g., a battery discharges) and Monte Carlo analyses for statistic distributions. Moreover, corner analyses have been conducted considering two main cases: the worst power case (minimum operating temperature = −50°C, maximum supply voltage= 3.63 V) and the worst speed case (maximum operating temperature = +130°C, minimum supply voltage = 2.97 V); in this way, also transistor size variations as well as mismatches, fabrication process errors and technological spread have been taken into account. All the simulations have confirmed the correct functionalities of the integrated interface solution guaranteeing a relative error on simulated period, with respect to the theoretical one, lower than 1%.

Then, the on-silicon complete version of the interface circuit has been fabricated, in a standard low-cost *CMOS* technology (AMS 0.35 μm) [48, 49]. Experimental measurements, performed on chip, have shown a good linearity confirming the expected front-end performances in a large frequency range. In particular, the integrated circuit, which has shown low power consumption (about 4 mW), for a single supply voltage (3.3 V), and good performances in a wide range of environmental temperature (from −20°C to +80°C), has been proved to be able to reveal with reduced relative error more than five decades of resistance variation (about from 500 kΩ to 100 GΩ) and, at the same time, to estimate the sensor parasitic capacitance in a more reduced range (about from 0 to 33 pF). The fabricated chip performances have been proved through the interfacing of a high-value resistance *MOX* sensor, monitoring both the resistance and the parasitic capacitance values during a fast thermal transient of the sensor. Moreover, the integrated front-end has been utilized for the detection of hydrogen, by means of the commercial Figaro TGS2600 resistive gas sensor, fluxing two different gas mixtures, constituted by hydrogen and nitrogen, and for different sensor operating temperatures.

More in detail, in order both to easily set the front-end sensitivity and, sometimes, to reduce the parasitic capacitance effects, the only external passive component in this integrated solution design is the integrating capacitor C_1, while the *OTA* frequency stability has been obtained by a compensation with $R_M = 8\,\mathrm{k\Omega}$ and $C_M = 2\,\mathrm{pF}$ values. Concerning the transistor level implementation, it has been designed so that the following uncertainty sources can be considered negligible: the asymmetric sensor excitation voltage; the non-zero voltage and current input offsets of the inverting integrator; the non-zero input offset voltage of $COMP_2$; the time delays introduced by $COMP_1$ and $COMP_2$.

Fig. 3.38 shows the fabricated chip photo (total chip size: $3 \times 3\,\mathrm{mm}^2$, see Fig. 3.38a) of the complete interface (interface area: $1.3\,\mathrm{mm} \times 0.65\,\mathrm{mm}$, see Fig. 3.38b). In Table 3.11, *OTA* main simulated (post-layout results) and measured characteristics have been reported, showing a good agreement among them and, in particular, a low input voltage offset and a good *SR* value.

Electrical on-chip measurements of the complete *CMOS* integrated interface, performed through a digital electronic system based on a programmable logic device (*PLD*) with a time resolution of about 50 ns, have confirmed the very high dynamic range and a good agreement with the theoretical expectations. In particular, in order to achieve experimental results on the chip as sensor interface, R_{SENS} and C_{SENS} have been firstly replaced with sample resistors and capacitors, respectively. The system sensitivity has been set to about 360 μs/MΩ by choosing a suitable value for the integrator capacitor ($C_1 = 100\,\mathrm{pF}$). In particular, when the sensor parasitic capacitance can be neglected ($C_{SENS} \approx 0\,\mathrm{pF}$), the value of sensor resistance R_{SENS} can be estimated starting from the output period T_C according to Eq. 3.18. In this case, the period of $COMP_1$ output square wave signal, obtained with different

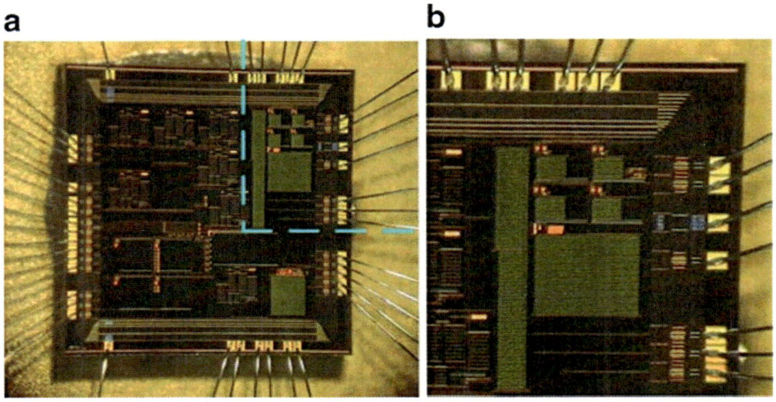

Fig. 3.38 (**a**) Photo of the fabricated chip (the interface is highlighted by the white dashed line); (**b**) Zoom of the area related to the interface

Table 3.11 *OTA* main characteristics (simulated and measured)

OTA parameter	Post-layout simulated value	On-chip measured value
Voltage supply	3.3 V	3.3 V
Power dissipation	992 μW	about 1 mW
GBW	65.8 MHz	61 MHz
Output dynamic range	Full	Full
Open loop DC voltage gain	66 dB	64 dB
Slew-Rate	40 $V/\mu s$	38 $V/\mu s$
Input voltage offset	100 μV	lower than 1 mV
Input equivalent noise	169 nV/sqrt(Hz) @ 1 kHz	210 nV/sqrt(Hz) @ 1 kHz

resistors at room temperature (30°C) and powering the integrated circuit at 3.3 V, has been measured and compared with the theoretical one, showing a reduced percentage relative error. Period values and relative percentage errors are listed in Table 3.12, showing good linearity for both simulation and experimental data, while Fig. 3.39 shows these experimental results in terms of resistance estimation starting from measured periods (the estimated resistance value $\langle R_{SENS} \rangle$ and its relative error related to the sample resistor values). Such estimations are averaged values of about 100 consecutive measurements for each R_{SENS} sample and, in this sense, the interface has revealed good reproducibility and, consequently, good precision. However, the quite high value of the relative error for the estimations (about 10%) that worsens interface accuracy is mainly due to circuit offsets and component uncertainty. However, this error can be partly compensated and, then reduced, employing a suitable linearization algorithm (e.g., the *WLMS* method). More precisely, with low resistance values, time T_C, so T_{C1}, T_{C2}, T_{C3}, T_{C4}, become small and a very good time-resolution instrument is needed to perform the time estimation. On the contrary, with very high resistance values, the main source of uncertainty is due to the noise, in particular concerning $COMP_2$; in fact,

Table 3.12 Chip measurement results compared with theoretical values and post-layout simulations (period estimations)

$R_{SENS}(\Omega)$ $[C_{SENS} = 0\,\text{pF}]$	Operating temperature = 30°C Theoritical period (s)	Simulated period (s)	Relative error in simulation results (%)	Measured period (s)	Relative error in measurement results (%)
500 k	180 μ	180.87 μ	+0.48	170.28 μ	−5.4
1 M	360 μ	360.42 μ	+0.12	340.80 μ	−5.33
5 M	1.8 m	1.7945 m	−0.31	1.6362 m	−9.10
10 M	3.6 m	3.5860 m	−0.39	3.270 m	−9.15
50 M	18 m	17.913 m	−0.48	16.30 m	−9.45
100 M	36 m	35.819 m	−0.50	32.45 m	−9.86
500 M	180 m	179.06 m	−0.52	162.04 m	−9.98
1 G	360 m	358.10 m	−0.53	324.95 m	−9.74
5 G	1.8	1.7904	−0.53	1.6340	−9.22
10 G	3.6	3.5808	−0.53	3.2440	−9.89
50 G	18	17.904	−0.53	16.7723	−6.82
100 G	36	35.808	−0.53	34.066	−5.37

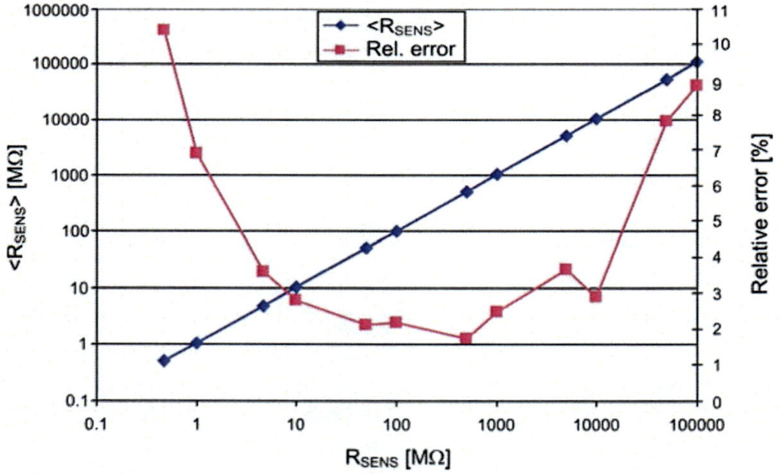

Fig. 3.39 On chip measurement results: estimated R_{SENS} and its relative error compared with sample resistance vs. sample R_{SENS} (with $C_{SENS} = 0$ pF and operating temperature = 30°C)

the integrator output voltage V_R is a quasi-constant signal for very high R_{SENS} values and, when it is around the threshold value (e.g., $V_{DD}/2$), some spurious commutations (i.e., jitters) can occur due to the presence of noise.

When a pure sensor resistor is considered, Eqs. 3.18 and 3.20 produce very similar results, but, if the sensor capacitance C_{SENS} is taken into account, Eqs. 3.21 and 3.22 have to be necessarily adopted so to properly estimate R_{SENS} and C_{SENS} values, starting from the *EX-OR* output squared signal. In this case, the capacitors which represent C_{SENS} are in parallel with the resistor emulating R_{SENS} whose

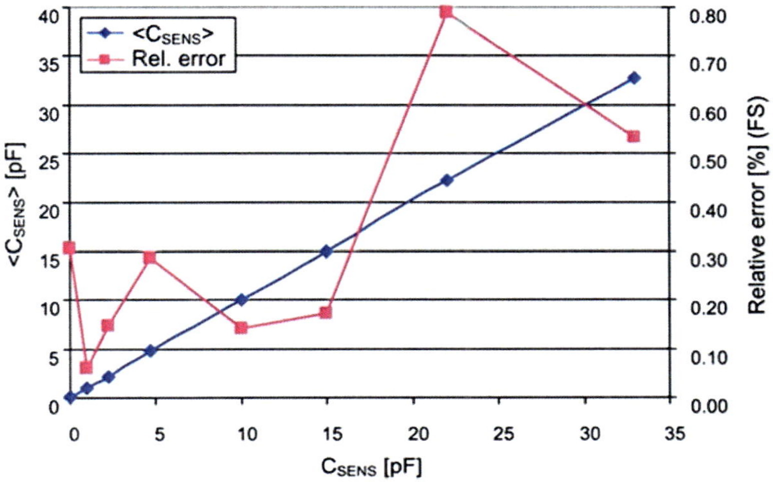

Fig. 3.40 Linearity error (in percentage of full scale - FS) of the C_{SENS} estimation vs. sample capacitors ($R_{SENS} = 100\,\mathrm{M\Omega}$)

estimation maintains its value (within 0.4%, even if C_{SENS} varies from 0 to 33 pF). For an example, if only a pure resistor of 100 MΩ is considered to emulate R_{SENS}, Eq. 3.18 produces an R_{SENS} estimation of 102.0 MΩ while from Eqs. 3.21 and 3.22 $R_{SENS} = 101.9\,\mathrm{M\Omega}$ and $C_{SENS} = 0.1$ pF; on the contrary, if a capacitor of 33 pF is in parallel to the 100 MΩ resistor, then, Eq. 3.18 produces an R_{SENS} estimation of 60.4 MΩ while from the same equations $R_{SENS} = 101.8\,\mathrm{M\Omega}$ and $C_{SENS} = 32.8$ pF. Anyway, regarding sensor capacitance estimation, Fig. 3.40 shows the linearity error for the C_{SENS} estimation with respect to the utilized sample capacitors, ranging from 0 to 33 pF, and with a $100\,\mathrm{M\Omega} R_{SENS}$. The system is able to estimate the parasitic capacitance C_{SENS} with a reduced linearity error for different R_{SENS} values, ranging from 1 MΩ to 10 GΩ.

In order to characterize the integrated circuit behaviour at different environmental temperatures, a climatic chamber (Perani UCI 50/40) has been used. Only the integrated circuit have been placed inside the chamber, whereas the *PLD*-based system for the time measurement has been kept outside. The room temperature has been changed from −20°C to +80°C (30°C has been used as the reference) while keeping the relative humidity to about 20–30%. Concerning these experimental results related to both R_{SENS} and C_{SENS} estimations, the main issues are related to the low resistances at low temperatures and with the high resistances at high temperatures. In the first case, the delay of components plays a determinant role, while, in the last case, the input bias current of the amplifiers increases, leading to the saturation of the integrator. Fig. 3.41 shows the system performances at some different temperature values. In particular, Y-axis reports the relative error between the linearized estimated R_{SENS} value at a certain temperature and the expected value at the reference temperature (30°C), therefore it is a quantification of the thermal error.

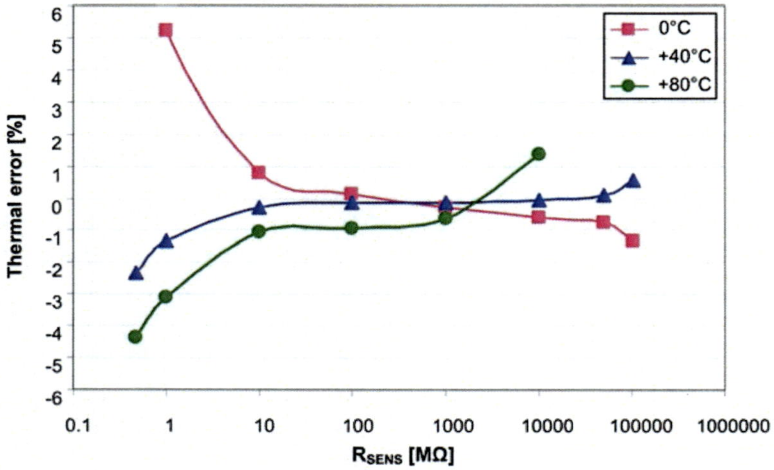

Fig. 3.41 Thermal behaviour of the proposed device. Each curve shows the relative deviation (percentage) of the estimated R_{SENS} with respect to the value at 30°C

It should be noticed that the proposed integrated circuit allows an estimation within about 1% of thermal error over three decades of resistance values (from 10 MΩ to 10 GΩ) and over a temperature range from 0°C to 80°C; while, if a reduced thermal range is considered (from 20°C to 60°C), the thermal error is below 5% over five decades of resistance values (from 470 kΩ to 50 GΩ). On the contrary, concerning C_{SENS} estimations, the chip performances are not influenced by the temperature, even if the capacitance evaluation is available only within a limited range.

In order to better demonstrate the suitability of the proposed interface to sensor applications, a *MOX* sensor has been used and a fast transient has been induced by quickly changing the power applied to the sensor heater. The aim of this test is to prove that the proposed interface is able to track the sensor resistance behaviour even with high resistance values and with wide and quick changes of the resistance itself. The sensor used in this experiment is a TiO_2–based *MOX* sensor, used in a chamber with controlled temperature and relative humidity (25°C, 25%). The sensor working temperature, which depends on the power issued to the heater, has been set to change quickly from 350°C to 470°C, corresponding to a power variation from about 400 to 600 mW. As shown in Fig. 3.42, the proposed system is able to track and, then estimate, a fast transient of both the resistive, which decreases of about two decades in about 10 s, and capacitive sensor components.

The fabricated chip has been finally tested and completely characterized through a suitable experimental apparatus, whose scheme is shown in Fig. 3.43, with a commercial resistive gas sensor. Experimental measurements have been conducted, firstly, by varying the sensor internal heater current (in dry air) and, successively, fixing an operating temperature, by means of the same sensor, to detect into a closed chamber (a chemical reactor) the presence of hydrogen mixed with nitrogen, with different concentrations (0, 40 and 80 ppm). In particular, the sensor heater,

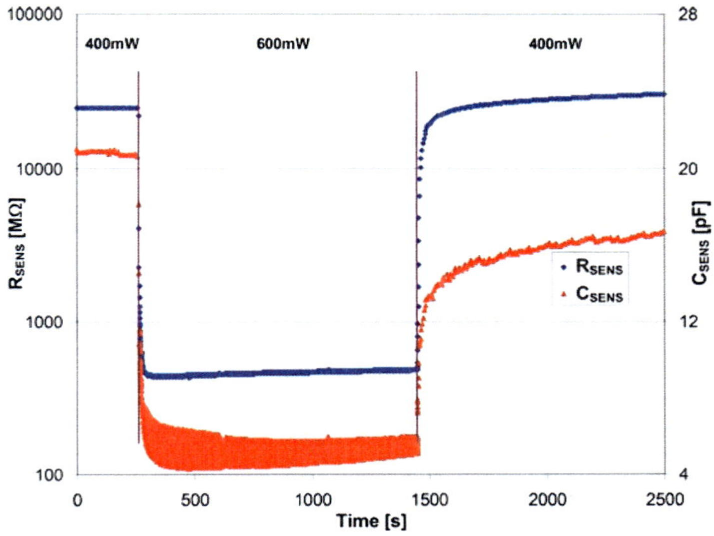

Fig. 3.42 TiO_2 *MOX* sensor response to a thermal transient in terms of estimated C_{SENS} and R_{SENS}

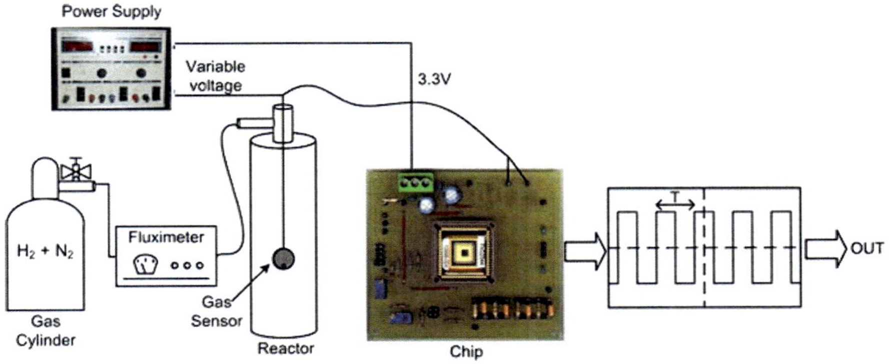

Fig. 3.43 Block scheme of the proposed experimental setup

so the gas sensor operating temperature, has been powered with external different voltages so to achieve a set of working temperatures, while R_{SENS} and C_{SENS} have been monitored through the fabricated chip when the chosen target gas (with different and well defined concentrations) has been fluxed so providing a sensor variation. In order to properly control the hydrogen concentration, a gas flux-meter and a chemical reactor thermally controlled with a thermocouple device have been employed; then, a digital oscilloscope has been used as frequency-meter so to measure the output square wave period T_C.

The R_{SENS} estimation is obtained using Eq. 3.18, neglecting the parasitic capacitance effect ($C_{SENS} = 0\,F$). The sensor employed for the gas detection is the

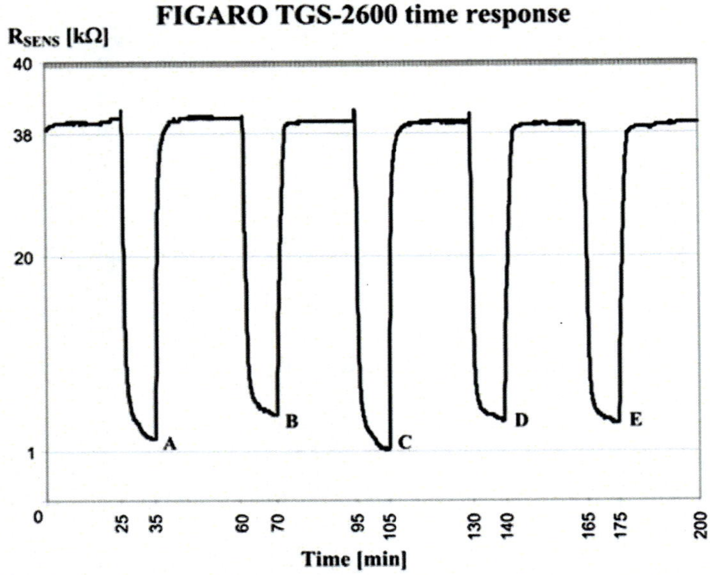

Fig. 3.44 Measured time response of the estimated gas sensor resistance vs. time (minimum sampling time = 1s, see Table 3.13)

commercial low cost device Figaro TGS2600 [50]. The sensitivity of the fabricated circuit has been increased and suitably set to about 7 ms/kΩ by means of the choice of the external capacitance $C_1 = 2.2\mu F$, so to use the proposed system with the low resistance values shown by the chosen sensor (typically from about 1 up to 100 kΩ). In such a way, it is possible to reveal small sensor variation under the presence of reduced ppm of hydrogen and the related time intervals to be measured are on the order of tens of milliseconds, therefore estimable, with reduced error, by the digital oscilloscope preserving, at the same time, a good value of the relative error in sensor resistance estimation. More in detail, the heater current value has been set to 42 mA (corresponding to a heater power consumption of about 216 mW) and a mixture of N_2 and H_2 has been fluxed at different concentrations for 10 min, alternating it with a 25 min dry air flux, repeating this cycle in several measurement sessions. Fig. 3.44 shows the typical system time response, considering the estimated sensor resistance for different H_2 concentrations, as detailed in Table 3.13 (B, D, E = 40 ppm; A, C = 80 ppm) where the mean values of the estimated sensor resistance have been reported, calculated over all the experimental measurements. Moreover, Table 3.13 shows also the related measured interface mean output period $<T_C>$, acquired with a minimum sampling time equal to 1 s, versus the specific gas concentration, so to show the order of magnitude of the revealed periods T_C. In agreement with the Figaro TGS2600 data-sheet [50], the sensor resistance value presents a large difference when gas concentration varies from 0 (dry air) to 40 ppm, whereas a smaller resistive variation between 40 and 80 ppm of H_2 concentrations can be observed.

Table 3.13 Experimental results achieved through the gas sensor FIGARO TGS2600 and related sensor resistance R_{SENS} estimation from T_C measurement (see Fig. 3.44)

Measurement time [min]	H_2 concentration [ppm]	Estimated sensor resistance $\langle R_{SENS} \rangle$[kΩ]	$\langle T_C \rangle$[ms]
0–25 35–60 70–95 105–130 140–165 175–200 Cleaning	(Dry air only)	38.31	270.01
60–70 (B) 130–140 (D) 165–175 (E) $N_2 + H_2$mixture	40	1.21	8.24
25–35 (A) 95–105 (C) $N_2 + H_2$mixture	80	1.05	7.61

3.3.2 Evolutions of AC-Excited Sensor Based Solutions

In order to reduce errors due to non-idealities of active components, in particular the various asymmetries and the noise in the "zero-comparator", and to simplify the time measurement method, a possible evolution of the circuit previously described is here explained [51]. This circuit, reported in Fig. 3.45, allows always to estimate both sensor elements (R_{SENS} and C_{SENS}) through an AC excitation voltage and *OAs* as active blocks.

The main improvement of this topology relies in the fact that the estimation of both resistance and capacitance values can be done through the evaluation of two square wave signals, generated by the comparators, and the measurement of only two times. In particular, referring to the PSpice simulated timing diagram shown in Fig. 3.46, considering the period of squared signal T_1 (which is the same for both generated signals) and the overlapping time T_2 between the signals, a straightforward analysis gives the following two expressions for R_{SENS} and C_{SENS} estimation:

$$R_{SENS} = \left[\frac{1}{2C_1} \frac{G_1 G_2}{(1 - G_1 G_2 - G_2)} \right] (T_1 - 2T_2) \tag{3.25}$$

$$C_{SENS} = \left[\frac{T_1(1 + G_2 - G_1 G_2) + 4T_2(G_1 G_2 - 1)}{2G_1 G_2 (T_1 - 2T_2)} \right] C_1 \tag{3.26}$$

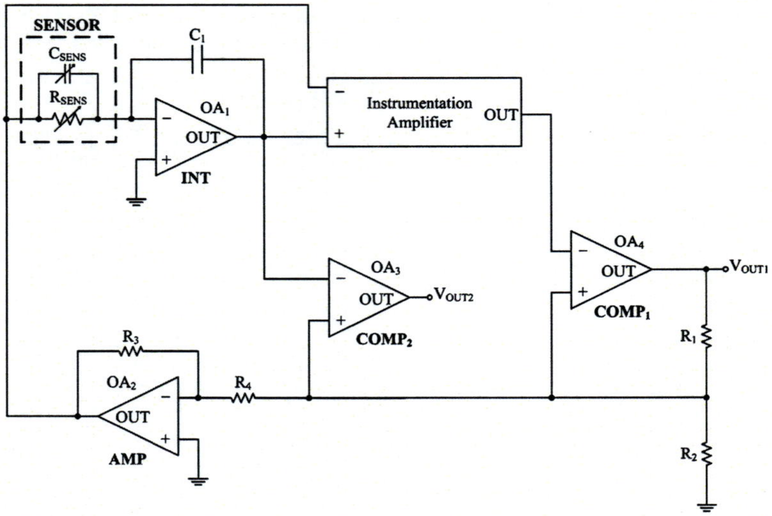

Fig. 3.45 Block scheme of the modified proposed front-end (*INT* = inverting integrator, *COMP*$_1$ = inverting hysteresis comparator, *COMP*$_2$ = voltage comparator, *AMP* = inverting voltage amplifier)

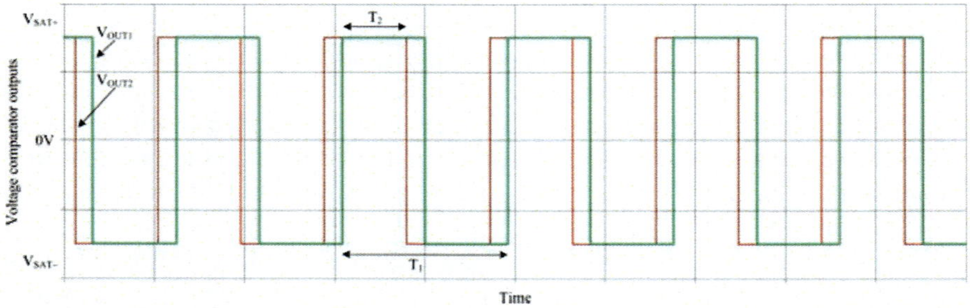

Fig. 3.46 Square wave voltages generated by two comparators and their relationship

being G_1 the voltage gain of the inverting voltage amplifier (typically higher than 1), G_2 the voltage gain of the instrumentation amplifier (in this interface solution, it has to be lower than 1; in this case, this block performs a differential-to-single-ended voltage conversion with an attenuation factor) and C_1 the capacitor of the inverting integrator. It is important to highlight that the two times T_1 and T_2 can be easily determined through the use of an *AND* digital gate which, receiving at the input terminals the two square wave signals (V_{OUT1} and V_{OUT2}, see Fig. 3.45), generates another squared voltage whose period and duty-cycle are exactly T_1 and T_2/T_1, respectively. In order to confirm the validity of the proposed front-end, both PSpice simulations and experimental measurements have been performed.

Table 3.14 Experimental measurements on fabricated prototype *PCB* with sample resistors

Sample resistor (expected R_{SENS}) [Ω]	Measured period T_1 [s]	Estimated R_{SENS} [Ω]	Relative error [%]
1 M	280 μ	1.076 M	+7.60
10 M	2.76 m	10.608 M	+6.08
100 M	27.2 m	104.544 M	+4.54
1 G	276 m	1.0608 G	+6.08
10 G	2.76	10.6081 G	+6.08

The fabricated prototype, developed with commercial components (e.g., LF411 by Texas Instruments), has shown good performances. Since only commercial sample resistors have been utilized emulating sensor resistance, without the presence of other external capacitors as C_{SENS} parasitic component ($C_{SENS} = 0\,\mathrm{pF}$), the overlapping time T_2 has not been here measured. In this way, the expression of R_{SENS} is now the following one (independent from T_2 period):

$$R_{SENS} = \frac{G_1 G_2}{4C_1(1 - G_1 G_2)} T_1. \tag{3.27}$$

Starting from Eq. 3.27 and considering $C_1 = 132\,\mathrm{pF}$, $G_1 = 3.3$, $G_2 = 0.208$, R_{SENS} values have been properly estimated showing a reduced relative error for about four frequency decades, as shown in Table 3.14.

In order to reduce the number of active blocks, a further improvement has been also done, so implementing an interface whose block scheme is reported in Fig. 3.47 [51]. The circuit implements a square wave oscillator whose output period is proportional to sensor resistance value. Also in this case, through the use of an AC sensor excitation voltage, the circuit is able to reveal both the resistive and the capacitive parasitic elements of the gas sensor. More in detail, through the evaluation of T_1 signal period, which is the same for both generated signals at V_{OUT1} and V_{OUT2} terminals and the related overlapping time T_2, as shown in Fig. 3.48, it is possible to estimate both R_{SENS} and C_{SENS} values, by simple calculations (for ideal conditions), as follows:

$$R_{SENS} = \frac{2T_2 - T_1}{2C} \frac{R_1 R_3}{R_2 R_3 - R_1 R_4}, \tag{3.28}$$

$$C_{SENS} = C\frac{R_4}{R_3} - \frac{T_1}{4R_{SENS}}. \tag{3.29}$$

Experimental measurements have been achieved through the fabricated *PCB* developed with suitable commercial components (OPA350, powered at ±2.5 V supply voltage) and using sample resistors to emulate sensor resistance. Concerning the

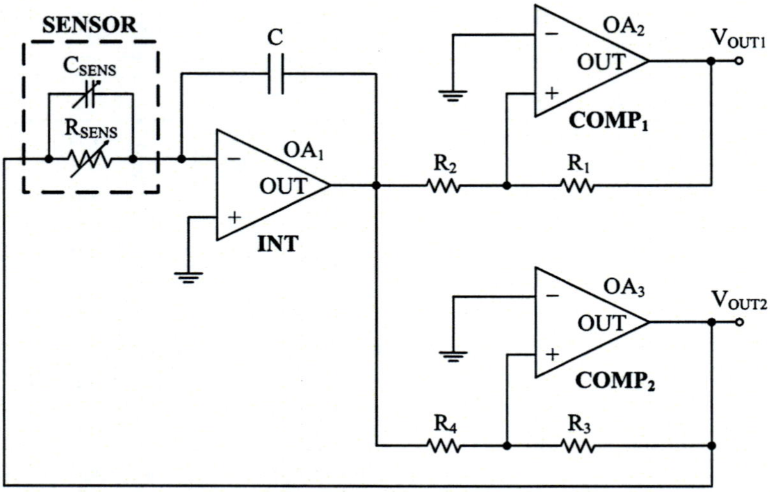

Fig. 3.47 The proposed interface at block level, with a reduced number of *OA* (*INT* = inverting voltage integrator, $COMP_1$ and $COMP_2$ = non-inverting hysteresis comparators)

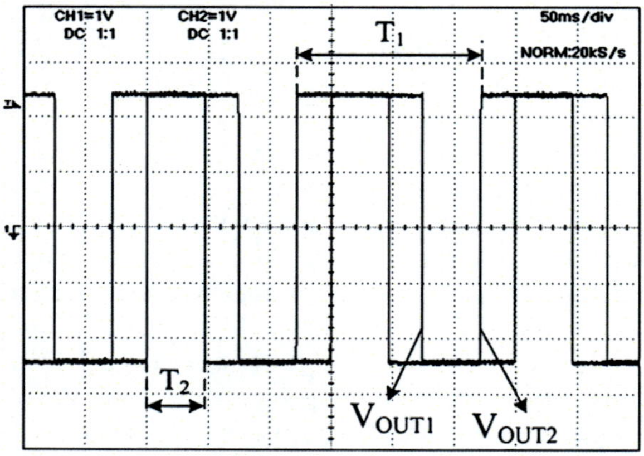

Fig. 3.48 Measured output signal waveforms and their relationship

elements shown in Fig. 3.47, we have set: $R_1 = 10\,\mathrm{k}\Omega$, $R_3 = 2.7\,\mathrm{k}\Omega$, $R_2 = R_4 = 1\,\mathrm{k}\Omega$, $C = 100\,\mathrm{pF}$. In this way, the interface sensitivity has been fixed to about $0.11\,\mathrm{ms/M}\Omega$. The conducted measurements have shown high linearity, reduced percentage error (lower than 10%) and are in a satisfactory agreement with theoretical expectations, as shown in Table 3.15, where the measured times T_1 and T_2 as a function of the sensor resistance have been reported. Table 3.16 shows the estimated values of sensor components (R_{SENS} and C_{SENS}), showing low relative errors (lower than 10%) for a wide frequency range.

Table 3.15 Experimental results: measured times vs. R_{SENS}

R_{SENS} (Measured sample resistance) [Ω]	Theoretical T_1 [s]	Measured T_1 [s]	Relative error [%]	Theoretical T_2 [s]	Measured T_2 [s]	Relative error [%]
101.90 k	11.02 μ	12.50 μ	11.84	2.76 μ	3.40 μ	23.19
0.99 M	107.10 μ	113.00 μ	5.51	26.80 μ	28.00 μ	4.48
9.96 M	1.08 m	1.10 m	1.85	269.30 μ	275.00 μ	2.12
99.75 M	10.80 m	11.00 m	1.85	2.70 m	2.65 m	−1.85
1.02 G	110.30 m	111.00 m	0.63	27.60 m	27.00 m	−2.17
10.01 G	1.08	1.06	−1.85	270.60 m	260.00 m	−3.92

Table 3.16 R_{SENS} and C_{SENS} estimations from time measurements

R_{SENS} (Measured sample resistance) [Ω]	Estimated R_{SENS} [Ω]	Relative error for R_{SENS} estimation [%]	Estimated C_{SENS} value = 10 pF) [F]	Relative error for C_{SENS} estimation [%]
101.90 k	104.31 k	+2.40	7.51 p	−24.90
0.99 M	1.04 M	+5.10	10.37 p	+3.70
9.96 M	10.07 M	+1.10	10.13 p	+1.30
99.75 M	102.48 M	+2.70	10.62 p	+6.20
1.02 G	1.04 G	+2.00	10.85 p	+8.50
10.01 G	9.88 G	−1.30	10.64 p	+6.40

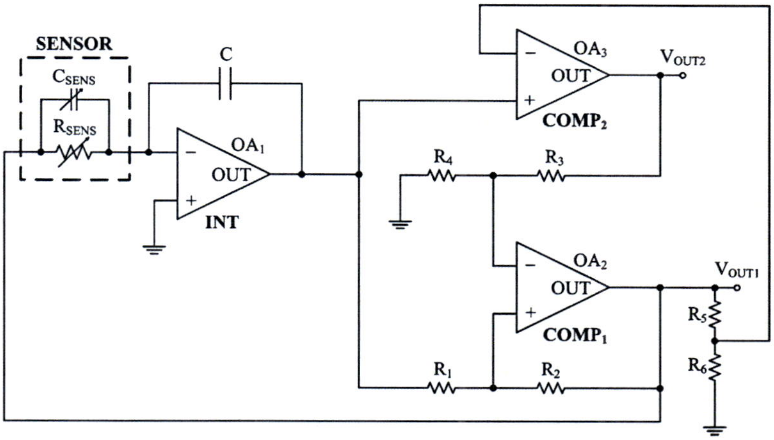

Fig. 3.49 The proposed interface at block level (*INT* = inverting integrator, *COMP*$_1$ and *COMP*$_2$ = voltage comparators)

In order to avoid ground noise disturbs, a further modified version of the previous circuit has been developed, shown in Fig. 3.49. The interface employs always three *OAs* implementing a square wave oscillator whose output period is proportional to sensor resistance value and always using an AC sensor excitation voltage. Also in

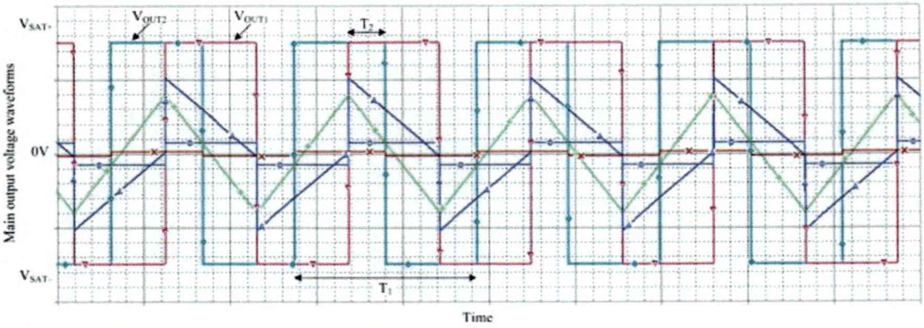

Fig. 3.50 Output signal waveforms at each terminals and their relationship (PSpice simulated timing diagram)

this case, through the evaluation of T_1 signal period (at V_{OUT1} or at V_{OUT2} terminals) and the related overlapping time T_2, it is possible to estimate both R_{SENS} and C_{SENS} values (for ideal conditions), as follows:

$$R_{SENS} = \frac{1}{C}\left(T_2 - \frac{T_1}{2}\right)\left(\frac{1-\dfrac{R_1}{R_1+R_2}}{\left(\dfrac{R_1}{R_1+R_2}\dfrac{R_6}{R_5+R_6}\right)-\dfrac{R_1}{R_1+R_2}-\dfrac{R_4}{R_3+R_4}-\dfrac{R_6}{R_5+R_6}}\right), \tag{3.30}$$

$$C_{SENS} = C\left(\frac{R_1}{R_1+R_2}+\frac{R_4}{R_3+R_4}1-\frac{R_1}{R_1+R_2}\right)+$$

$$-\frac{T_1}{4}\frac{2C}{2T_2-T_1}\left(\frac{\left(\dfrac{R_1}{R_1+R_2}\dfrac{R_6}{R_5+R_6}\right)-\dfrac{R_1}{R_1+R_2}-\dfrac{R_4}{R_3+R_4}-\dfrac{R_6}{R_5+R_6}}{1-\dfrac{R_1}{R_1+R_2}}\right). \tag{3.31}$$

Experimental measurements have been performed through the fabricated *PCB* developed with suitable commercial components (OPA350, powered at ±2.5 V supply voltage) and using sample resistors to emulate sensor resistance. The two times T_1 and T_2 have been measured through the use of an *AND* digital gate which, receiving at the input terminals the two square wave signals (V_{OUT1} and V_{OUT2}), generates another squared voltage whose period and duty-cycle are exactly T_1 and T_2/T_1, respectively, as shown in Figs. 3.50 and 3.51. These measurements are in a satisfactory agreement with theoretical expectations, as shown in Table 3.17, where the measured times T_1 and T_2 as a function of the sensor resistance have been reported, while Table 3.18 shows the estimated values of sensor components (R_{SENS} and C_{SENS}), showing low relative errors (lower than 10%) for a wide frequency range.

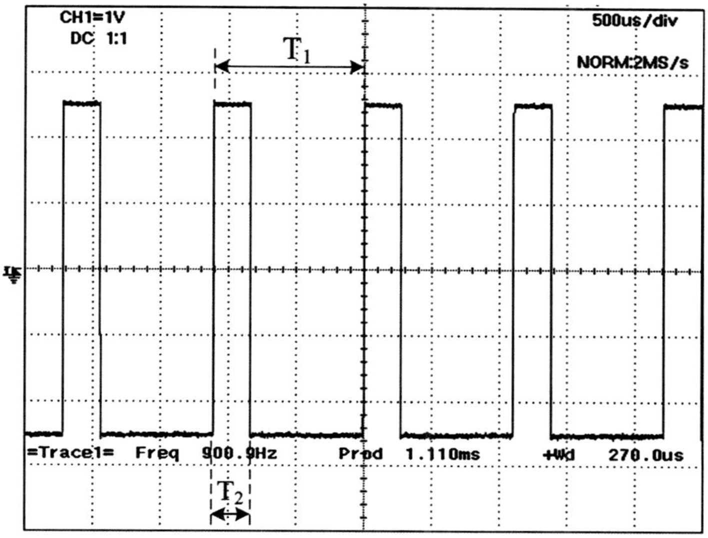

Fig. 3.51 Revealed signal waveform at the output terminal of AND digital block for the measurement of T_1 and T_2 times

Table 3.17 Experimental results: measured times vs. R_{SENS}

R_{SENS} (Measured sample resistance) [Ω]	Theoretical T_1 [s]	Measured T_1 [s]	Relative Error [%]	Theoretical T_2 [s]	Measured T_2 [s]	Relative Error [%]
99.30 k	19.913 μ	21.40 μ	+7.47	2.983 μ	3.6 μ	+20.68
1.01 M	199.135 μ	208 μ	+4.45	29.834 μ	32 μ	+7.26
10.01 M	1.991 m	2.040 m	+2.46	298.336 μ	300 μ	+0.56
99.7 M	19.913 m	20.40 m	+2.45	2.983 m	3 m	+0.57
1.024 G	199.135 m	204 m	+2.44	29.834 m	30 m	+0.57
10 *G* ± 5%	1.991	1.940	−2.56	298.336 m	280 m	−6.15

Table 3.18 R_{SENS} and C_{SENS} estimations from time measurements

R_{SENS} (Measured sample resistance) [Ω]	Estimated R_{SENS} [Ω]	Relative error for R_{SENS} estimation [%]	Estimated C_{SENS} (expected value = 10 pF) [F]	Relative error for C_{SENS} estimation [%]
99.30 k	101.815 k	+2.53%	7.238 p	−27.62%
1.01 M	1.032 M	+2.18%	9.420 p	−5.80%
10.01 M	10.325 M	+3.15%	10.389 p	+3.89%
99.7 M	101.815 M	+2.12%	10.184 p	+1.84%
1.024 G	1.047 G	+2.25%	10.389 p	+3.89%
10 G ± 5%	9.895 G	−1.05%	10.768 p	+7.68%

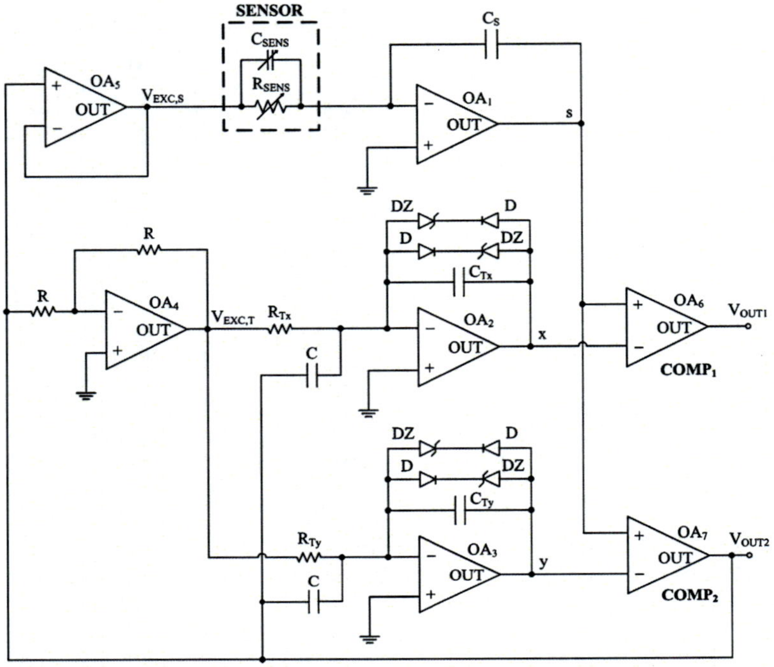

Fig. 3.52 The schematic circuit at block level of the proposed interface

3.3.3 *Fast Uncalibrated AC-Excited Sensor Interfaces with Reduced Measurement Times*

Also for the AC-excited sensor based solutions, a drawback of the *R-T* conversion is related to the long measuring times for the evaluation of high-valued sensor resistances. For this reason, an architecture, always exploiting the *R-T* conversion technique but based on suitable moving thresholds, has been developed so to estimate both sensor resistive and parasitic capacitance components, reducing the higher measuring time at about tens of milliseconds also when very high values of the sensor resistance occur [52–54]. Simulations and experimental measurements conducted with commercial resistors (values between $1\,\text{M}\Omega$ and $100\,\text{G}\Omega$) have confirmed the validity of this proposal.

More in detail, Fig. 3.52 shows the block scheme of the proposed interface solution. It consists of an oscillator circuit directly derived on the circuit shown in Fig. 3.27. By this solution, the two main limits of the previous circuits, that is the long measuring time for high resistive values and the noise at the zero-comparator level, are contemporaneously overcome.

The novelty is in the comparator thresholds used to generate the output waveforms: in this improved solution, such thresholds are not given by fixed voltages, but

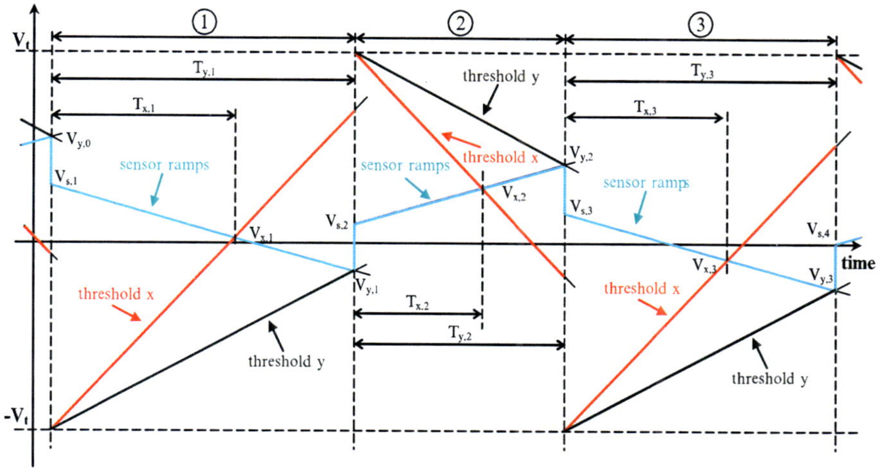

Fig. 3.53 Time diagram for the interface depicted in Fig. 3.52

are implemented by means of two ramp voltages, x and y, which move in the opposite direction with respect to the integrator ramp s (self-moving threshold). In such a way, timings related to the oscillating circuit do not depend on the sensor resistive value only, but also on the moving threshold slopes. In fact, the working principle is always based on the integration of a constant current flowing through the sensor and generating a ramp. Such a ramp is compared with two threshold ramps with known slope and the maximum measuring time is limited by the slower threshold.

This front-end is constituted by seven *OAs*; in particular, referring to Fig. 3.52, OA_1 is the sensor integrator, OA_2 is the faster threshold ramp generator (the first one crossing the sensor ramp), OA_3 is the slower threshold ramp generator (OA_2 and OA_3 are integrators, similar to OA_1), OA_4 is an inverting amplifier to switch the threshold ramp slope with respect to the integrator ramp, OA_5 is a buffer to decouple the circuit from the capacitive effect of the cables connecting the sensor and OA_6 and OA_7 are the two voltage comparators which generate square wave signals ($COMP_1$ and $COMP_2$). The two capacitors C are needed to generate the charge transfer effect on OA_2 and OA_3 during the slope commutation which determines the step on the threshold ramps (bootstrapping effect). The series of a diode D and a Zener diode DZ are needed to make the threshold ramps starting always from the same value $\pm V_t$ after the step due to the charge transfer. In particular, the V_t value is equal to the sum of the direct voltage of the diode D and the reverse voltage of the Zener diode DZ. The two moving thresholds (threshold ramps) have different slopes and the maximum measurement time depends on the time T_y taken by the slower threshold ramp y, starting from V_t, to reach the sensor ramp s (voltage signal at s node, see Fig. 3.52), as depicted in Fig. 3.53. It can be designed to be as little as necessary by means of a suitable choice of C_{Ty} and R_{Ty} values.

However, the information obtained measuring the time interval T_y allows the estimation of the resistive component R_{SENS}, but only under the hypothesis that the parasitic capacitance C_{SENS} is neglected. On the contrary, if the parasitic capacitance cannot be neglected, the only measure of the time interval T_y leads to a significant error in the resistance estimation. The use of the faster threshold ramp x allows the estimation of the resistive component R_{SENS} without being affected by the parasitic capacitance C_{SENS}, which is furthermore estimated.

Always, referring to the timing diagram shown in Fig. 3.53, this front-end allows to estimate the sensor resistive value and parasitic capacitance by means of the commutation time measurements. More in detail, the sensor is supplied by a voltage $V_{EXC,S}$ which derives from the $COMP_2$ output voltage V_{OUT2} through the buffer OA_5. In the same way, the threshold ramp generators are supplied with a voltage $V_{EXC,T}$ which derives from the $COMP_2$ output voltage V_{OUT2} through the inverting amplifier OA_4. If we suppose V_{OUT2} to commutate between the two values $\pm V_{EXC}$, then $V_{EXC,S}$ and $V_{EXC,T}$ voltages commutates, with opposite phase, between the same values $\pm V_{EXC}$. This hypothesis can be considered valid, for example, if rail-to-rail components are used to implement the comparators and the amplifiers. Odd and even cycles have been defined when the sensor ramp at s node is decreasing and increasing, respectively. In this way, the circuit behaviour can be easily analyzed considering both odd and even cycles (e.g., cycle number 1 and cycle number 2), achieving the circuit parameter relationships. In fact, through a straightforward analysis, the following expression, allowing to estimate the sensor resistance R_{SENS} during a generic cycle k (that means there are not matter if the considered cycle is odd or even), can easily be obtained:

$$R_{SENS,k} = \frac{C_{Tx}R_{Tx}C_{Ty}R_{Ty}(T_{y,k} - T_{x,k})}{C_S(C_{Ty}R_{Ty}T_{x,k} - C_{Tx}R_{Tx}T_{y,k})}. \tag{3.32}$$

Therefore, for the sake of simplicity, the evaluation of current-cycle R_{SENS} value can be performed by simply measuring time intervals T_x and T_y in that cycle k and then using the Eq. 3.32. In addition, considering some circuital simplification, such as $C_T = C_{Tx} = C_{Ty}$, and $R_{Tx} = \alpha \cdot R_{Ty} = \alpha \cdot R_T$, with $0 < \alpha < 1$, the following equation can be used to reckon R_{SENS} in a generic cycle k:

$$R_{SENS,k} = \alpha \frac{C_T R_T}{C_S} \frac{T_{y,k} - T_{x,k}}{T_{x,k} - \alpha T_{y,k}}. \tag{3.33}$$

The estimation of the parasitic capacitance C_{SENS} takes advantage of the knowledge for the sensor ramp s both the final value (equal to V_y) at the end of one cycle and the initial value V_s at the beginning of the next cycle. In fact, the step that the signal s performs during the slope commutation is due to the effect of the parasitic capacitance, when the sensor supply voltage switches instantaneously between two values (e.g., from $+V_{EXC}$ to $-V_{EXC}$). In general, considering the commutation between cycle $k-1$ and cycle k, the following expression can be obtained:

$$C_{SENS,k} = -C_S \frac{V_{s,k} - V_{y,k-1}}{V_{EXC,S,k} - V_{EXC,S,k-1}} \tag{3.34}$$

being $V_{s,k}$ the initial voltage value of the sensor ramp s (after the step due to the charge transfer effect on OA_1 caused by C_{SENS}) and $V_{y,k-1}$ the final voltage value of the threshold y (equal to the final voltage value of the sensor ramp s before the k next-cycle step due to C_{SENS}). Also in this case, if all the simplifications described above are applied, the following equation can be used to estimate the parasitic capacitance value C_{SENS} in a generic cycle k:

$$C_{SENS,k} = \frac{C_S}{2C_T R_T}\left(2C_T R_T \frac{V_t}{V_{EXC,S}} - \frac{1-\alpha}{\alpha}\frac{T_{x,k}T_{y,k}}{T_{y,k}-T_{x,k}} - T_{y,k-1}\right). \quad (3.35)$$

It should be highlighted that the moving threshold approach limits also the noise problem due to the lack of hysteresis on zero-comparator. In fact, if the threshold value is fixed and the integrator ramp is slow, spurious commutations of the zero-comparator output can happen, when the integrator ramp is next to the threshold value. With the proposed approach, even in case of very slow integrator ramp, the threshold on $COMP_1$ moves with a certain speed, thus limiting the possibility of spurious commutations of the $COMP_1$ output. It should be furthermore noticed that, in case of very fast sensor ramp s (low sensor resistance value), a great accuracy is required for the time measurement. In fact, time interval T_x can be very small as well as the difference between T_y and T_x. Therefore, the slope of ramp thresholds x and y should be chosen with a great care. More in detail, the slope of ramp y fixes the maximum measurement time, whereas the slope of ramp x should be chosen in order to measure, with a sufficient resolution, the time interval T_x, as well as the difference between T_y and T_x. Referring to Eq. 3.33, resolution in time measurement can affect both the numerator and the denominator of the fraction. With low values of R_{SENS}, term $T_y - T_x$ is very small, whereas with high values of R_{SENS}, term $T_x - \alpha T_y$ can suffer from resolution problems. Starting from similar considerations, a small value of α allows the estimation of low values of R_{SENS}, whereas a big value is more suitable if high R_{SENS} values must be estimated.

The feasibility of the proposed approach has been firstly verified and tested through PSpice and Matlab simulations, so to evaluate its performances. The circuit has been initially simulated using different values of R_{SENS} and without the parasitic capacitance C_{SENS}. PSpice simulations have been conducted using low input bias current operational amplifiers OPA350 and fast-response comparators TLC3702 (all from Texas Instruments) with a ± 3.3 V power supply. Fig. 3.54 shows the absolute value of the resistive estimation error as a function of the value of R_{SENS} and α if the timing resolution is 100 ns and the measurement time is about 40 ms. With these values, the estimation error is expected to be below 1% for $4\,\text{M}\Omega < R_{SENS} < 100\,\text{G}\Omega$ (both in simulations and in prototype board, α value has been set to about 0.2). Fig. 3.55 shows the effect of timing resolution on the estimation error (Matlab simulations).

The component values have been chosen to achieve a maximum measuring time in the order of 100 ms (considering a timing resolution of 100 ns) with a slope ratio between the two ramps of about $8 (1/\alpha \cong 8)$. The initial value of the ramps ($\pm V_t$) has been set using a clamp realized with 2.4 V Zener diodes as DZ and

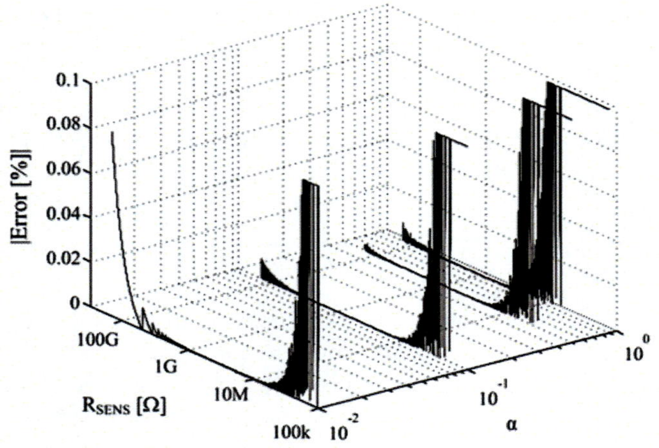

Fig. 3.54 R_{SENS} estimation error due to resolution in timing measurement (set to 100 ns) as a function of α

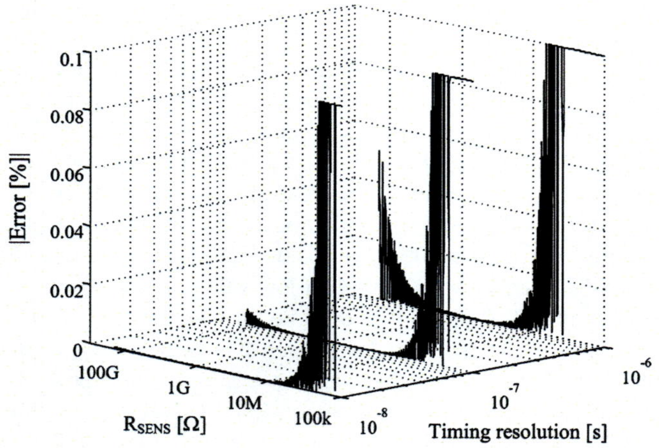

Fig. 3.55 R_{SENS} estimation error as a function of the timing resolution (α set to 0.2)

Schottky diodes with about 0.4 V forward voltage as D (see Fig. 3.52). It is evident that device non-idealities affect the final clamping voltage, that however, reaches a stable value around 2.5 V. Capacitors C have been chosen of the same value of C_{Tx} and C_{Ty} (470 nF), to generate at the output of the threshold ramp generators a voltage step, due to the charge transfer, equal to $2V_{EXC}$. In such a way, it is guaranteed that the threshold ramp steps are big enough to reach the initial point $\pm V_t$ in every possible situation. Table 3.19 shows the results obtained from such simulation tests for different values of sensor resistance, where the estimation of R_{SENS} have been performed using the *WLMS* line, which is better than the usual

Table 3.19 PSpice simulation results: R_{SENS} estimations from time measurements

R_{SENS} [MΩ]	$\langle R_{SENS}\rangle$ *WLMS* [MΩ]	Std./$\langle R_{SENS}\rangle$ [%]	Error *WLMS* [%]	Error Matlab [%]
1.00E − 01	1.287E − 01	87,2	28.7	100
1.00E + 00	1.003E + 00	5,3	0.3	9.7
1.00E + 01	9.619E + 00	2,1	−3.8	0.03
1.00E + 02	1.021E + 02	0,4	2.1	<0.01
1.00E + 03	1.019E + 03	1,7	1.9	<0.01
1.00E + 04	1.038E + 04	8,2	3.8	0.03
1.00E + 05	9.671E + 04	17,4	−3.3	0.5

Table 3.20 Experimental results from the prototype *PCB* ($C_{SENS} = 0$ pF)

R_{SENS} [MΩ]	$\langle R_{SENS}\rangle$ *WLMS* [MΩ]	err. *WLMS* [%]
1.00E + 01	9.975E + 00	−0.3
1.00E + 02	.021E + 02	2.0
1.00E + 03	1.049E + 03	4.6
1.00E + 04	1.010E + 04	1.0
1.00E + 05	9.132E + 04	−9.5

LMS one, when a wide range of variation is considered. Except for the lowest resistance value, where the limit is the time resolution in the measurement (about 100 ns) and the aforementioned problem about threshold overshoots, as the high value of standard deviation clearly highlights, the relative linearity error is below 4% over more than five decades of resistance variation, from 1 MΩ up to 100 GΩ. The differences between Matlab and PSpice results shown in Table 3.19 depend on component behaviour. Timing resolution is not the only source of non-ideality; in the case of high values of R_{SENS}, the current flowing in the integrator is very small, comparable with the bias current of the *OA*. In addition, PSpice takes into account the non-idealities due to non-ideal component offset and propagation delay, while Matlab simply applies equations describing the ideal behaviour.

A discrete component prototype has been developed and tested so to practically validate the proposed circuit, using the same commercial active devices and circuit settings adopted in the previously described simulations. Experimental results, which are in a good agreement with PSpice simulations in a wide resistance value range, have been obtained using commercial resistors simulating the R_{SENS} behavior of the sensor. An accurate calibration of the circuit has been performed to make possible the experimental test (resistance estimations) using the realized prototype *PCB*. In particular, the initial values of the ramps ($\pm V_t$), the delay T_d, and the threshold slope ratio α have been estimated before executing the measurements. Table 3.20 shows, in this sense, the achieved experimental results regarding R_{SENS} estimation.

The problem related to the diode limiter circuit, which implied a high T_d delay, makes the circuit unsuitable for the estimation of resistance values smaller than 10 MΩ. However, in the range between 10 MΩ and 10GΩ, the circuit shows a

Table 3.21 Experimental results from the prototype for the C_{SENS} estimation with different values of R_{SENS}

	$R_{SENS} = 10\,M\Omega$		$R_{SENS} = 100\,M\Omega$		$R_{SENS} = 1\,G\Omega$	
C_{SENS} [pF]	$\langle C_{SENS}\rangle$ *LMS* [pF]	err. *LMS* [% FS]	$\langle C_{SENS}\rangle$ *LMS* [PF]	err. *LMS* [% FS]	$\langle C_{SENS}\rangle$ *LMS* [PF]	err. *LMS* [% FS]
0	−2.06	−4.39	−0.75	−1.59	−0.89	−1.90
1	0.86	−0.29	0.46	−1.15	−0.52	−3.22
2.2	2.07	−0.27	1.69	−1.08	1.08	−2.38
4.7	4.98	0.59	4.75	0.11	4.47	−0.50
10	10.88	1.88	10.87	1.85	11.06	2.26
15	16.13	2.40	15.63	1.33	17.27	4.82
22	23.60	3.40	23.34	2.86	25.93	8.36
47	45.44	−3.31	45.90	−2.33	43.51	−7.44

WLMS error smaller than 5%, in agreement with the simulation results. With very high resistance values ($>10\,G\Omega$), the main issue is related to the threshold slope ratio α. In fact, when the R_{SENS} value is very high, the denominator of Eq. 3.33 becomes very small and can determine high R_{SENS} estimation errors if the ratio α is not well known. In addition, when very-high resistances are considered, the current flowing in the sensor is very small and the behavior of the circuit can be influenced by the non-linearity and non-ideality of the components (e.g. the input bias current of the *OA*).

Furthermore, the parasitic capacitance estimation feature has been experimentally tested using sample capacitors (from 1 pF to 47 pF, accuracy 10% for 1 pF and 2.2 pF, 2% otherwise) in parallel with the resistors used for the previous experiment. Table 3.21 reports the results of such test, when different values of R_{SENS} have been used. In this case, due to the limited range of the input range, the classic least mean square (*LMS*) linearization has been adopted and the linearity error has been reckoned with respect of the full scale (*FS*) value of 47 pF. The performances of the capacitance estimation depend on the R_{SENS} value, because, with high values of R_{SENS} value, the denominator of the second term of Eq. 3.35 is close to zero.

3.4 Voltage-Mode Approach in Capacitive Sensor Interfacing

Capacitive sensors have been proved to be good transducers for signal conditioning electronics with reduced current consumption. In fact, they have a high impedance up to reasonably high measurement frequencies and high signal levels. Generally, capacitive sensor interface topologies have to respect the following constraints: high dynamic range, good linearity and high precision, low input noise and offset, long-term temperature stability, reduced silicon area, low effect of parasitic capacitances and calibration and compensation of the transducer characteristics. These constraints have to be satisfied by interface circuits which, if designed with

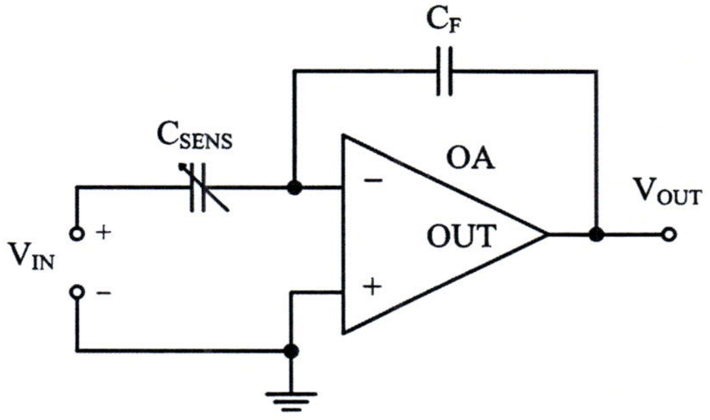

Fig. 3.56 Block scheme of a charge-pump circuit topology based on *OA*

LV and *LP* characteristics, can be utilized in portable, remote and wireless integrated systems for industrial, biomedical, automotive and consumer applications, where a great need of reliable and miniature sensor systems has emerged.

The simpler circuits suitable for the read-out of the capacitive sensors are based on a Capacitance-to-Voltage (*C-V*) conversion. In this case, the implemented solutions usually need to be designed with a high accuracy, because the interfacing circuit is directly contacted to the detected capacitive sensor. This can be simply done by one of the bridge configurations shown in Chap. 2 (see Fig. 2.26). Alternatively, a charge-pump configuration (or charge pre-amplifier), based on *OA* in an inverting topology, as reported in Fig. 3.56, allows to convert the sensor capacitance into an output voltage V_{OUT}, according to the following relationship:

$$V_{OUT} = -\frac{C_{SENS}}{C_F} V_{IN}. \tag{3.36}$$

In this case, the offset voltage of the utilized *OA* and the charge injection error of a switch should be taken into account so to reduce the error in sensor capacitance estimation.

A better solution of the charge-pump circuit is the topology shown in Fig. 3.57 [55]. In this scheme, C_{SENS} is the sensor capacitance to be detected, while C_R and C_F are the designed fixed capacitors. V_{R1} is the common-mode voltage and V_{R2} is a reference voltage. The signals ck_1 and ck_2 represent two non-overlapping phase clocks. When the signal ck_2 is logically high, the voltage V_{R2} will charge the capacitor C_{SENS}, whereas the capacitor C_F stores the offset voltage of the *OA*. The *MOS* transistors M_1 and M_{C1} are switched on and the output voltage is V_{R1}. When the signal ck_1 is logically high, capacitor C_F is connected to the output and the voltage V_{R2} charges the capacitor C_R. Thus, by following the principle of charge conservation, the output voltage V_{OUT} will be expressed as follows:

$$V_{OUT} = \frac{C_{SENS} - C_R}{C_F} (V_{R2} - V_{R1}). \tag{3.37}$$

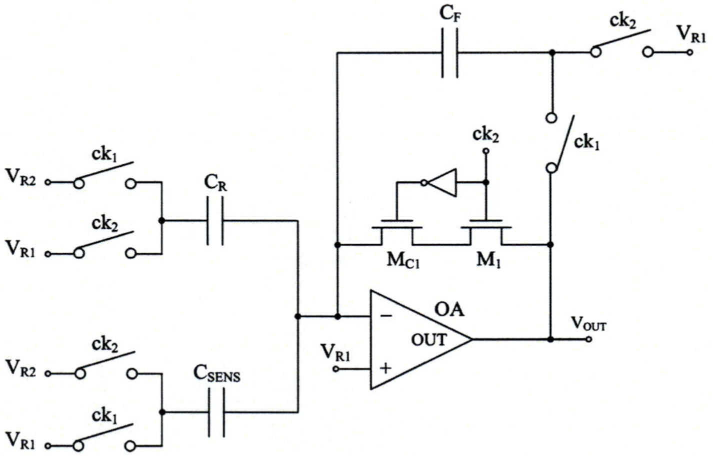

Fig. 3.57 Schematic of the *C-V* converter with the *OA* offset and switch charge injection insensitive properties

Equation 3.37 shows that *OA* offset does not affect the output voltage signal, even if an accurate design of the employed switches has to be operated. Moreover, it is important to remark that this kind of interface is able to reveal reduced capacitance variations, ranging from hundreds of fF up to tens of pF.

Finally, as an example for the *C-V* converter application, we consider here the accelerometer device which represents a typical case of differential capacitive sensor. Referring to the capacitive accelerometer described in Chap. 2, which detects an acceleration proportional to a relative displacement δ of a central electrode with respect to a central position $x_0(\delta = x/x_0)$, the *C-V* conversion can be operated through the use of a capacitive bridge which, in this case, detects and suitably converts differential sensor capacitance variations, as shown in Fig. 3.58.

More in detail, considering that an accelerometer typically presents capacitance variations of the two capacitors, C_1 and C_2 (differential capacitors) having an initial capacitance value C_0 for null acceleration, by connecting the two capacitors (C_1 and C_2) in a passive capacitive bridge configuration with other two capacitors having a value C_0, as shown in Fig. 3.58, it is possible to achieve at the bridge output terminals the following voltage signal:

$$V_{OUT} = V_{IN}\left[\frac{1}{\frac{1}{j\omega C_0}+\frac{1}{j\omega C_1}}\frac{1}{j\omega C_1} - \frac{1}{\frac{1}{j\omega C_0}+\frac{1}{j\omega C_2}}\frac{1}{j\omega C_2}\right]$$
$$= V_{IN}\left[\frac{C_0}{C_0+\frac{C_0}{1+\delta}} - \frac{C_0}{C_0+\frac{C_0}{1-\delta}}\right] = V_{IN}\frac{2\delta C_0^2}{4C_0^2-\delta C_0^2} \cong \frac{V_{IN}}{2}\delta, \quad (3.38)$$

assuming that the relative displacement $\delta \ll 1$ and $C_1 = C_0(1+\delta)$ and $C_2 = C_0(1-\delta)$.

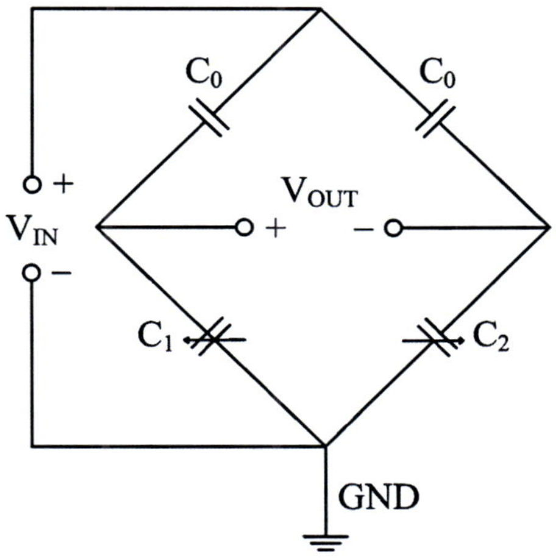

Fig. 3.58 The capacitive bridge configuration: C_1 and C_2 have a grounded common electrode which is equivalent to connect the mobile electrode, fixed to the seismic mass, at the reference ground

In terms of acceleration measurement, when an acceleration a_x on x axis occurs, the seismic mass M of the accelerometer undergoes a displacement δ which corresponds to the capacitance variations of the two capacitors C_1 and C_2 and the equation which allows to estimate the acceleration along the x direction, starting from the measurement of the capacitive bridge output voltage V_{OUT}, can be expressed as follows:

$$a_x = -2\frac{V_{OUT}}{V_{IN}}\frac{k_{el}x_0}{M}, \tag{3.39}$$

since that

$$\delta \cong 2\frac{V_{OUT}}{V_{IN}} = -\frac{M \cdot a_x}{k_{el}x_0}, \tag{3.40}$$

being k_{el} the elastic constant of the device. Eqs. 3.38, 3.39 and 3.40 show that the acceleration a_x is proportional to the relative displacement δ.

A commercial integrated device, performing the accelerometer functions and operating a *C-V* conversion, which can be considered for this kind of application, is the *ADXL50* of Analog Devices (see Chap. 2) [56]. In Fig. 3.59, a possible example of an external schematic circuit for the utilization of this active commercial component is reported. It is provided of a reference voltage which fixes the voltage threshold from which the output signal is generated. This last, suitably amplified, gives the signal which control, as an applicative example, the charge explosion of the airbag for passengers safety.

Alternatively, capacitive sensors are also often interfaced with read-out electronic circuits that perform a Capacitance-to-frequency (*C-f*) conversion [57–64], so that a digitized signal is produced without implementing the analog-to-digital converter. In this case, some of those previously described circuits, performing the *R-f* conversion, can be also employed to perform a *C-f* conversion. The output frequency of these transducers could not be higher than a 100 kHz frequency band (considering the typical *BW* of *OA*). Generally, in this case, the sensor capacitance

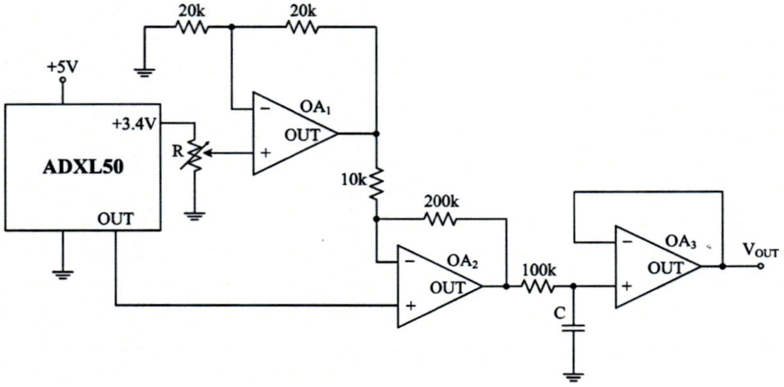

Fig. 3.59 An example of an external conditioning circuit for the commercial integrated accelerometer *ADXL50* utilization in automotive airbag applications (i.e., an accelerometer signal processing)

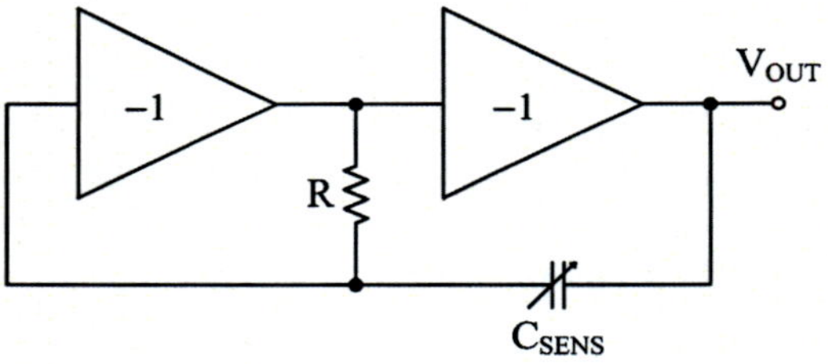

Fig. 3.60 A simple *RC*-based oscillator circuit

is charged and discharged by a constant current and the frequency of the signal revealed at the output of the designed system allows to determine the value of the capacitive sensor, this frequency being inversely proportional to the sensor capacitance value. Successively, an automatic storage of the oscillation frequency can be also performed, using a digital frequency counter.

Moreover, some *RC*-based oscillators, as basic capacitive sensor interfaces, for frequency output sensing circuit, which needs one resistor and one capacitor, have been proposed in the literature [65]. An example is shown in Fig. 3.60 where a ring oscillator, whose output frequency shifts because of C_{SENS} sensor capacitance change, so performing a *C-f* conversion, is depicted.

Fig. 3.61 shows another example of capacitive sensing circuits operating a *C-f* conversion through a high frequency ring oscillator, in a standard integrated *CMOS* technology [66].

Other capacitance read-out circuits are based on switched-capacitor (*SC*), continuous-time current generator (*CTCG*) and continuous-time voltage generator (*CTVG*), etc.. Usually, the *CTVG* sensing has superior noise performances, when compared to the other two, therefore is more suitable for high precision capacitive sensor interfacing.

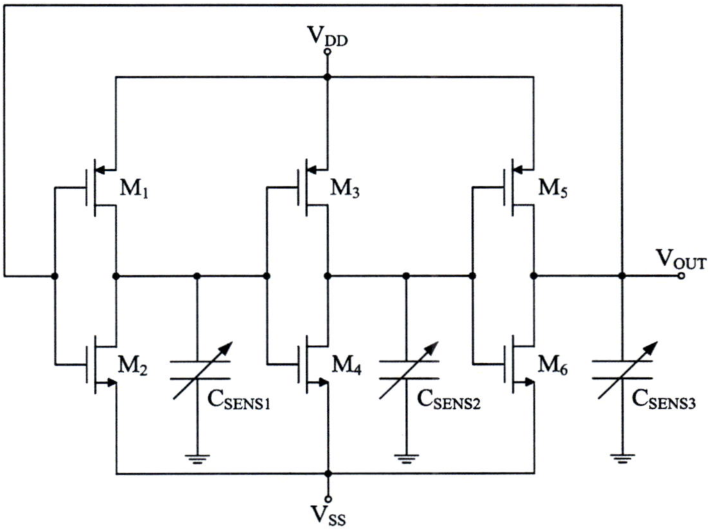

Fig. 3.61 Example of capacitive sensing oscillating circuit

Fig. 3.62 Astable-multivibrator as simple basic capacitive sensor interface for the *C-f* conversion

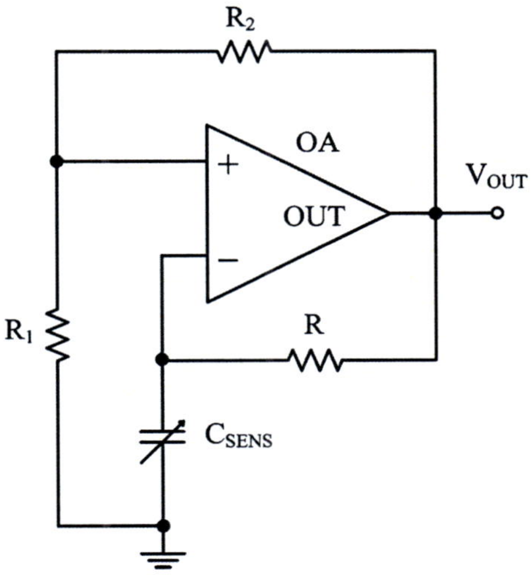

Starting from these considerations, the simpler *VM* capacitive sensor interface circuit, implemented by an *OA*, is the astable multivibrator configuration reported in Fig. 3.62 (also considered for resistive sensor interfaces, see Fig. 3.24) where, in this case, the capacitor represents the sensor [43]. This solution is based on an oscillating architecture generating a square wave signal whose output oscillation frequency, allowing to determine the sensing element value of the capacitive sensor, can be expressed by the following relation (considering $R_1 = R_2$):

$$f \cong \frac{1}{2.2RC_{SENS}}. \tag{3.41}$$

Ideally, the circuit is able to operate for a large span of capacitance variation, which correspond to a frequency span of the same number of decades.

Precise values of resistance (in particular of R) have to be utilized and non-linear effects (among which the temperature) have to be taken into account and verified so to be negligible in the frequency determination, therefore in the capacitance estimation. Moreover, the sensitivity, for this kind of capacitive sensor interface, is very low, being in the order of hundreds of fF/Hz.

3.5 Temperature Sensor Interfaces: Circuits for Temperature Control

In order to have an optimal operating condition for the employed resistive sensor (typically for gas detection), a complete electronic interface must include also a suitable temperature control system. This can be generally formed by a couple of resistances, one of which is the temperature sensor and the other is the heating element [67]. The "in situ" heating [40] is an important requirement in many integrated sensors, especially those for gas detection and flow rate measurements. The heating characteristics are fundamental to determine sensitivity and selectivity of the gas sensor array. Anyway, the problem of the temperature control through the heater element can be reduced to the control of a simple resistance.

A simple low-cost temperature sensor interface can be easily designed in a standard technological process, employing only resistive passive element. It makes use of a temperature sensitive resistive Wheatstone bridge, as shown in the Fig. 3.63. The passive bridge is fabricated, for an example, using polysilicon resistor layers

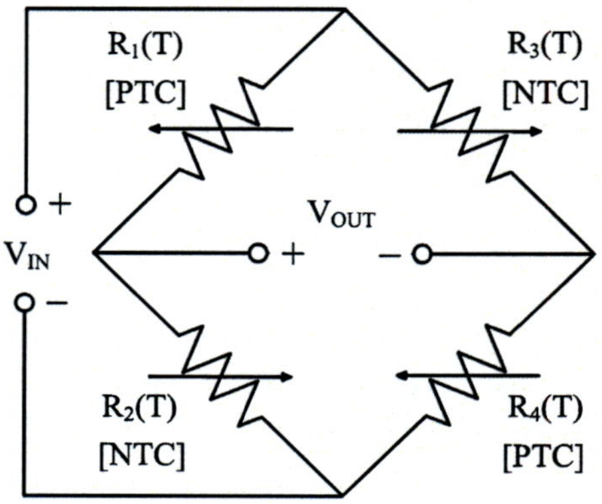

Fig. 3.63 The Wheatstone bridge as temperature sensor interface

of positive first order temperature coefficients (*PTC*) c_1[+ppm/°C] to fabricate opposite resistors R_1 and R_4, and of negative temperature coefficients (*NTC*) c_2[−ppm/°C] to implement R_2 and R_3.

In this manner, the output voltage of the bridge is proportional to temperature variation as follows:

$$V_{OUT} \cong \frac{1}{2} V_{IN} (c_1 - c_2)(T - T_0) \tag{3.42}$$

being

$$R_1 = R_4 = R_0[1 + c_1(T - T_0)] \tag{3.43}$$

and

$$R_2 = R_3 = R_0[1 + c_2(T - T_0)]. \tag{3.44}$$

The output voltage V_{OUT} is proportional to absolute temperature and is independent from the values chosen for the resistances. The circuit can be also powered at low supply voltage (e.g., $V_{IN} = 1$ V) and its output sensitivity is in the order to few mV/°C. As for resistive sensors, the introduction of a differential instrumentation voltage amplifier at the output nodes allows to improve the circuit sensitivity, even if introduces some problems such as offset, noise, etc..

Nevertheless, noise calculation is fundamental to determine the system resolution. An accurate evaluation of the noise brings us to the conclusion that noise is generally given only by external resistances if a suitable differential amplifier is developed. A proper design allows to obtain a high linearity for a large temperature range variation, while a 0.01°C resolution, one order of magnitude lower than commercial digital thermometers, can be also achieved even if for a reduced range of temperature variation (about 0 − 40°C).

A possible application of this simple basic sensor interface concerns the use of thermal conductivity sensors suitable for the absolute humidity measurements, by quantifying the difference between the thermal conductivity of dry air and that of air containing water vapour. These devices (or thermal conductivity absolute humidity sensors), typically composed by two matched *NTC* thermistors, are used in a DC powered resistive bridge circuit configuration (see Fig. 3.63) whose output voltage is directly proportional to absolute humidity: one element is hermetically encapsulated in dry nitrogen (e.g., R_2 as a dry nitrogen sealed thermistor) and the other is exposed to the environment (e.g., R_1 as an ambient air thermistor utilized for examination and detection). When current flows through the thermistors, resistive heating increases their temperature (typically to >200°C). The heat dissipated from the sealed thermistor is greater than the exposed thermistor due to the difference in the thermal conductivity of the water vapour as compared to dry nitrogen. Since the heat dissipated yields different operating temperatures, the thermistor resistance difference, so the bridge output voltage, is proportional to the absolute humidity. Typically, the simple resistor network provides a voltage output variation of about 0–13 mV corresponding to the humidity range of 0–130 g/m^3 at 60°C. In addition, the required calibration is performed by placing the sensor in moisture-free air or nitrogen and adjusting the output voltage of the bridge to zero [68].

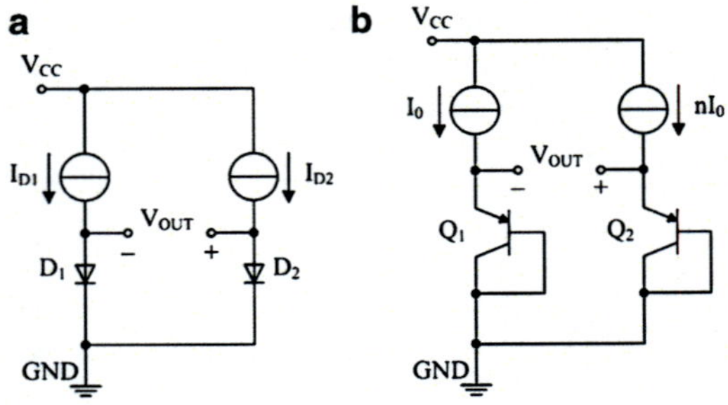

Fig. 3.64 The *PTAT* sensors: a basic principle scheme of a circuit which generates a PTAT signal. It is implemented by either two matched diodes **(a)** or two diode-connected bipolar transistors **(b)** which are powered by two different current generators

Temperature sensors make use also of bipolar technology integrated in a chip: they normally sense the difference of two base-emitter voltages, biased by different currents, to detect temperature variation, since that, concerning a semiconductor-based electronic devices, the current densities depend on the operating temperature of the same device. Thus, the junction diode, which represents the main basic element for the junction-based devices, can be utilized as a temperature sensor, exploiting its temperature-dependent characteristics. In particular, if the diode is supplied with a constant current level, when the temperature increases, a reduction of the voltage at the diode terminals can be observed and this behaviour can be seen as a resistance decrease. Since that the integrated circuit design allows to implement "matched" devices that show identical characteristics, it is possible to exploit the diode sensibility to the temperature variation implementing circuit solutions which provide the so-called *PTAT* signals. In fact, in order to easily evaluate the temperature through the read-out of a voltage signal, for an example, it is possible to inject two different currents into two equal and "matched" diodes, as in the circuits shown in Fig. 3.64a, or diode-connected bipolar transistors, as reported in Fig. 3.64b. It is important to highlight that, also in this case, in order to improve the sensitivity of these circuits, they need a suitable differential amplifier connected at the output terminals [67].

In particular, referring to Fig. 3.64a, the operating principle of the basic *PTAT* measurement technique can be described as follows: if the two injected currents into the diodes have a temperature-independent and accurate ratio ($n = I_{D2}/I_{D1}$, this condition is easily achievable in the integrated circuits), the output voltage of the circuit, revealed at the diode terminals, is given by:

$$V_{OUT} = \Delta V_D = V_{D2} - V_{D1} = \frac{kT}{q}\left[\ln\left(\frac{I_{D2}}{I_0}\right) - \ln\left(\frac{I_{D1}}{I_0}\right)\right] = \frac{kT}{q}\ln\left(\frac{I_{D2}}{I_{D1}}\right). \tag{3.45}$$

In the same manner, considering matched transistors as shown in Fig. 3.64b, the difference between the emitter to base voltages of these transistor is:

$$V_{OUT} = \Delta V_{EB} = V_{EB2} - V_{EB1} = \frac{kT}{q}\ln(n) \tag{3.46}$$

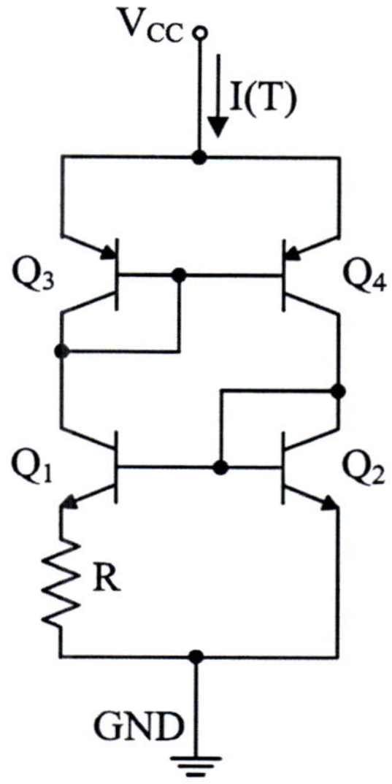

Fig. 3.65 The principle scheme of the commercial device *AD590*

being n the ratio between the transistor emitter currents, k the Boltzmann constant and q the absolute value of the electron charge. Therefore, the resulting output voltage signal is accurately proportional to the absolute temperature T. Moreover, considering that $k/q = 86\mu V/K$, temperature sensitivity is not high, so this solution can be utilized for the thermal compensation in the integrated circuits.

The *PTAT* current principle is exploited in some commercial integrated temperature sensors as, for example, the *AD590* produced by Analog Devices [69], whose principle scheme is reported in Fig. 3.65. This sensor can be considered as a temperature-dependent current generator, powered by a constant supply voltage, and typically, its sensitivity is about 1 $\mu A/K$. Fig. 3.66 shows the typical sensor *I-V* characteristics highlighting that the device can be considered a current generator only for biasing voltages higher than about 4 V. Moreover, in Fig. 3.67 a simple application example of this commercial sensor is reported, where an output voltage signal is achieved through the use of an external load resistor. On the other hand, the device can be connected to a Current-to-Voltage converter implemented by an *OA* (i.e., a transresistance amplifier configuration) so to achieve an output voltage level proportional to the generated current and, therefore, depending on the absolute temperature [67].

Another commercial integrated device which can be utilized as a temperature sensor is the *LM35* produced by National Semiconductor [70]. In fact, this discrete active component, produced and commercialized by National Semiconductors, is based on *PTAT* signal principle, but provides directly a voltage signal proportional to the temperature which has to be revealed. In Fig. 3.68 the principle internal scheme of this temperature sensor is shown.

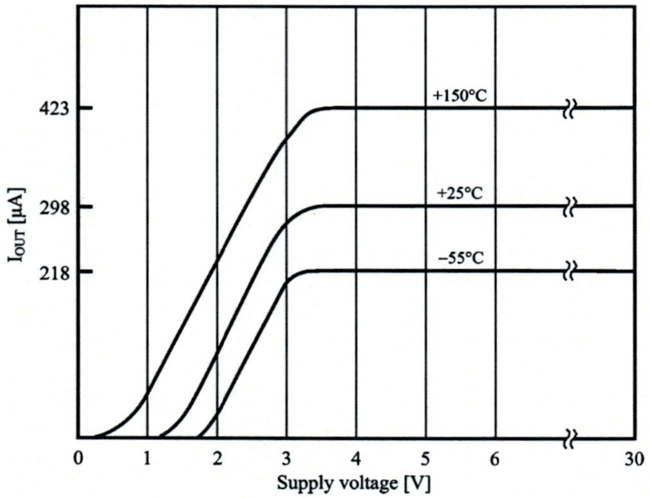

Fig. 3.66 The *AD590* sensor *I-V* typical characteristics for different temperatures: the output current is independent from the applied voltage and proportional to the absolute temperature for supply voltages higher than about 4 V

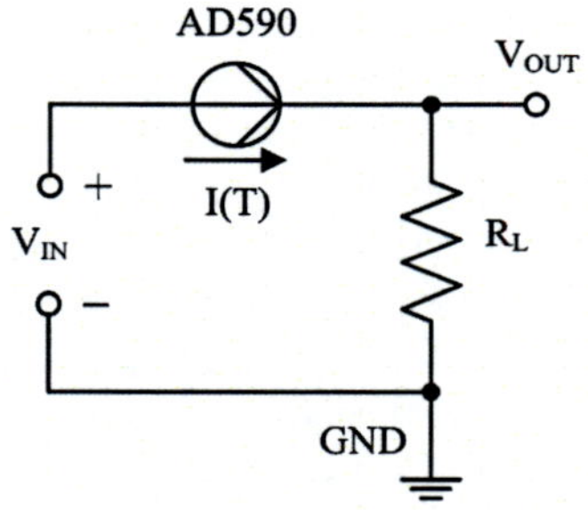

Fig. 3.67 A simple example scheme for the utilization of the *AD590* temperature sensor ($V_I N > 4\,V$)

In this case, the generated output voltage V_{OUT}, depending on the temperature T, can be expressed by the following relationship:

$$V_{OUT} = T\left[\frac{R_4 + R_5 + R_6}{R_5}\frac{k}{q}\ln\left(\frac{R_2}{R_1}\right)\right], \tag{3.47}$$

being k the Boltzmann constant and q the absolute value of the electron charge. Typically, internal components have been designed so to have a sensibility of about 10 mV/K in a temperature range of about $1-40°$C, while its best resolution is in the order of about 0.5°C at 25°C (typical value, as for all the other related commercial *ICs*) [67].

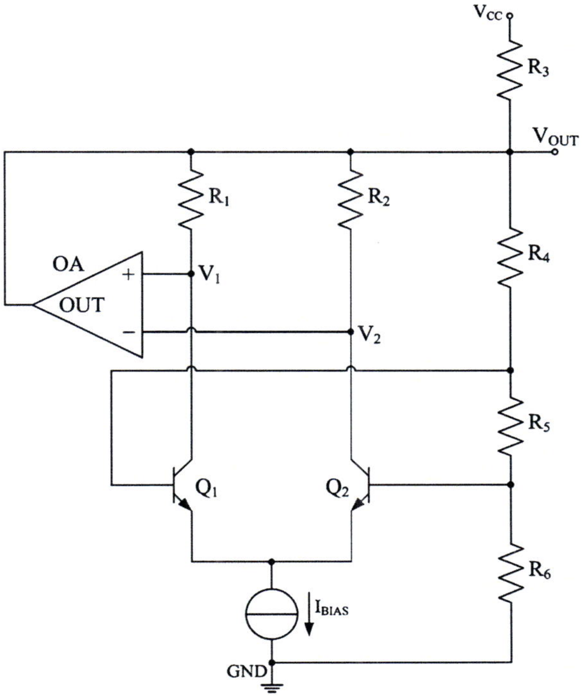

Fig. 3.68 The principle scheme of the commercial device *LM35*

3.5.1 *An Integrated Temperature Control System for Resistive Gas Sensors*

The temperature control is often necessary in gas sensor systems, therefore an automatic temperature regulation system must include a temperature sensor, a thermal actuator and an electronic interface [71–74]. In this Section we describe an example of an integrated temperature control system for resistive gas sensors, utilizing a constant powered heater element, developed for an Italian Research Project (Italian PRIN project 2003/091427) [11–15]. Fig. 3.69 shows a simplified block scheme of the complete gas sensing microsystem, where a suitable temperature control sub-system has been designed. In particular, this microsystem consists of: a semiconductor gas microsensor array based on *MOX* thin films deposited on micromachined substrates, fabricated and fully characterized in their functional and electrical performance (with particular attention to drifts and noise); a mixed-signal integrated front-end with embedded customized *A*/*D* converter, developed in a standard *CMOS* technology, for the read-out and the digitalization of the data coming from the sensors, able to reveal wide-range resistive variations; an electronic sub-section contains a micro-heater (R_{HEAT}) for a smart control of both the chip temperature and the gas sensors, as shown in Fig. 3.70; a final data processing system that uses pattern recognition algorithms to allow information extraction

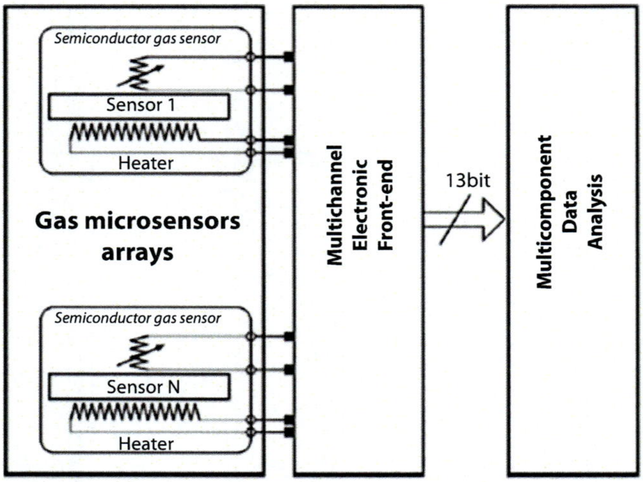

Fig. 3.69 A complete gas sensing microsystem with internal temperature control sub-system

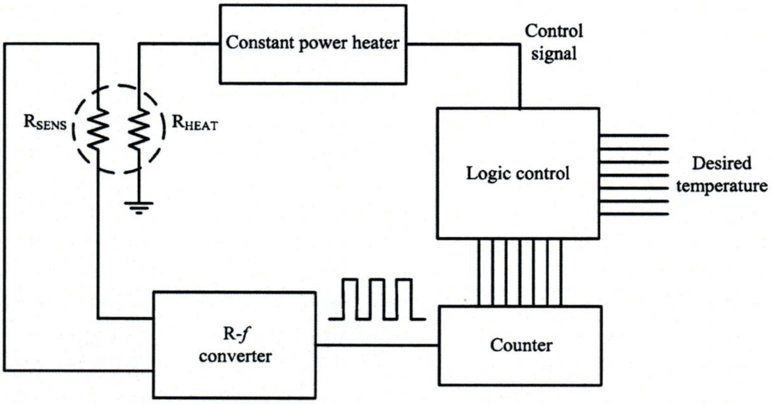

Fig. 3.70 Developed temperature control sub-system block scheme, based on a Resistance-to-frequency converter (R_{SENS} = temperature sensor resistance; R_{HEAT} = heater element resistance)

from the data acquired from the front-end. As regards the thermal characteristics, so the features of the temperature control sub-system, sensor specifications are the following: $R_{SENS} = 1\,\mathrm{k}\Omega$ (at 21°C); $R_{HEAT} = 100\,\Omega$ (at 21°C); supply voltage = 3.3 V; required heater power = 25 mW; operating temperature of the resistive gas sensor = 200°C.

The current, or more generally, the power delivered to the heater resistance must be such that the temperature has to remain constant. This task is made easy thanks to the presence of the Logic Control block (see Fig. 3.70) that allows, through a *R-f* conversion (i.e., a solution based on a oscillator circuit), the measurement and

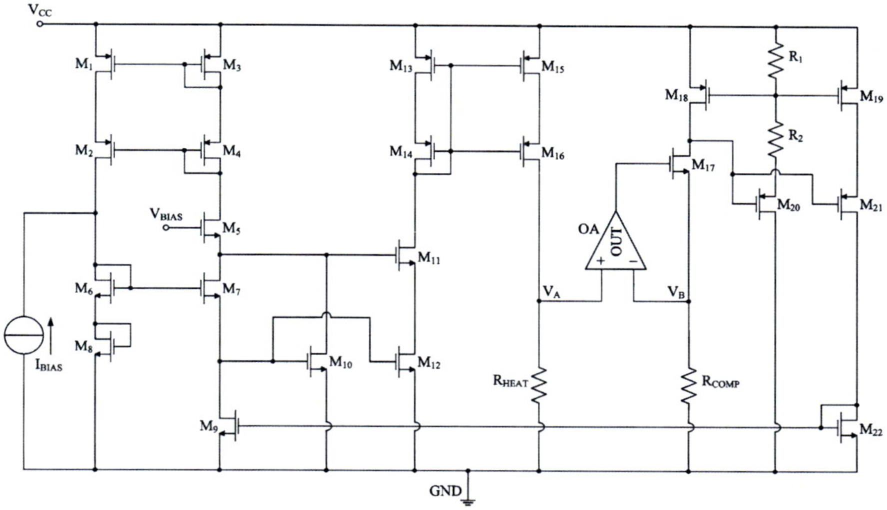

Fig. 3.71 Heater schematic circuit at transistor level

control of the heater resistance and, therefore, of the gas sensor temperature. The suitable control logic evaluates the frequency and, consequently, the resistance, generating a feedback signal to maintain the gas sensor at the proper operating temperature. The power delivered to the heater is electrically controlled by a digital sub-system. This solution has been demonstrated to be the more interesting and affordable considering the 3.3 V total supply voltage. The whole temperature control system shows a 60 mW total power consumption, so it can be considered suitable for portable sensor applications.

This temperature control system is driven by a constant supply voltage, which generates a constant power signal that can be controlled through an external voltage. A constant current heater system has not been considered because it needs a current generator which has to be very much stable, especially with temperature, insensible to load variations and has to know exactly the current injected into the heater. On the contrary, the developed circuit delivers a constant power that is independent from the variation in the heater resistor R_{HEAT}. In particular, since the gas microsensor performances depend on the substrate temperature, the sensor temperature has to be revealed and proper controlled. The designed integrated power generator scheme (heater circuit), at transistor level, is shown in Fig. 3.71 and represents a modified version of that presented in [71].

The circuit allows to achieve good performances in terms of R_{SENS} measurement, since that it does not suffer of V_{CC} drifts; in fact, the heater resistance R_{HEAT} and the compare resistance R_{COMP} have been connected to ground, so to be insensitive to supply variations. The circuit delivers a constant power independent from the variation in the heater resistor R_{HEAT}. The cascade stages provide better power supply rejection and reduced channel length modulation effects while maintaining

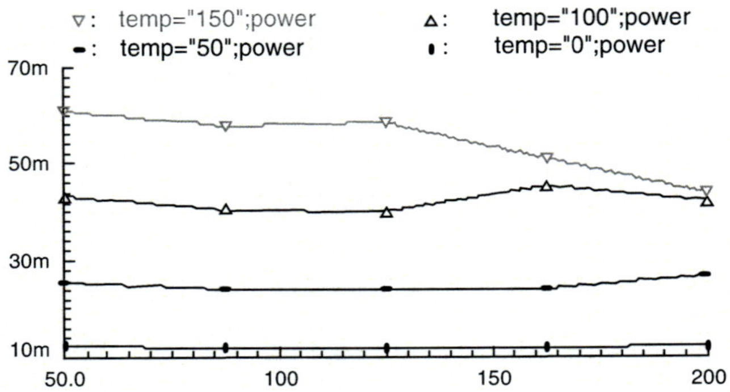

Fig. 3.72 Power vs. R_{HEAT} (at different temperatures, power level equal to 25 mW)

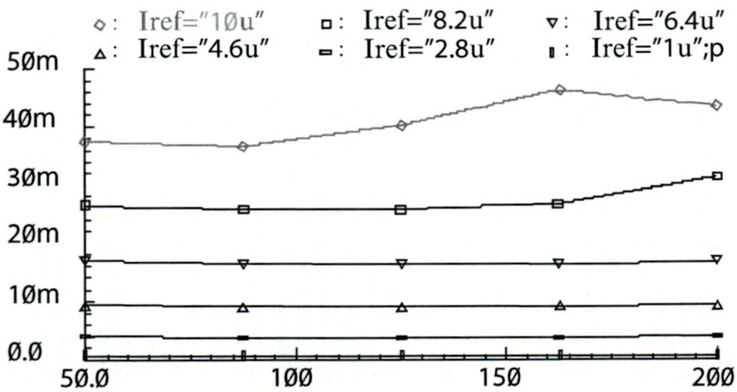

Fig. 3.73 Dissipated heater power vs. R_{HEAT} (at different I_{BIAS} current values)

the same frequency response. In particular, the two pairs M_6, M_8 and M_7, M_{10} constitute a trans-linear loop through a negative feedback that ensures the following relation:

$$4I_{BIAS}^2 = I_{D7,SAT} I_{D10,SAT}. \tag{3.48}$$

Transistor M_5 acts as a current buffer for transistors M_9 and M_{10} to isolate them from the supply voltage, while V_{BIAS} level, applied to the gate of M_5, sets the required power on R_{HEAT} through the injection of a suitable current in the branch. The drain currents of M_9 and M_{10} are added and mirrored through the cascoded current mirror M_1–M_4. The current which flows into R_{HEAT} is equal to I_{D10}, suitably mirrored by M_{13}–M_{16}. By comparing V_A and V_B voltages, M_{17} drives the current into M_{19} and M_{20}, which act directly on the translinear loop. Typical R_{HEAT} and R_{COMP} values are about 100 Ω, while the power, which gives the required heating level to the considered chemioresistive gas sensor, is about 25 mW [40].

Fig. 3.72 shows the simulated power provided by the heater circuit vs. heater resistance at different temperatures, while Fig. 3.73 shows the same power vs. heater

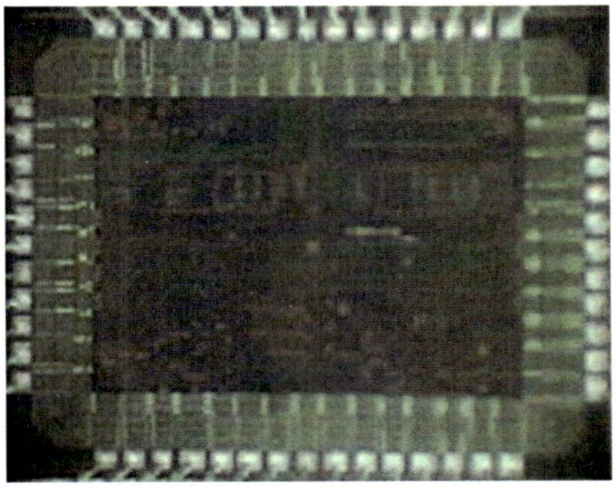

Fig. 3.74 Chip photo

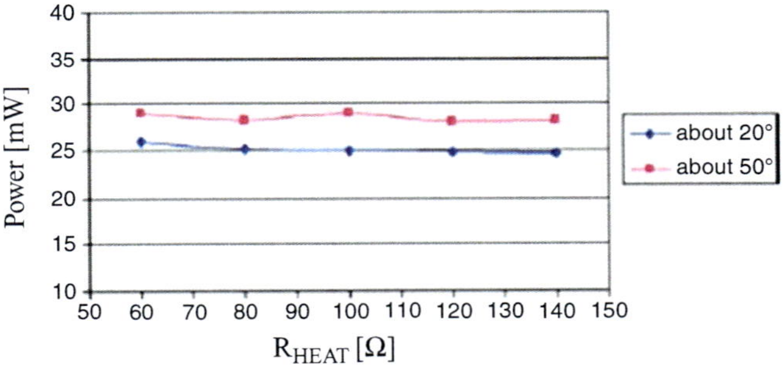

Fig. 3.75 On-chip measurements for the heater at two different temperatures

resistance at different reference currents I_{BIAS}. The typical I_{BIAS} value related to 25 mW power level is 8.2 μA. The V_{BIAS} level that ensures 25 mW power on heater resistance is about 0.8 V. Small variations of this value do not affect the amount of power dissipated on the heater resistance.

Fig. 3.74 shows the photo of fabricated chip implemented in a standard *CMOS* technology (AMS 0.35μm). Power measurements on the heater excellently agree with post-layout simulations. In Fig. 3.75 the power delivered by the heater versus different R_{HEAT} values at two specified temperatures (about 20°C and 50°C) is reported [14, 15, 71]. The digital subsystem has been implemented through a XILINX FPGA board which allows to maintain the temperature at the desired level. In the system design specifications, we have considered a temperature resolution of 0.5°C, to which corresponds a 3.7 KHz of frequency variation. This imposes a time acquisition window larger than 300 μs. Then, a 1 ms value has been chosen for the acquisition window.

Since the maximum frequency is lower than 5 MHz, a 13 bit digital counter has been chosen. With this specification, we have managed temperatures higher than 0°C (R_{SENS} = 500 Ω). If the temperature is lower than the minimum acceptable value, an alarm signal will warn. The maximum testing temperature

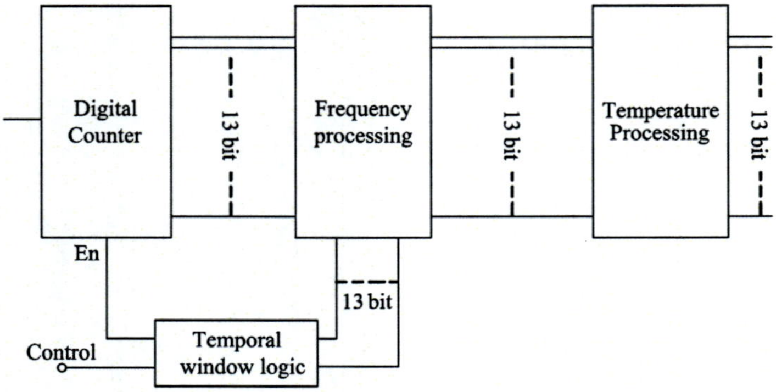

Fig. 3.76 Digital sub-system block scheme

value is defined by the fusion point of the components. If the temporal window dimension is reduced to 500 μs, a rise in temperature testing range occurs, so a lower sensibility is obtained. Fig. 3.76 shows the complete digital Frequency-to-Temperature conversion block scheme. The frequency processing block computes, by the information provided from Digital Counter and Temporal window logic, the frequency value. Moreover, the "Temporal window logic" block manages the temporisation of the digital counter. The digital counter dimension is the same (13 bit), so we can count 8192 different configurations. The frequency value is defined by the ratio between the output data of the digital counter and the temporal window dimension.

References

1. G. Ferri, V. Stornelli, *Complementi di Microelettronica analogica* (Aracne, Roma, 2006). ISBN 8854803200
2. A. D'Amico, C. Di Natale, G. Ferri, A ±0.6 V-supply CMOS topology to improve Wheatstone bridge performance, in *Proceedings of AISEM (Associazione Italiana Sensori e Microsistemi)* Conference, Roma, Feb 1999, pp. 403–407
3. A. Morgenshtein, L. Sudakov-Boreysha, U. Dinnar, C.G. Jakobson, Y. Nemirovsky, Wheatstone bridge readout interface for ISFET-REFET applications. Sensor. Actuat. B **98**, 18–27 (2004)
4. T. Sahm, L. Mädler, A. Gurlo, N. Barsan, U. Weimar, A. Roessler, S. E. Pratsinis, High performance porous metal oxide sensors via single step fabrication, in *Proceedings of Eurosensors*, Barcelona, Sept 2005
5. A. Fort, M. B. Serrano-Santos, R. Spinicci, N. Ulivieri, V. Vignoli, Electronic noses based on metal oxide gas sensors: the problem of selectivity enhancement, *Proceedings of IMTC*, Como, May 2004, pp. 599–604
6. H.T. Chueh, J.V. Hatfield, A real-time data acquisition system for a handheld electronic nose. Sensor. Actuat. B **83**, 262–269 (2002)
7. G. Sberveglieri, Recent developments in semiconducting thin-film gas sensors. Sensor. Actuat. B **23**, 103–109 (1995)
8. S. Capone, P. Siciliano, Gas sensors from nanostructured metal oxides. Encyclopedia Nanosci. Nanotechnol. 3, 769–804 (2004)

9. A. Flammini, D. Marioli, A. Taroni, A low-cost interface to high value resistive sensors varying over a wide range, in *Proceedins of IEEE Instrumentation and Measurement Technology Conference*, Veil, 2003, pp. 726–731
10. D. Barrettino, M. Graf, W.H. Song, K. Kirstein, A. Hierlemann, Hotplatebased monolithic CMOS microsystems for gas detection and material characterization for operating temperatures up to 500°C. IEEE J. Solid-St. Circ. **39**, 1202–1207 (2004)
11. M. Grassi, P. Malcovati, A. Baschirotto, A 0.1% accuracy 100Ω–20MΩ dynamic range integrated gas sensor interface circuit with 13 + 4 bit digital output, in *Proceedings of IEEE European Solid-State Circuits*, France, pp. 351–354, 2005
12. M. Grassi, P. Malcovati, A. Baschirotto, A high-precision wide-range front-end for resistive gas sensors arrays. Sensor. Actuat. B **111–112,** 281–285 (2005)
13. P. Malcovati, M. Grassi, F. Borghetti,V. Ferragina, A. Baschirotto, Design and characterization of a 5-decade range integrated resistive gas sensor interface with 13-bit A/D converter, in *Proceedings of IEEE International Conference on Sensors*, Irvine, 2005, pp. 472–475
14. A. Baschirotto, S. Capone, A. De Marcellis, C. Di Natale, V. Ferragina, G. Ferri, L. Francioso, M. Grassi, N. C. Guerrini, P. Malcovati, E. Martinelli, P. Siciliano, V. Stornelli, A portable integrated wide-range gas sensing system with smart A/D front end, *Proceedings of IMCS, Brescia,* July 2006
15. A. Baschirotto, S. Capone, A. D'Amico, C. Di Natale, V. Ferragina, G. Ferri, L. Francioso, M. Grassi, N. Guerrini, P. Malcovati, E. Martinelli, P. Siciliano, A portable integrated wide-range gas sensing system with smart A/D front-end. Sensor. Actuat. B **130**(1), 164–174 (2008)
16. A. De Marcellis, G. Ferri, V. Stornelli, A. Depari, A. Flammini, D. Marioli, A novel op-amp based front-end for high valued resistive sensors, in *Proceedings of AISEM (Associazione Italiana Sensori e Microsistemi) Conference*, Roma, Feb 2008
17. A. Depari, A. Flammini, D. Marioli, E. Sisinni, E. Comini, A. Ponzoni, A 10 ms-readout interface for experimental resistive sensor characterization. Proc. ISOEN **1137**, 220–223 (2009)
18. A. Depari, A. Flammini, D. Marioli, E. Sisinni, A fast-readout interface circuit for high-value and wide-range resistive chemical sensors, in *Proceedings of IEEE I2MTC*, 2010, pp. 116–120
19. A. Depari, A. Flammini, D. Marioli, E. Sisinni, A new interface circuit for high-value wide-range resistive chemical sensor dynamic characterization. Proc. Eurosensors **1**(1), 1327–1330 (2009)
20. M. Grassi, P.Malcovati, A.Baschirotto, A high-precision wide-range front-end for resistive gas sensor arrays, in *Proceedings of Eurosensors*, Rome, Sept 2004
21. G. Ferri, N. Guerrini, V. Stornelli, C. Catalani, A novel CMOS Temperature Control system for resistive gas sensor array, *Proceedings of ECCTD*, Cork, 2005, pp. 351–354
22. G. Ferri, V. Stornelli, A. De Marcellis, A. Flammini, A. Depari, Novel CMOS fully integrated interface for wide-range resistive sensor arrays with parasitic capacitance estimation. Sensor. Actuat. B **130**, 207–215 (2008)
23. A. De Marcellis, A. Depari, G. Ferri, A. Flammini, D. Marioli, V. Stornelli, A. Taroni, Uncalibrated integrable wide-range single-supply portable interface for resistance and parasitic capacitance determination. Sensor. Actuat. B **132**, 477–484 (2008)
24. A. De Marcellis, A. Depari, G. Ferri, A. Flammini, D. Marioli, V. Stornelli, A. Taroni, A CMOS integrable oscillator-based front-end for high dynamic range resistive sensors. IEEE Trans. Instr. Meas. 57(8), 1596–1604 (2008)
25. A. Flammini, D. Marioli, A. Taroni, A low-cost interface to high-value resistive sensors varying over a wide range. IEEE Trans. Instr. Meas. **53**(4), 1052–1056 (2004)
26. L. Fasoli, F. Riedijk, J. Huijsing, A general circuit for resistive bridge sensors with bitstream output. IEEE Trans. Instr. Meas. **46**(4), 954–960 (1997)
27. K. Mochizuki, K. Watanabe, A high-resolution, linear resistance-to-frequency converter. IEEE Trans. Instr. Meas **45**, 761–764 (1996)
28. S. Middlehoek, P.J. French, J.H. Huijsing, W. Lian, Sensors with digital or frequency output. Sensor. Actuat **15**, 119–133 (1988)

29. I. Hotovy, V. Rehacek, P. Siciliano, S. Capone, L. Spiess, Sensing characteristics of NiO thin films as NO_2 gas sensors. Thin Solid Films **418**(1), 9–15 (2003)
30. A. Martucci, M. Pasquale, M. Guglielmi, M. Post, J.C. Pivin, Nanostructured silicon oxide nickel oxide sol–gel films with enhanced optical carbon monoxide gas sensitivity. J. Am. Ceram. Soc. **86**(9), 1638–1640 (2003)
31. G. Sberveglieri, E. Comini, G. Faglia, M.T. Atashbar, W. Wlodarski, Titanium dioxide thin films prepared for alcohol microsensor Applications. Sensor. Actuat. B **66**, 139–141 (2000)
32. S. Bianchi, E. Comini, M. Ferroni, G. Faglia, G. Sberveglieri, Metal oxide nanowires for optical gas sensing, in *Proceedings of Eurosensors*, Barcellona, Sept 2005
33. T. Sahm, L. Madler, A. Gurlo, N. Barsan, U. Weimar, A. Roessler, S. E. Pratsinis, Direct formation of highly porous gas-sensing films by in situ thermophoretic deposition of flame-made Pt/SnO_2 nanoparticles, *Proceedings of Eurosensors*, Barcellona, Sept 2005
34. G. Neri, A. Bonavita, G. Rizzo, C. Baratto, G. Faglia, G. Sberveglieri, Towards enhanced performances in gas sensing: nanocrystalline oxides application, in *Proceedings of Eurosensors*, Barcellona, Sept 2005
35. E. Comini, M. Ferroni, V. Guidi, P. G. Merli, V. Morandi, G. Sberveglieri, Low temperature CO detection by goldactivated nanosized titania, in *Proceedings of Eurosensors*, Barcellona, Sept 2005
36. A. Fort, S. Rocchi, M. Serrano-Santos, R. Spinicci, N. Ulivieri, V. Vignoli, A high-performance measurement system for simultaneous mass and resistance variation measurements on gas sensing polymer films, *Proceedings of IMTC*, 2004, pp. 599–604
37. I. Sayago, M.C. Horrillo, S. Baluk, M. Aleixandre, M.J. Fernandez, L. Ares, M. Garcia, J.P. Santos, J. Gutierrez, Detection of toxic gases by a tin oxide multisensor. IEEE Sens. J. 2(5), 387–393 (2002)
38. A. Depari, M. Falasconi, A. Flammini, D. Marioli, S. Rosa, G. Sberveglieri, A. Taroni, A new hardware approach to realize low-cost electronic noses, *Proceedings of IEEE Sensors*, Irvine, 2005, pp.239–242
39. G. Ferri, V. Stornelli, W. Cappucci, C. Cantalini, Integrated CMOS interfaces for wide-range resistive gas sensors. Sensor. Actuat. B **B118**(2), 269–275 (2006)
40. S.S.W. Chan, P.C.H. Chan, A resistance-variation tolerant constant-power heating circuit for integrated sensor applications. IEEE J. Solid-St. Circ. **34**(4), 432–437 (1999)
41. G. Ferri, V. Stornelli, A. De Marcellis, A. Flammini, V. Depari, Novel CMOS integrable wide-range resistive sensor interface, *Proceedings of IMCS*, Brescia, July 2006
42. A. Depari, A. Flammini, D. Marioli, S. Rosa, A. Taroni, A low-cost circuit for high-value resistive sensors varying over a wide range. IOP Measur. Scien. Tech. **17**(2), 353–358 (2006)
43. A. Sedra, K.C. Smith, *Microelectronic Circuits*, 5th edn. (Oxford University Press, New York, 2007). ISBN 0195142527
44. A. Depari, A. Flammini, D. Marioli, A. Taroni, A. De Marcellis, G. Ferri, V. Stornelli, A new CMOS integrable oscillating circuit for high-value wide-range resistive sensors, in *Proceedings of IMTC*, Varsavia, May 2007, pp. 1–4
45. A. Depari, A. Flammini, D. Marioli, A. Taroni, A. De Marcellis, G. Ferri, V. Stornelli, An uncalibrated wide-range single-supply integrable front-end for resistance and capacitance estimation, in *Proceedings of Transducers and Eurosensors*, Lyon, June 2007, pp. 2031–2034
46. W. Sansen, *Analog Design Essentials* (Springer, Dordrecht, 2006). ISBN 0387257462
47. F. Maloberti, *Analog Design For CMOS VLSI Systems* (Kluwer Academic Publishers, Dordrecht, 2001). ISBN 1441949194
48. C. Di Carlo, A. De Marcellis, V. Stornelli, G. Ferri, A. Flammini, A. Depari, Integrated CMOS resistance-to-period converter with parasitic capacitance evaluation, *Proceedings of IEEE ISCAS*, Taipei, May 2009
49. G. Ferri, C. Di Carlo, V. Stornelli, A. De Marcellis, A. Flammini, A. Depari, N. Jand, A single chip integrated interfacing circuit for wide-range resistive gas sensor arrays. Sensor. Actuat. **B B 143, n. 1**, 218–225 (2009)
50. Internet resource: http://www.figarosensor.com, datasheet TGS2600

51. A. De Marcellis, G. Ferri, V. Stornelli, An oscillator topology as wide range resistive-capacitive sensor interface, in *Proceedings of AISEM (Associazione Italiana Sensori e Microsistemi) Conference*, Rome, 2008
52. A. Depari, A. Flammini, D. Marioli, E. Sisinni, A. De Marcellis, G. Ferri, V. Stornelli, A new interface for resistive chemical sensors with low measuring time, in *Proceedings of IEEE IMTC*, Singapore, May 2009
53. A. Depari, A. Flammini, D. Marioli, E. Sisinni, A. De Marcellis, G. Ferri, V. Stornelli, A new, fast readout, interface for high-value resistive chemical sensors, in *Proceedings of ISOEN*, Brescia, April 2009
54. A. Depari, A. Flammini, D. Marioli, E. Sisinni, A. De Marcellis, G. Ferri, V. Stornelli, A new and fast-readout interface for resistive chemical sensors. IEEE Trans. Instrum. Meas. **59**(5), 1276–1283 (2010)
55. C.T. Chiang, C.S. Wang, Y.C. Huang, A monolithic CMOS autocompensated sensor transducer for capacitive measuring systems. IEEE T. Instrum. Meas. **57**(11), 2472–2486 (2008)
56. Internet resource: http://www.analog.com, datasheet ADXL50
57. G. Ferri, P. De Laurentiis, A novel low voltage low power oscillator as a capacitive sensor interface for portable applications. Sensor. Actuat. A **76**, 437–441 (1999)
58. N. Yadzi, A. Mason, K. Najafi, K.D. Wise, A generic chip interace for capacitive sensors in low power multiparameter microsystems. Sensor. Actuat. A **84**, 351–361 (2000)
59. G. Amendola, G.N. Lu, I. Badadjian, Signal processing electronics for a capacitive microsensor. Analog. Integr. Circ. S. **29**(1–2), 105–120 (2001)
60. S. Chatzandrolis, D. Tsoukalas, Capacitance to frequency converter suitable for sensor applications using telemetry. Analog. Integr. Circ. S. **27**(1–2), 31–38 (2001)
61. K. Watanabe, H. Matsumote, K. Fujiwara, Switched-capacitor frequency-to-voltage and voltage-to-frequency converters. IEEE Trans. Circ. Syst. **CAS-33, n. 8**, 836–838 (1986)
62. H. Matsumoto, K. Watanabe, Switched-capacitor frequency-to-voltage and voltage-to-frequency converters based on charge balancing principle, in *Proceedings of IEEE International Symposium on Circuits and System*, 1988, pp. 2221–2224
63. D. Yin, Z. Zhang, J. Li, A simple switched-capacitor-based capacitance-to-frequency converter. Analog. Integr. Circ. S. **1**(4), 353–361 (1991)
64. F. Krummenacher, A high-resolution capacitance-to-frequency converter. IEEE J. Solid-St. Circ. **SSC-20, n. 3**, 666–670 (1985)
65. M. Suster, W.H. Ko, D.J. Young, An optically powered wireless telemetry module for high-temperature MEMS sensing and communication. J. Microelectromech. S. **13**(3), 536–541 (2004)
66. T.G. Constandinou, J. Georgiou, C. Toumazou, Micropower front-end interface for differential capacitive sensor systems. IET Electron. Lett. **44**(7), 470–472 (2008)
67. A. D'Amico, C. Di Natale, *Introduzione ai sensori* (Aracne, Roma, 2008). ISBN 9788854816633
68. Internet resource: http://www.sensorsmag.com/sensors/humidity-moisture/choosing-a-humidity-sensor-a-review-three-technologies-840
69. Internet resource: http://www.analog.com, datasheet AD590
70. Internet resource: http://www.national.com, datasheet LM35
71. G. Ferri, V. Stornelli, A high precision temperature control system for CMOS integrated wide range resistive gas sensors. Analog. Integr. Circ. S. **47**(3), 293–301 (2006)
72. G.C.M. Meijer, G. Wang, F. Fruett, Temperature sensors and voltage references implemented in CMOS technology. IEEE Sens. J. 1(3), 225–234 (2001)
73. G. Wang, G.C.M. Meijer, The temperature characteristics of bipolar transistors fabricated in CMOS technology. Sensor. Actuat. A **87**, 81–89 (2000)
74. J. Sun, C.Y. Yeong, H.H. Wang, A low voltage CMOS current source with temperature compensation, in *Proceedings of IEEE SSMSD*, 2003

Chapter 4
The Current-Mode Approach in Sensor Interfaces Design

In this chapter, considering the Second Generation Current Conveyor (*CCII*) as the main active block in Current-Mode (*CM*) approach, as an alternative to *OA* utilized in *VM* design, some *CM* interface solutions both for resistive and capacitive sensors will be described at system level. The presented circuits have been implemented as discrete element prototype *PCBs*, using commercial components and, sometimes, in the case of integrated circuit design, with *LV LP* characteristics, in a standard *CMOS* technology.

4.1 Introduction to Current-Mode Resistive Sensor Interfaces

Actually, the *CM* design represents a new challenge in *LV LP* microelectronics, then also in sensor interface development. In particular, *CM* electronic front-ends for resistive sensors may take advantage of the use of the main *CM* block, the *CCII* (see Appendix 1 for more details) [1–3]. The simplest topology of resistive sensor interface is, as well-known, the Wheatstone bridge. As mentioned before, its sensitivity (and resolution) can be increased by using a differential voltage amplifier connected to its outputs. This can be also done in *CM* approach by using a *CCII*-based (instrumentation) voltage amplifier (a possible example of this circuit is described in Appendix 2, where also offset and noise compensations are considered).

Fig. 4.1 shows a *CCII*-based analog interface, designed for piezoresistive pressure sensors [4], but, obviously, can be utilized in other DC-excited resistive sensor applications. The advantage of this *CM* circuit in the sensor interface is the capability to perform the offset compensation. The output voltage is linearly proportional to the (piezo) resistive variation. The only feature to be considered is the design of *CCIIs* having negligible parasitic impedances. In particular, in this case, it is assumed that the piezoresistor is modelled by the resistance R_{SENS}, whose variation is $R_{SENS} = R_0(1 + x)$, being R_0 the resistance at reference pressure and

A. De Marcellis and G. Ferri, *Analog Circuits and Systems for Voltage-Mode and Current-Mode Sensor Interfacing Applications*, Analog Circuits and Signal Processing, DOI 10.1007/978-90-481-9828-3_4, © Springer Science+Business Media B.V. 2011

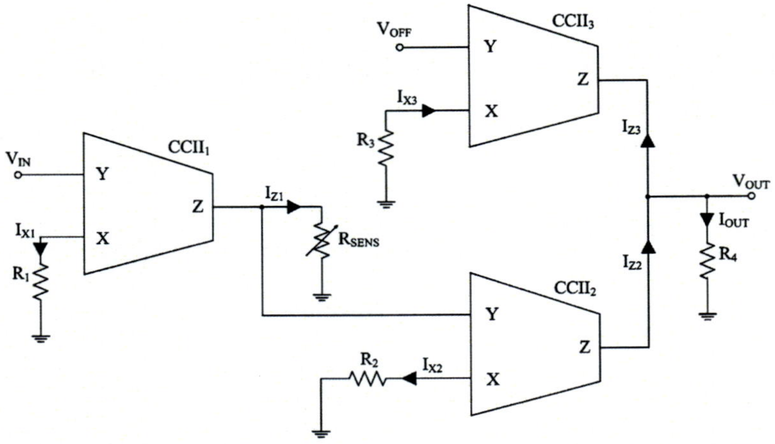

Fig. 4.1 *CM* electronic interface for (piezoresistive) pressure sensors based on *CCII*

x the relative sensor variation. A current source configuration is obtained through $CCII_1$. The current I_{Z1} is equal to I_{X1} and therefore is set by V_{IN} and R_1 values. The change in piezo value affects the voltage at Z_1 node as follows:

$$V_{Z1} = \frac{V_{IN}}{R_1} R_0(1 + x). \tag{4.1}$$

Referring to Fig. 4.1, the output current of $CCII_3$ also follows the input current, equal to the ratio between V_{OFF} and R_3 since Y_3 is a high impedance node, while the output current of $CCII_2$ is given by the ratio between V_{Z1} and R_2. The two currents I_{Z2} and I_{Z3} are then added to obtain the output current I_{OUT}. Consequently, in this manner, a linear dependence between I_{OUT} and the piezo variation x is obtained.

Therefore, the output voltage signal can be expressed by:

$$\begin{aligned} V_{OUT} &= I_{OUT} R_4 = (I_{Z2} - I_{Z3}) R_4 \\ &= \left(\frac{R_4 R_0 V_{IN}}{R_1 R_2} \right) x + \left(\frac{R_4 R_0 V_{IN}}{R_1 R_2} - \frac{R_4 V_{OFF}}{R_3} \right), \end{aligned} \tag{4.2}$$

where the first term is linearly proportional to the relative resistance variation x, while the second one can be set to zero by a suitable choice of V_{IN}, V_{OFF}, R_1, R_2, R_3 and R_4. In particular, the second term in Eq. 4.2 might allow to easily cancel the offset without reducing the speed of the interface, even if time-varying errors, such as drift and 1/f noise, might not be compensated. This interface can be easily implemented in a standard *CMOS* technology, designing an *CCII*, at transistor level, with *LV LP* characteristics (see Appendix 1), so to achieve a complete integrated system suitable for portable resistive sensor applications.

4.2 The AC Excitation Voltage for Resistive/Capacitive Sensors

In this Paragraph, we will describe some *CM* solutions, based on either sinusoidal or square-wave oscillators, which can be utilized as analog interface circuits for both resistive and capacitive sensors.

4.2.1 Wien Oscillators as Small Range Resistive/Capacitive Sensor Interfaces

It is not unusual to find circuit transformation methods that allow to obtain a *CM* solution from its *VM* counterpart. Then, as for *OAs*, also *CCIIs* have been employed with success in the implementation of oscillators which operate the conversion of the sensor parameter (resistance or capacitance) into a period (or frequency) [5–14]. For these sinusoidal oscillators, low parameter variations are generally considered.

Concerning this approach, a simple circuit is constituted by the design of a Wien oscillator, whose solution using a *VCVS* as amplifier block in a Wien bridge is shown in Fig. 4.2. The design equations for this oscillator are the following:

$$f_0 = \frac{1}{2\pi\sqrt{R_1 R_2 C_1 C_2}}, \tag{4.3}$$

$$K = 1 + \frac{R_1}{R_2} + \frac{C_2}{C_1}, \tag{4.4}$$

being f_0 the oscillation frequency and K the *VCVS* voltage gain, representing also the oscillation condition.

We can also consider R_1 as a part of the active block; in this case the *VCVS* is replaced by a *VCCS*, whose gain is K/R_1. In this sense, two *CCII*-based Wien oscillator configurations can be designed, as shown in Fig. 4.3 [1].

In this case, the oscillation frequency and conditions are the same, provided that the following relationships are verified:

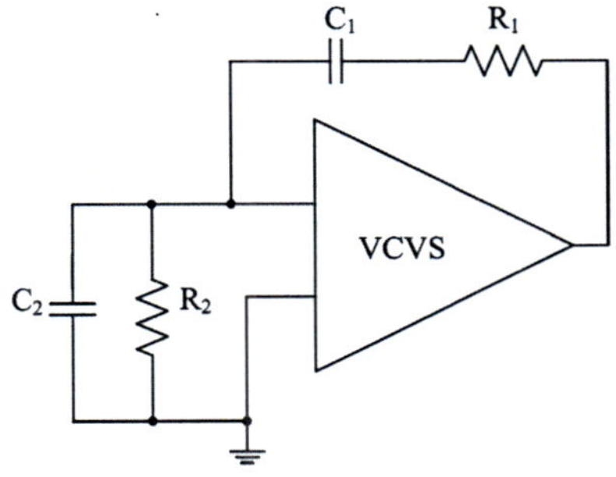

Fig. 4.2 Wien oscillator block scheme

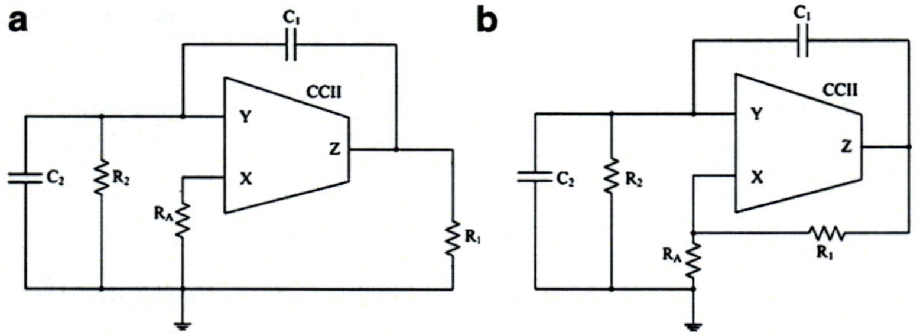

Fig. 4.3 *CCII*-based Wien oscillators

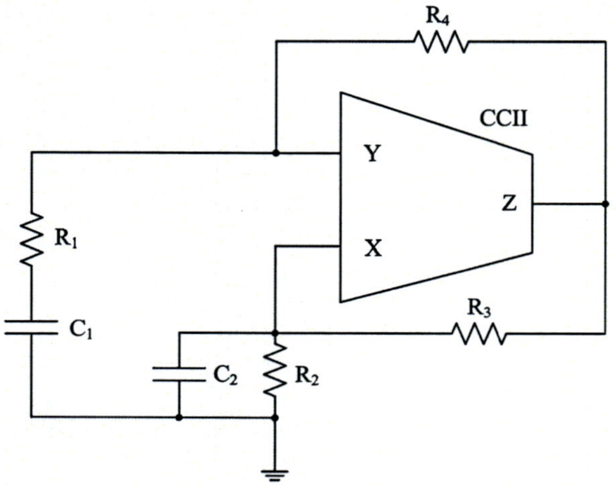

Fig. 4.4 *CCII*-based Wien oscillator with only grounded capacitances

$$R_A = \frac{R_1}{K} \tag{4.5}$$

for the topology of Fig. 4.3a, while for the other topology reported in Fig. 4.3b:

$$R_A = \frac{R_1}{K-1}. \tag{4.6}$$

In order to have all the capacitors grounded, in Fig. 4.4 a possible solution of *CCII*-based oscillator has been reported, whose oscillation conditions are the following:

$$R_1C_1 = R_2C_2, \tag{4.7}$$

$$2\frac{R_3}{R_4} = \frac{R_2}{2R_1} - 1, \tag{4.8}$$

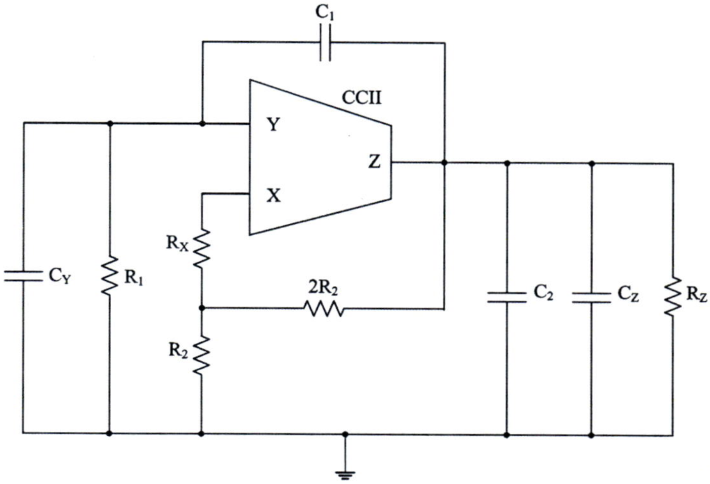

Fig. 4.5 Proposed *CCII* based oscillator

while the oscillation frequency, which can be easily varied modifying the resistance value, is given by:

$$f_0 = \frac{1}{2\pi R_1 C_1}. \tag{4.9}$$

Moreover, it has to be considered the fact that without changing the *CCII* topology it is possible to set oscillation frequencies from about 1 kHz up to 10 MHz.

Another integrable solution of a *CCII*-based Wien oscillator is shown in Fig. 4.5 (including the main *CCII* parasitic components) [15]. This front-end consists of a single block oscillating circuit performing an R-f conversion. This solution is typically suitable for resistive sensor interfacing, with a low dynamic range variation (about two decades).

A routine analysis gives the following expression for the sinusoidal oscillation frequency of the output signal:

$$f_0 = \frac{1}{2\pi\sqrt{R_1 R_2 C_1 (C_2 + C_Z)}}. \tag{4.10}$$

Generally speaking, it is important that an oscillator circuit presents the advantage that, for an example, a grounded resistor can be utilised to control the oscillation condition without affecting the oscillation frequency. Moreover, typically, it is highly desirable to have an oscillating circuit with all the passive components grounded [10]. With this aim, different topologies of *CCII*-based oscillators showing also other particular features have been presented, like that proposed in [11] which uses only positive *CCIIs* and four (or two) grounded resistors and two (or four) grounded capacitors, showing independent oscillation frequency and condition.

Another possible oscillator can be implemented starting from two Dual Output *CCIIs* and only grounded resistances and capacitances, achieving a circuit particularly attractive for integrated applications [12, 13]. Other simple *CM* oscillators can be implemented also using both first and second generation current conveyors [14]. In the literature, there are also different topologies of oscillators which utilize only one *CM* active building block: those based on Differential Voltage *CCII* (*DVCCII*) [16] and on Fully Differential *CCII* (*FDCCII*) [17].

4.2.2 *Astable Multivibrator as Wide Range Resistive/Capacitive Sensor Interface*

In this Section, a *CM LV* astable multivibrator, implemented with a single *CCII*, performing a controlled square wave generation [18], is presented. The circuit can be used in capacitive (or resistive) sensor interface. This solution shows a linear relation between (sensor) capacitance and oscillation period, in an operating frequency range up to about 50 MHz. Since the utilized *CCII* has been implemented at transistor level, in a standard *CMOS* technology, the proposed oscillating circuit, owing to its topological simplicity, has been completely integrated so to achieve a *LV* front-end solution suitable for portable sensor applications.

As in *VM* approach, the interface is based on an inverting Schmitt trigger (i.e., an inverting hysteresis comparator) implemented, in this case, through the use of a *CCII* block, as shown in Fig. 4.6a. A regenerative feedback takes part of the output voltage from the Z node and apply it to the Y node [19]. The Schmitt trigger, whose transcharacteristic is reported in Fig. 4.6b, has the following threshold voltages at Y node:

$$V_{TH+} = \frac{R_1 - R_S}{R_2 + R_1} V_{SAT+}, \tag{4.11}$$

$$V_{TH-} = \frac{R_1 - R_S}{R_2 + R_1} V_{SAT-}, \tag{4.12}$$

where V_{SAT+} and V_{SAT-} are the saturation voltages that the *CCII* is able to reach at its output node V_{OUT}. In addition, it is important to consider that, referring to Eqs. 4.11 and 4.12, this circuit is able to operate also as a zero comparator (i.e., with a null hysteresis, having a constant threshold voltage fixed to zero) simply by choosing $R_1 = R_S$. The developed *CCII*-based astable multivibrator, which has been used as sensor interface, is obtained by substituting the voltage generator at V_{IN} node of the inverting Schmitt trigger with a capacitor, representing a capacitive sensor, C_{SENS}, as shown in Fig. 4.7.

In this case, supposing that V_{OUT} has an initial value of V_{SAT+}, V_C (the voltage on C_{SENS} capacitor) has an initial value of V_{TH-} and, if the *CCII* is ideal, C_{SENS} is charged by the X node voltage through the resistance R_S. Considering the voltage

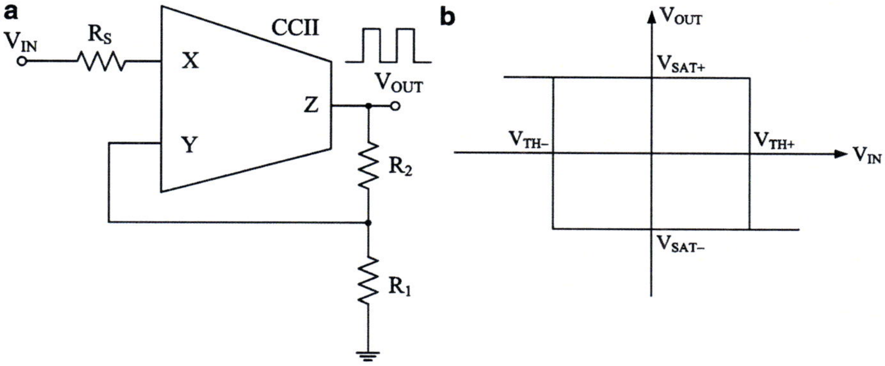

Fig. 4.6 (**a**) *CCII*-based inverting Schmitt trigger; (**b**) Transcharacteristic of the inverting Schmitt trigger

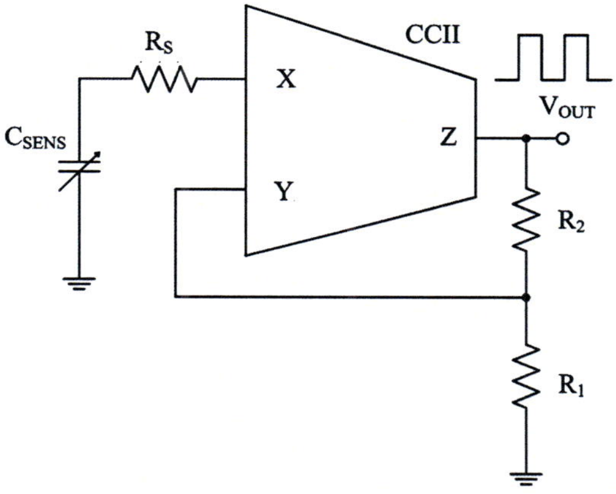

Fig. 4.7 *CCII*-based astable multivibrator employed as capacitive sensor interface

signals shown in Fig. 4.8, the voltage V_C, as a function of time t, can be expressed as follows:

$$V_C(t) = V_{X+} - (V_{X+} - V_{TH-})e^{-\left(\frac{t}{R_S C_{SENS}}\right)}. \tag{4.13}$$

This condition is valid until V_C reaches V_{TH+}. The time taken by V_C to reach V_{TH+} starting from V_{TH-} is:

$$T_1 = R_S C_{SENS} \ln\left(\frac{V_{TH-} - V_{X+}}{V_{TH+} - V_{X+}}\right). \tag{4.14}$$

When V_C reaches V_{TH+}, V_{OUT} switches to V_{SAT-}; analogously V_C, as a function of time, is now given by:

$$V_C(t) = V_{X-} - (V_{X-} - V_{TH+})e^{-\left(\frac{t}{R_S C_{SENS}}\right)}. \tag{4.15}$$

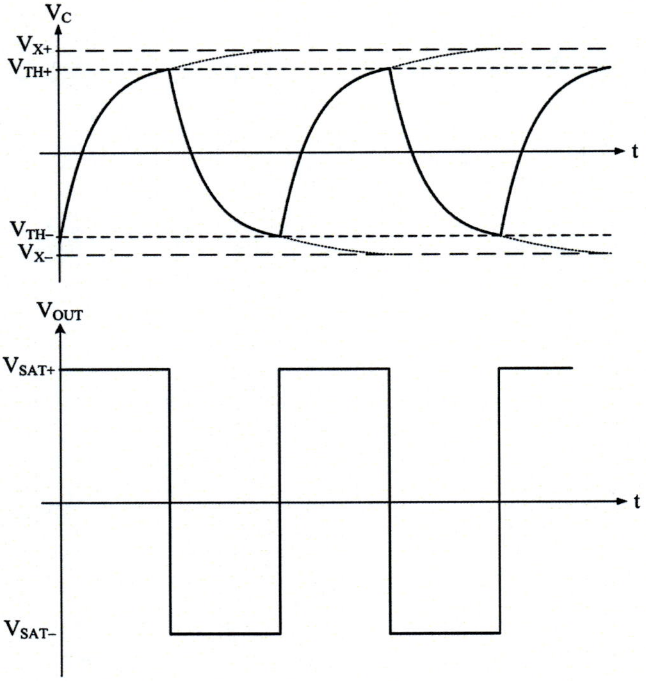

Fig. 4.8 Behaviour and relationship of the voltages V_c, on the C_{SENS} capacitor, and V_{OUT}, at the circuit output terminal

Once again, this condition will be valid until V_C reaches V_{TH-}, so the time taken by V_C to reach V_{TH-} starting from V_{TH+} is now:

$$T_2 = R_S C_{SENS} \ln\left(\frac{V_{TH+} - V_{X-}}{V_{TH-} - V_{X-}}\right). \tag{4.16}$$

In this case, when V_C reaches V_{TH-}, V_{OUT} switches to V_{SAT+}. As a consequence, the output square-wave signal period is ideally given by the sum of eq.s (4.14) and (4.16), as follows:

$$T = T_1 + T_2 = R_S C_{SENS} \ln\left(\frac{V_{TH-} - V_{X+}}{V_{TH+} - V_{X+}} \cdot \frac{V_{TH+} - V_{X-}}{V_{TH-} - V_{X-}}\right). \tag{4.17}$$

Then, the period T can be varied by changing either C_{SENS} or R_S (in the latter case, for the resistive sensor interfacing, series parasitic resistance at X node must be carefully considered). Moreover, from the expression of the oscillation period, it is evident that, in order to obtain a 50% duty cycle, it is necessary to design a *CCII* so that $V_{SAT+} \approx -V_{SAT-}$. Since for this solution the chosen total supply voltage is relatively low (1.5 V), it is mandatory to design and employ a *CCII* with a rail-to-rail

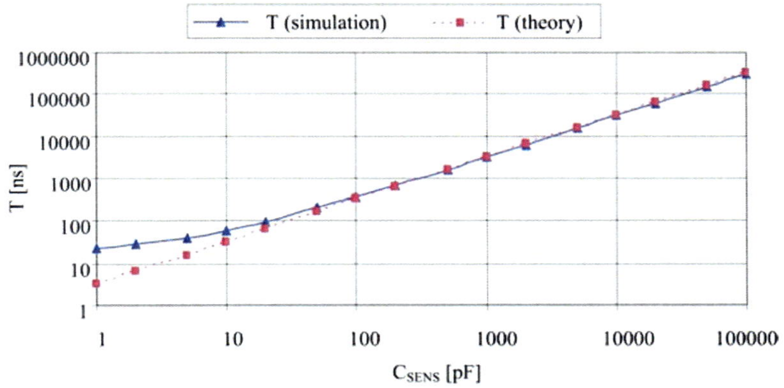

Fig. 4.9 Oscillation period T as a function of capacitance C_{SENS}: theory vs. simulation results

operative range. Therefore, in this case, the utilized integrated *CCII* topology is that reported in Fig. A1.6 (see Appendix 1) which is able to guarantee both the respect of a 50% duty cycle and a low voltage operation. External passive component values have been chosen, for PSpice simulations, so to neglect the contribution of the parasitic effects in the *CCII*, in particular: $R_S = 1\,\text{k}\Omega$, $R_1 = 3\,\text{k}\Omega$, $R_2 = 6\,\text{k}\Omega$. The supply voltage are $V_{DD} = 0.75\,\text{V}$, $V_{SS} = -0.75\,\text{V}$. In this conditions, considering $C_{SENS} = 500\,\text{pF}$, V_{OUT} shows a total voltage amplitude of about 1.25 V and an oscillation period $T = 1.631\,\mu\text{s}$. The relative error between the theoretical and the simulated period is lower than 1%.

The wide linear relation between the capacity C_{SENS} and period of oscillation T allows the circuit to be suitable for generic capacitive sensor interfacing. In this sense, CADENCE simulations have been performed sweeping the capacitive value C_{SENS} from 1 pF to 100 nF, as reported in Fig. 4.9. The relationship between C_{SENS} and T shows a very good linearity in the range [100 pF,100 nF], with a sensitivity of about 3 ns/pF. The lower limit for the oscillation period is determined by the *CCII* bandwidth, which is of about 20 ns (50 MHz).

4.2.3 *Uncalibrated Solution for High-Value Wide-Range Resistive/Capacitive Sensors*

Generally, oscillating circuits (i.e., square waveform generators) can be typically implemented by an active device as a switching current source to charge and discharge a grounded timing capacitor, so employing a Schmitt trigger and an integrating cell (typically, these solutions are based on a passive or active integrator). These solutions are able to reveal, with a good linearity, a number of capacitance variations, but are unsuitable for low valued capacitive sensors because of both

CCII bandwidth limitations and its parasitic impedances which strongly affect the capacitive measurements, also because *CCII* parasitic component values depend numerically on the particular operating point to which the *CCII* is working. In particular, referring to the astable multivibrator previous described, since both Z and Y nodes are biased with a high voltage level, X node parasitic impedances (i.e., its resistive and capacitive components) can assume very high values affecting directly the excitation voltage of the sensor and its estimation.

Therefore, a wide range capacitive (or resistive) sensor interface, which overcome these problems, has been developed [20, 21]. Its main operation is based on a current differentiation rather than a voltage integration, as in the previous described *CM* solutions, as well as in the *R-T* converters developed in the *VM* approach. In particular, this oscillator, performing an impedance-to-period (*C-T* or *R-T*) conversion and being suitable for the integration on chip in a standard *CMOS* technology with *LV LP* characteristics, allows to neglect the Z and Y nodes saturation effects in the square waveform generation, so in capacitive sensor behaviour estimation, utilizing only resistive loads on X node whose values can be chosen sufficiently higher than parasitic resistances. In fact, the capacitive sensor is connected at Z node, so is not strongly affected by its parasitic capacitance and there are not limitations for wide variation ranges (higher than 6 decades) and high frequency (i.e., small period) values since it is possible to easily set the interface working range through several external parameters (only resistances) which allow also to set the desired sensitivity of the read-out circuit. As a consequence, this circuit configuration can be employed as a suitable solution for capacitive (or resistive) sensor analog front-end which allows to reveal, with a good linearity and accuracy, variations of floating capacitive sensors having a baseline or changing their value in the range [pF, μF] as well as variations of grounded resistive sensors ranging in [MΩ, GΩ].

More in detail, the proposed front-end, whose schematic circuit at block level is shown in Fig. 4.10, is formed by six resistors, a capacitor and two positive *CCIIs*: the first, $CCII_1$, is a voltage-to-current converter, while the second, $CCII_2$, is a hysteresis current comparator, based on a *CM* Schmitt trigger.

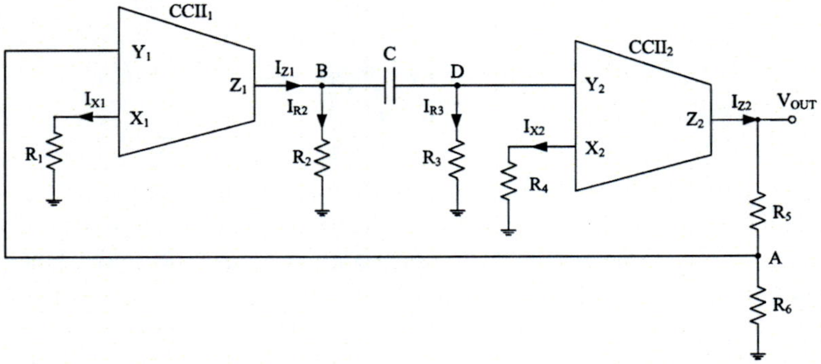

Fig. 4.10 Block scheme of the proposed novel interface

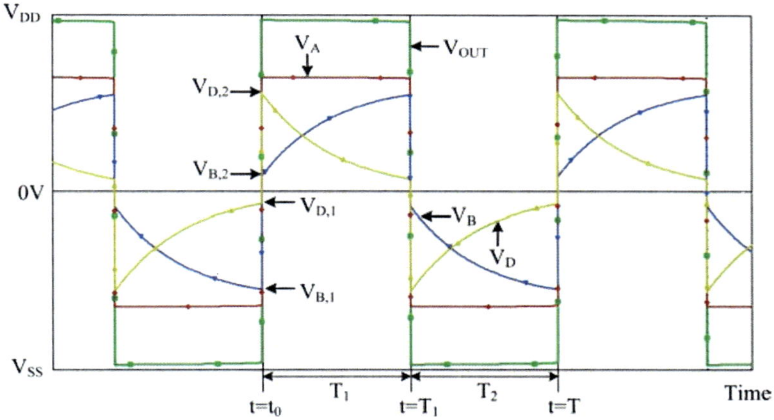

Fig. 4.11 Time responses evaluated at main interface nodes

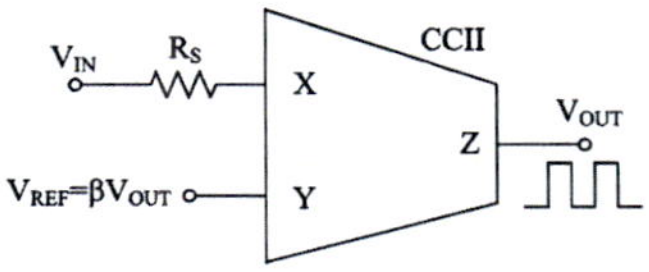

Fig. 4.12 Inverting *CM* Schmitt Trigger simplified scheme (inverting bistable circuit, $\beta = R_1/(R_1 + R_2)$)

Fig. 4.11 shows the voltage signals at each node of the interface under the hypothesis of a constant C during the measuring operation. Referring to this figure, the whole interface works as follows: the saturated output current I_{Z2} of the $CCII_2$ comparator, converted into a saturation voltage ($V_{OUT} = \pm V_{SAT}$) through R_5 and R_6, represents both the periodic signal (from which it is possible to measure the period, proportional to sensor capacitance C) and the input signal (V_A) of the voltage-to-current converter, reduced by the voltage divider implemented through R_5 and R_6. The latter gives an AC excitation current for the capacitive sensor C. In particular, the output signal of $CCII_1$ is a square-wave current signal (I_{Z1}) which is differentiated by the $C - R_3$ passive cell. Consequently, at D node an exponential signal (V_D) is generated, as shown in Fig. 4.11. This signal is converted into a current I_{X2}, through $CCII_2$, and compared with the saturation current I_{Z2} by the same hysteresis comparator $CCII_2$, so generating the square-wave voltage V_{OUT}, whose period T is proportional to C.

In this sensor interface, the hysteresis current comparator is based on a non-inverting bistable circuit (i.e., non-inverting *CM* Schmitt trigger), which is different from that previously described (i.e., inverting *CM* Schmitt trigger), and reported in Fig. 4.6a whose simplified scheme is in Fig. 4.12 [22]. In fact, in the latter, when considered in the astable configuration, since the input signal V_{IN}, applied at the low impedance X node through a series resistor, exponentially tends to $V_{REF} = \beta V_{OUT} = \beta V_{SAT}$ (being $\beta = R_1/(R_1 + R_2)$), the commutation occurs when V_{IN} reaches one of the threshold voltages $V_{TH} = ((R_1 - R_S)/(R_1 + R_2))V_{SAT}$, as

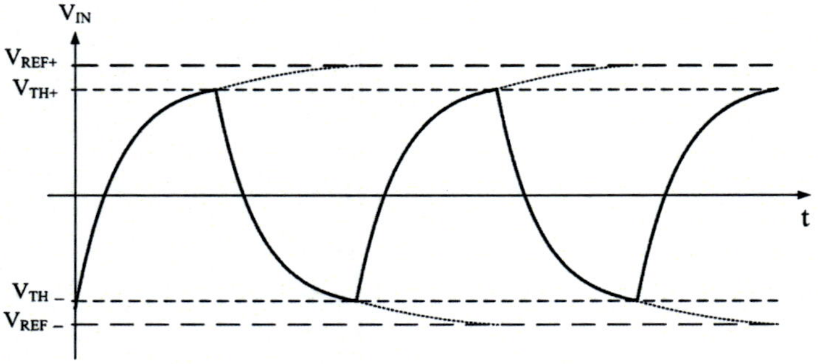

Fig. 4.13 Typical time response of the input signal V_{IN} in the inverting *CM* Schmitt trigger configured as astable multivibrator

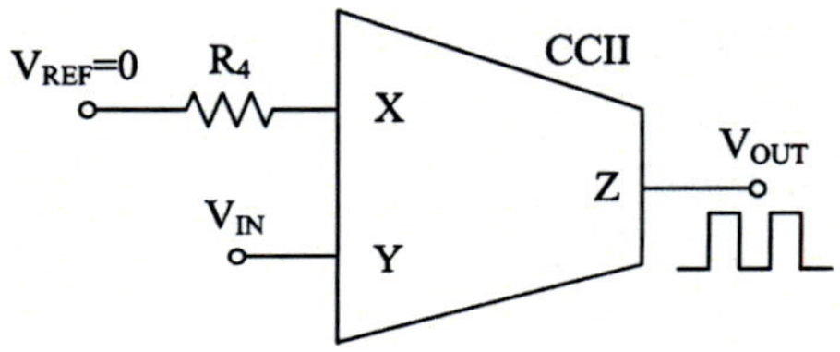

Fig. 4.14 A sketch of the proposed circuit (*CCII*$_2$, see Fig. 4.10): the simplified scheme of the non-inverting *CM* Schmitt trigger (non-inverting bistable circuit)

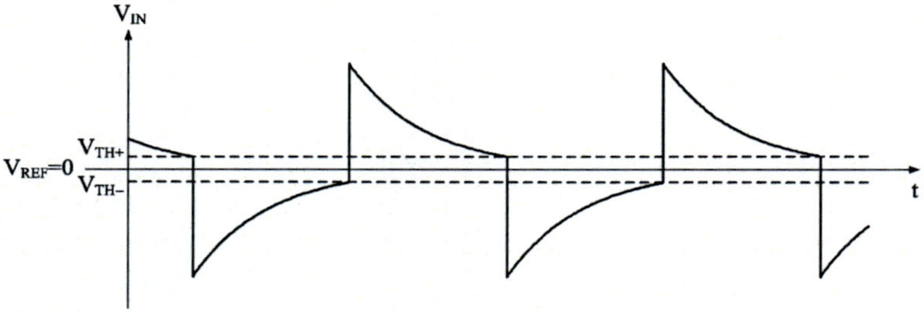

Fig. 4.15 Typical time response of the input signal V_{IN} of the non-inverting bistable circuit implemented in the proposed sensor interface (non-inverting *CM* Schmitt trigger configured as astable multivibrator)

shown in Fig. 4.13. On the contrary, in this interface solution, the hysteresis current comparator reported in Fig. 4.14 (referring to Fig. 4.10, this is only a part of the complete oscillating circuit which takes into account only the $CCII_2$), the input signal V_{IN} is applied to the high impedance Y node. Also in this case, V_{IN} tends to the reference voltage V_{REF} and the commutation occurs when V_{IN} reaches one of the threshold voltages $V_{TH} = (R_4/(R_5 + R_6))V_{SAT}$ but now, V_{REF} is applied to the low impedance X node through a series resistor and is always equal to zero, as shown in Figs. 4.14 and 4.15.

Both the inverting and non-inverting *CM* Schmitt triggers (i.e., bistable circuits) configured as astable multivibrators, represent exactly the *CM* counterparts of the

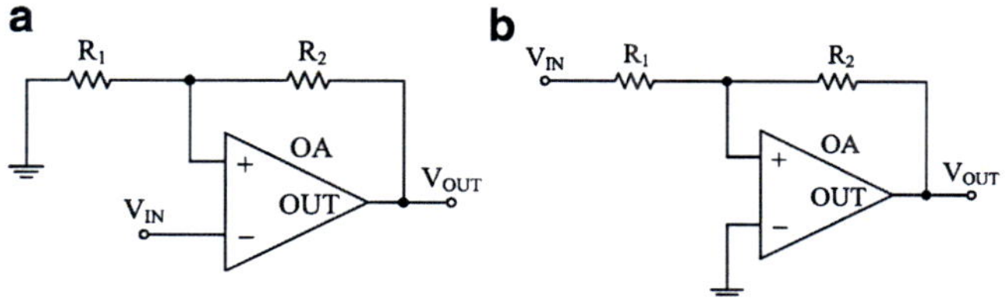

Fig. 4.16 *VM* bistable circuits (Schmitt triggers): (**a**) inverting topology; (**b**) non-inverting topology

two different well-known *VM* bistable circuits (both inverting and non-inverting). In particular, the *VM* inverting configuration reported in Fig. 4.16a [22] corresponds to *CM* inverting Schmitt Trigger (see Fig. 4.12), while the non-inverting configuration shown in Fig. 4.16b [22] corresponds to the second solution that is the *CM* non-inverting Schmitt Trigger described in Fig. 4.14.

Hence, while oscillators based on circuit of Fig. 4.12 performs an integrating operation, the here proposed interface is based on a current differentiation based on circuit of Fig. 4.14. In this way, nearby the commutation, X and Y voltage are about zero, where the parasitic components are typically calculated (i.e., *CCII* bias point) and can be properly used in theoretical calculations; this makes the circuit more accurate. On the contrary, in the inverting Schmitt trigger, the parasitic elements can change their value substantially, due to the fact that the all node operating points heavily changes since they work in the saturation conditions.

Starting from these considerations, referring to Figs. 4.10 and 4.11, the output voltage V_{OUT} can assume the two possible saturation values, V_{SAT+} or V_{SAT-}; consequently, V_A shows two constant values, to be applied at Y_1 node. Resistor R_1 converts this voltage, reported at X_1 node, into a constant current, I_{X1}. At the commutation time, the current I_{Z1}, equal to I_{X1}, flows through R_2 and R_3 proportionally to their values (if $R_2 = R_3$, then $I_{R2} = I_{R3}$). After the commutation, I_{R2} increases, while I_{R3} decreases. Starting the analysis from the condition $V_{OUT} = V_{SAT+}$, in order to have an output commutation, I_{X2} has to be equal to I_{Z2} (i.e., I_{SAT+} or I_{SAT-}). In particular, if we consider the commutation of V_{OUT} from V_{SAT-} to V_{SAT+}, we have:

$$I_{X2} = \frac{V_{X2}}{R_4} = \frac{V_{Y2}}{R_4} = I_{Z2} = I_{SAT-} = \frac{V_{OUT}}{R_5 + R_6} = \frac{V_{SAT-}}{R_5 + R_6}. \tag{4.18}$$

Consequently, the corresponding voltage commutation condition (i.e., the corresponding negative threshold voltage) is given by:

$$V_{Y2} = \frac{R_4}{R_5 + R_6} V_{SAT-} = V_{D,1} = V_{TH-}, \tag{4.19}$$

being $V_{D,1}$ the D node voltage value at the instant t_0^-, which immediately precedes the V_{OUT} commutation from V_{SAT-} to V_{SAT+}. During the commutation, the capacitor voltage V_C (equal to $V_D - V_B$) does not change, while V_{OUT} presents a variation equal to $2V_{SAT}$ (from V_{SAT-} to V_{SAT+}) and $I_{R2}+I_{R3} = 2I_{Z1}$. At the instant t_0^+, which immediately follows t_0, V_D starts to change its value, controlled by C discharge through R_2 and R_3.

Therefore, through a straightforward analysis, considering ideal *CCII* behaviour, it is possible to determine the expression for the period T of the generated output square wave signal, revealed at V_{OUT} node, as a function of the sensor capacitance C, as follows:

$$T = 2C(R_2 + R_3)\ln\left[\frac{2R_2R_3R_6 - R_1R_4(R_2 + R_3)}{R_1R_4(R_2 + R_3)}\right]. \tag{4.20}$$

More simply, if we consider $R_2 = R_3 = R$, Eq. 4.20 becomes:

$$T = 4CR\ln\left(\frac{RR_6 - R_1R_4}{R_1R_4}\right). \tag{4.21}$$

From Eq. 4.21, we have that circuit sensitivity can be opportunely set by choosing suitable values of resistances R_2 and R_3, especially.

This front-end topology has been also designed (employing the *CCII* internal topology reported in Appendix 1, see Fig. A1.6), as a complete integrated solution at transistor level in a standard *CMOS* technology (AMS 0.35 μm), with low voltage (±1 V) and low power (430 μW) characteristics. The proposed circuit properly works with integrable passive component values (resistance ≤ 100 kΩ and capacitance ≤ 100 pF), so it is suitable for integrated portable applications.

Simulation results have confirmed the circuit stability for working temperature drifts (the maximum difference of the obtained oscillation period with respect to its value at the room temperature, 27°C, is lower than 3% in the whole considered range of variation, equal to [−50°C; 110°C]), showing a good linearity in a wide oscillation period range, which can be independently adjusted through either capacitive (in the range pF − μF, about six decades, for capacitors higher than 10 pF) or resistive (in the range MΩ − GΩ, about three decades, for resistors higher than 500 kΩ) external passive components. More in detail, in order to verify the oscillation period variation with respect to the passive components C and R_3, time-domain simulations have been performed choosing $R_1 = R_2 = 100\,\text{k}\Omega$, $R_4 = 100\Omega$, $R_5 = 10\,\text{k}\Omega$, $R_6 = 5\,\text{k}\Omega$; the related results are reported in Figs. 4.17 and 4.18. As regards the capacitance dependence of the oscillation period, R_3 has been set to 100 kΩ, while C has been varied from 1 pF up to 10 μF. Then, for R_3 ranging from 10 kΩ to 1GΩ, C has been fixed both to 50 pF (integrable value) and 10 nF (external component) so to obtain the period variation with respect to R_3. Moreover, from Eq. 4.20, it is easy to note that, for the similar circuit passive components setting ($R_1 = R_3 = 100\,\text{k}\Omega$, $R_4 = 100\Omega$, $R_5 = 10\,\text{k}\Omega$, $R_6 = 5\,\text{k}\Omega$, $C = 50\,\text{pF}$ and 10 nF), R_2 variation provides the same effects on the oscillation frequency as R_3, but only for a reduced resistive range, from 10 to 100 kΩ, as depicted in

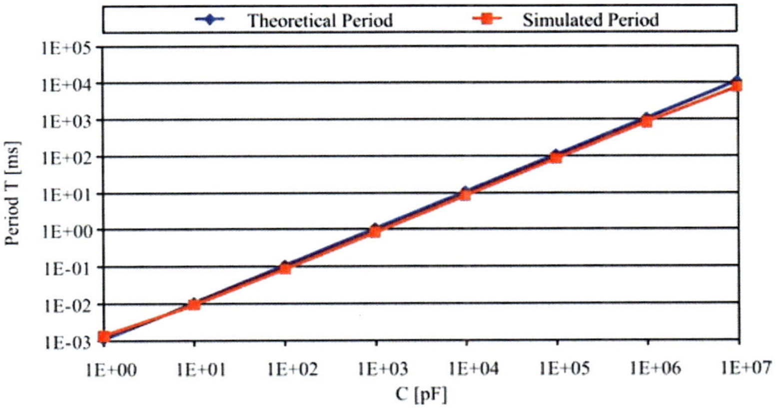

Fig. 4.17 Oscillation period T vs. C (simulation and theoretical values)

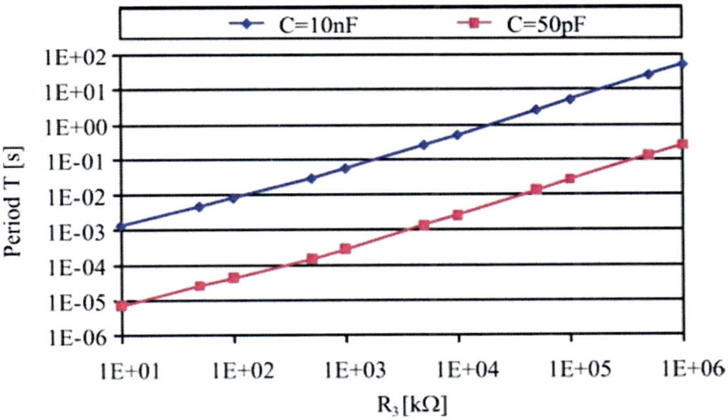

Fig. 4.18 Oscillation period T vs. R_3 (simulation values)

Fig. 4.19. This constraint is due to the presence of the parasitic resistance at $CCII_1$ Z node, whose finite value limits the resistive load R_2 (for R_2 variation, the linearity behaviour is reduced about in the range 20–100 kΩ).

In addition, experimental measurements have been performed implementing the circuit through a prototype *PCB* with the commercial component AD844 of Analog Devices [23] (supplied at ±15 V) as *CCII* and using commercial passive sample resistors and capacitors, emulating both capacitive and resistive sensor behaviour. In particular, Fig. 4.20 shows the obtained period variation with respect to R_2 ranging from 10 to 100 kΩ. As regard the capacitive dependence of the oscillation period, experimental results have confirmed the theoretical expectations, as reported

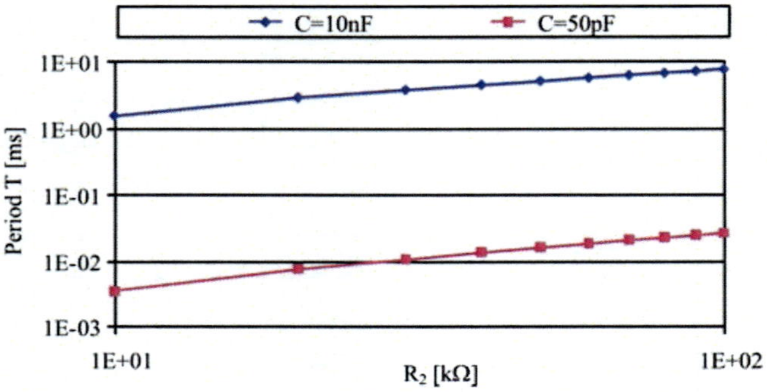

Fig. 4.19 Oscillation period T vs. R_2 (simulation values)

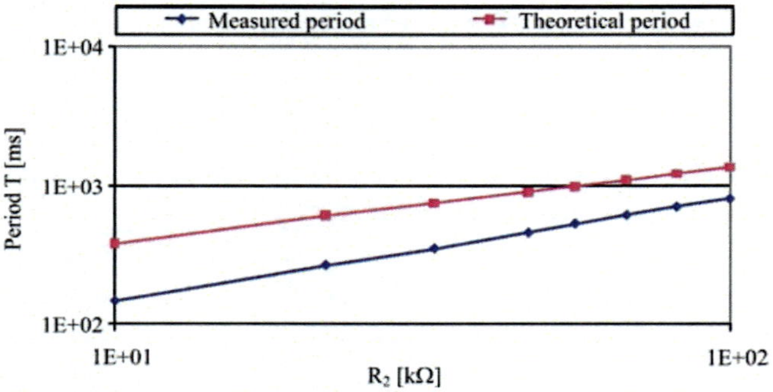

Fig. 4.20 Theoretical response (referred to ideal *CCII* behavior, see Eq. 4.20) and measurement results related to oscillation period of generated output square waveform vs. R_2

in Fig. 4.21 (the circuit sensitivity, considering ideal *CCIIs*, is about 10 μs/pF), showing a good linearity in an oscillation period range varying C from 10 pF up to 10 nF. This range covers a large number of commercial capacitive sensors (e.g., pressure and humidity sensors).

Further experimental measurements have been performed employing commercial sensors, in particular capacitive humidity (HCH-1000 Series by Honeywell) and resistive gas sensors (TGS 2600 Series by Figaro) [24, 25]. Fig. 4.22 shows the period variation versus the capacitive sensor variation (i.e., C-T conversion), when the *RH* has been changed in the range 10–80%, properly mixing dry air with wet air in a closed and controlled chamber. In this case, the *RH* reference measurements have been achieved by a commercial thermo-hygrometer (HTD-625 High Accuracy Thermo-Hygrometer) having a resolution equal to 0.1%*RH* and an accuracy of ±2%*RH*. On the contrary, as regard the resistive dependence of the oscillation period (i.e., R-T conversion), the achieved experimental results have been reported in Fig. 4.23, where the resistive gas sensor provides period variations for gas concentration changes ranging from 0 up to 150 ppm. In this case, the employed

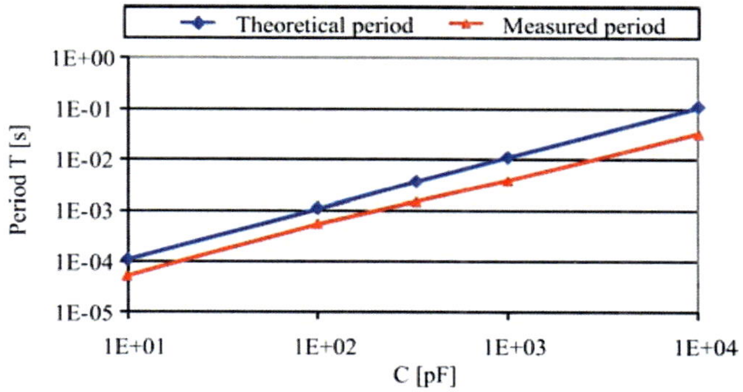

Fig. 4.21 Theoretical response (referred to ideal *CCII* behavior, see Eq. 4.20) and measurement results related to the oscillation period of the generated output square waveform vs. *C*

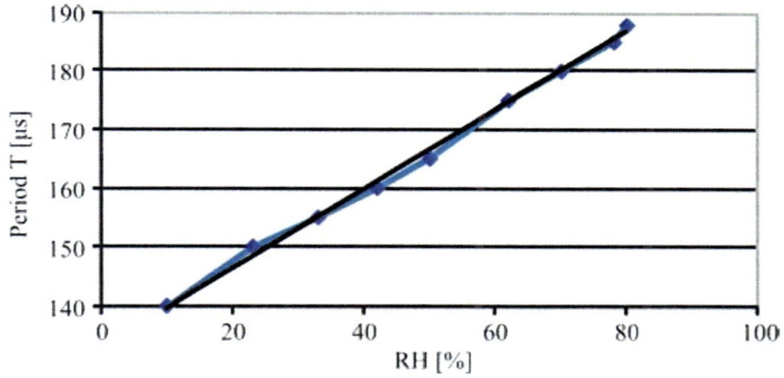

Fig. 4.22 Experimental measurements of *RH* detection through the commercial capacitive humidity sensor HCH-1000 Series by Honeywell

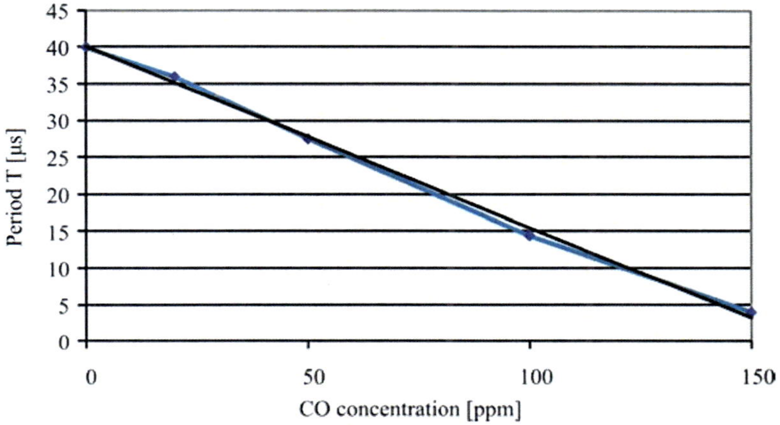

Fig. 4.23 Experimental measurements of CO detection through the commercial resistive gas sensor TGS 2600 Series by Figaro

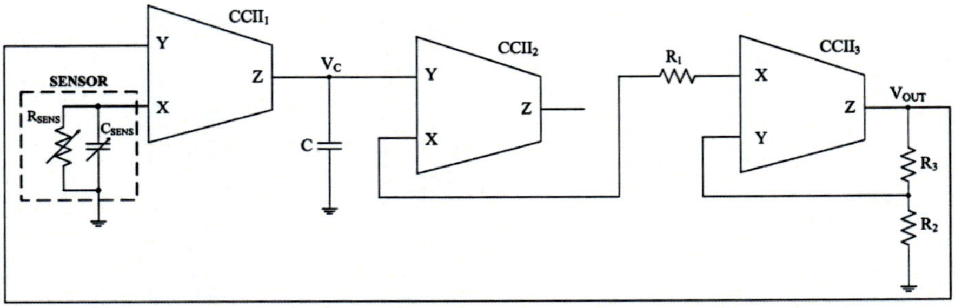

Fig. 4.24 Block scheme of the proposed interface ($CCII_1$ = voltage integrator, $CCII_2$ = buffer, $CCII_3$ = Schmitt trigger)

gas is the *CO*, fluxed into a closed chamber with controlled concentrations. Both experimental measurements show a good linearity in the oscillation period variation range.

4.2.4 Uncalibrated Solution for Small-Range Resistive Sensors

In this Section we present a *CM* interface circuit, whose block scheme is reported in Fig. 4.24, for AC-excited sensors showing a reduced resistive variation. This solution, which does not need any initial calibration, is based on an oscillating circuit performing an *R*-*T* conversion [26].

Referring to Fig. 4.24, the front-end is formed by three main blocks: a voltage integrator; a voltage buffer (that decouples input and output stages); a *CCII*-based hysteresis comparator (Schmitt trigger). The whole interface works as follows. The saturated output voltage of the Schmitt trigger ($V_{OUT} = \pm V_{SAT}$) represents both the periodic signal (from which it is possible to measure the period, proportional to sensor resistance) and the input signal for the voltage integrator, that gives the AC excitation voltage for the resistive sensor (due to the *CCII* voltage buffer operation). Output voltages $\pm V_{SAT}$, integrated by voltage integrator, generate a rising ramp when $V_{OUT} = +V_{SAT}$ and a falling ramp if $V_{OUT} = -V_{SAT}$. This triangular signal is compared with the voltage reference at Y node by the hysteresis comparator, so generating the square-wave voltage V_{OUT}, whose period T is proportional to R_{SENS}. Fig. 4.25 shows the main voltage signals in the circuit under the hypothesis of a constant R_{SENS} during the measuring operation. Through a straightforward analysis, considering ideal *CCII* behaviour, it is possible to determine the expression for the period T of generated output square wave signal, revealed at V_{OUT} node, as a function of the sensor resistance R_{SENS} as follows:

$$T = 4R_{SENS}C\left(\frac{R_2 - R_1}{R_2 + R_3}\right). \tag{4.22}$$

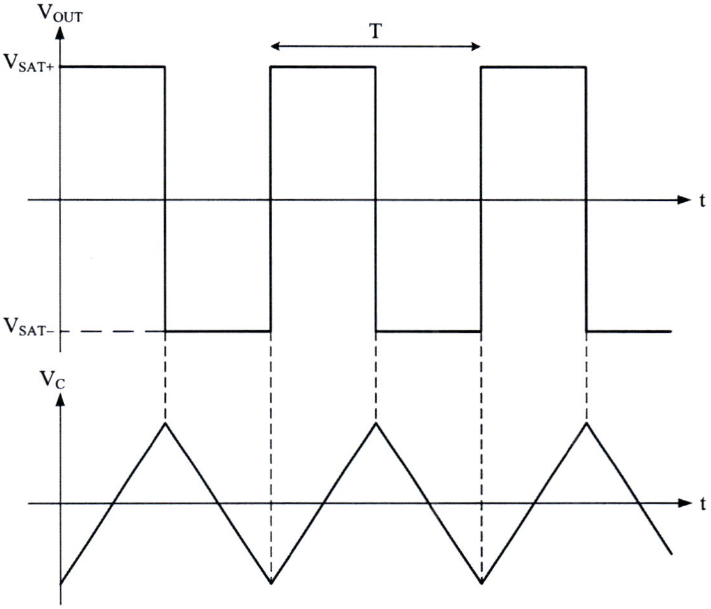

Fig. 4.25 Voltage behaviour at main interface internal nodes

From Eq. 4.22, circuit sensitivity can be opportunely set by choosing C, R_1, R_2 and R_3 values.

When the resistive sensor shows a parasitic capacitance, modelled in parallel to R_{SENS} as shown in Fig. 4.24, a straightforward computation gives the following expression for the output period:

$$T = 4R_{SENS}CG\left(1 - \frac{C_{SENS}}{2CG}\right) \tag{4.23}$$

being:

$$G = \frac{R_2 - R_1}{R_2 + R_3}. \tag{4.24}$$

From Eq. 4.23, it comes that C_{SENS} contribution is negligible if the factor $2CG$ is designed to be much higher than the same capacitance value. If either it is not possible to neglect the sensor parasitic capacitance or we want to estimate it, simple additional blocks as *EX-OR* gates must be added, according to the technique proposed in Chap. 3.

Experimental measurements on a prototype *PCB*, using the commercial component AD844 as *CCII* and sample resistances as R_{SENS} (ranging from 18 kΩ to 1.8 MΩ) are reported in Fig. 4.26 showing the measured periods compared with the theoretical ones, obtained with the following experimental values: $R_1 = 470\ \Omega$, $R_2 = 2.2\ \text{k}\Omega$, $R_3 = 4.7\ \text{k}\Omega$, $C = 47$ pF, $V_{EXC} = \pm 15$ V. In this case, the

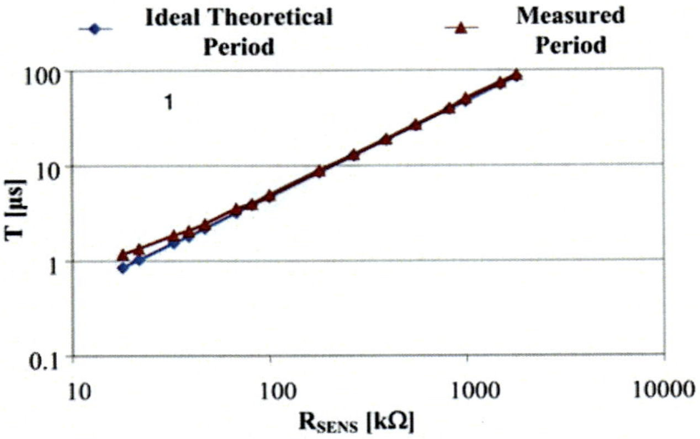

Fig. 4.26 Measured and theoretical output period T vs. R_{SENS}

Table 4.1 Measured and theoretical periods for sample values of R_{SENS} and C_{SENS}

R_{SENS} [Ω]	C_{SENS} [pF]	Measured period [μs]	Theoretical period [μs]
18 k	1	1.9	1.764
	5.6	1.4	1.598
	10	1.2	1.440
180 k	1	17.90	17.64
	5.6	14.60	15.98
	10	12	14.40
1.8 M	1	166	176.4
	5.6	126	159.8
	10	104	144.0

interface sensitivity has been set to about 50 μs/MΩ. Moreover, always with the same conditions except for the integrator capacitance value that has been here set to 100 pF, the presence of a non-zero sensor capacitance has been also investigated. In Table 4.1 the measured and simulated periods have been reported at different fixed sensor resistances, where sensor capacitance ranges from 1 up to 10 pF in three steps. As expected from Eq. 4.23, the higher the parasitic sensor capacitance is, the worst the estimated periods are, with respect to ideal theoretical ones.

4.3 Uncalibrated DC-Excited Resistive Sensor Interface

Finally, in this Paragraph we present a *CM* interface for DC-excited resistive sensors [27,28]. It is based on an oscillating circuit, suitable for either pure resistive sensors (or with negligible parasitic capacitances) or for resistive sensing elements which do not tolerate an AC excitation voltage (i.e., they provide bad responses and lower

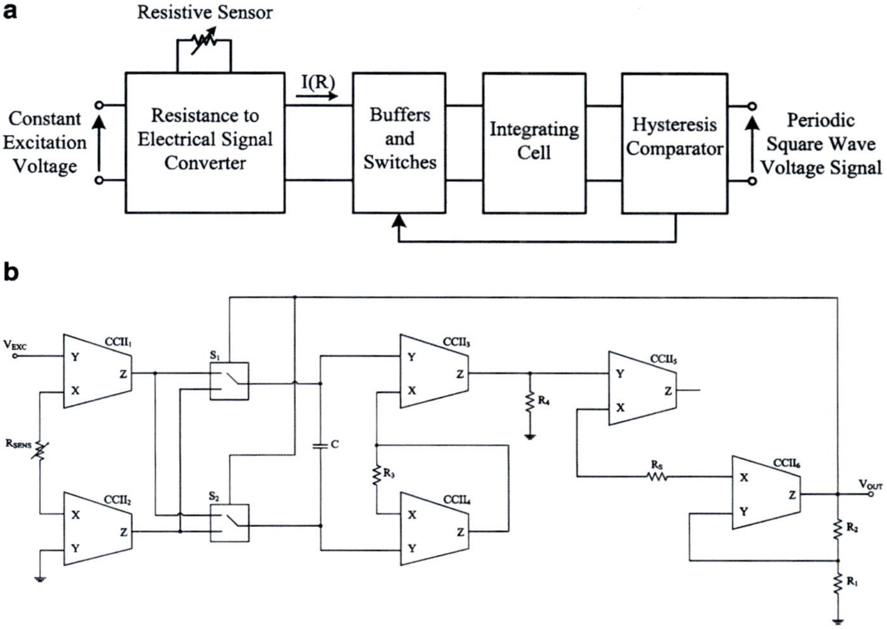

Fig. 4.27 (**a**) Basic block scheme of the proposed interface; (**b**) Detailed scheme of the proposed *CCII*-based *LV LP* front-end ($CCII_1, CCII_2$ = R-Iconverter, S_1, S_2 = switch stage, $CCII_3$, $CCII_4, CCII_5$ = instrumentationamplifier, $CCII_6$ = hysteresis comparator)

lifetimes). This low-cost fully-integrable front-end does not show any preliminary calibration and operates, once again, an R-T conversion.

In Fig. 4.27a the basic block scheme of the proposed interface is shown. A constant external excitation voltage drives a resistance to current converter, whose output signal is sent to a switch stage. This stage, driven by the last block of the system, is able to change the polarity of the current signal. Then, the current is integrated and handled by a hysteresis comparator which suitably controls the switch stage and gives a square-wave output signal, utilizing only a DC voltage for the sensor excitation. Starting from the proposed circuit, Fig. 4.27b shows the detailed block scheme of the designed *CCII*-based front-end. It shows four blocks: a resistance to current converter, containing the sensor resistance; a switch stage; a current-mode instrumentation amplifier; a hysteresis comparator (*CCII*-based Schmitt Trigger). The current signal coming from the switch stage depends on the sensor resistance value. This current alternatively charges and discharges the capacitor C, with a rate depending on the current value itself. In this way, the conversion from resistance to time is obtained.

Through a straightforward analysis, considering ideal *CCII* and switch behaviours, it is possible to determine the expression for the period T of the generated

Table 4.2 Measured and theoretical periods T revealed at the interface output node (system sensitivity equal to 1.2 μs/kΩ)

R_{SENS} [Ω]	Theoretical period [s]	Measured period [s]	Relative error [%]
10 k	0.0121 m	0.0133 m	9.92
100 k	0.1210 m	0.1230 m	1.65
1 M	1.2100 m	1.250 m	3.31
10 M	12.1000 m	12.400 m	2.48

output square wave signal as a function of the sensor resistance R_{SENS} as follows:

$$T = 2R_{SENS}C\left[\frac{k\,(V_{SAT+} - V_{SAT-})}{A \cdot V_{EXC}}\right], \tag{4.25}$$

where k is the ratio between $(R_1 - R_S)$ and $(R_1 + R_2)$, A is the voltage gain of instrumentation amplifier ($A = 2R_4/R_3$), V_{EXC} is the DC sensor excitation voltage, while V_{SAT+} and V_{SAT-} are the positive and negative saturation voltages at output terminal V_{OUT}, respectively.

The presented interface has been completely designed in a standard low cost AMS 0.35 μm *CMOS* technology at a ±0.75 V dual supply voltage and showing a total 0.7 mW power consumption (in particular, in this case, the schematic circuit of Fig. A1.5 has been utilized to implement the *CCIIs*). Post layout simulation results, in particular Monte Carlo and corner analysis, have shown a good immunity with respect to the supply voltage, technological and temperature variations, for R_{SENS} varying from 10 kΩ to 100 MΩ.

In addition, measurements on a prototype *PCB* (AD844 has been used as *CCII*, needing a higher supply voltage of ±15 V), for different values of the sensor resistance R_{SENS}, are reported in Table 4.2 where the following experimental values has been utilized: $R_1 = 10\,\text{k}\Omega$, $R_2 = 20\,\text{k}\Omega$, $R_3 = 30\,\text{k}\Omega$, $R_4 = 30\,\text{k}\Omega$, $R_S = 100\,\Omega$, $C = 100\,\text{pF}$, $A = 2$, $V_{EXC} = 0.6\,\text{V}$, $V_{SAT+} = -V_{SAT-} \cong 11\,\text{V}$. In this way, the interface sensitivity has been set here to about 1.2 μs/kΩ. Data have been achieved using high precision (1%) commercial sample resistors emulating the resistive gas sensor, ranging from 10 kΩ to 10 MΩ, and the relative percentage error between measured and theoretical periods is lower than 10% all over the considered resistive variation range. Then, the fabricated prototype board has been tested to detect the presence of hydrogen mixed with nitrogen, respectively in 40 and 80 ppm, into a closed chamber. In particular, a suitable laboratory experimental equipment, shown in Fig. 4.28, to detect different gas mixtures, has been utilized; in order to properly control the hydrogen concentration and the operating conditions, we have employed a gas flux-meter and a thermally controlled chemical reactor. For these experimental measurements the commercial Figaro TGS2600 [24] has been employed, at different operating temperatures, as resistive gas sensor ranging about from 10 kΩ to 10 MΩ.

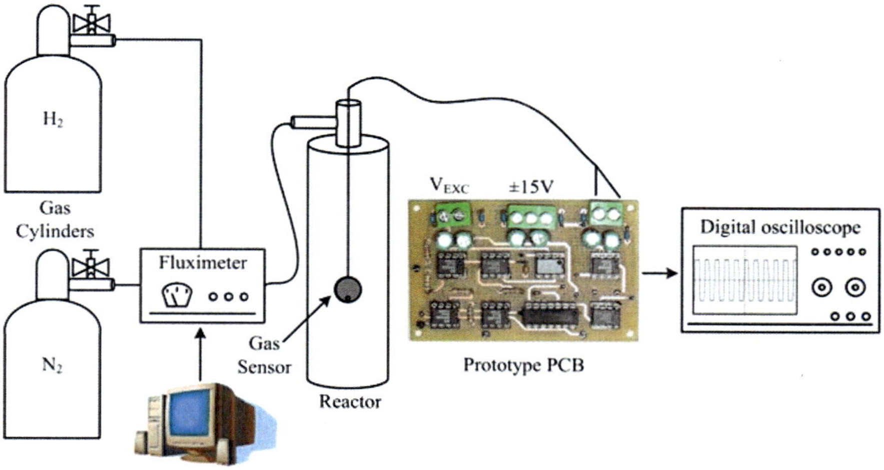

Fig. 4.28 Block scheme of the utilized experimental apparatus

Table 4.3 Measured and theoretical periods T determined at output node of the prototype board (system sensitivity equal to 7.1 μs/kΩ)

R_{SENS} [Ω]	Theoretical period [s]	Measured period [s]	Relative error [%]
500	3.55 μ	3.61 μ	1.81
1 k	7.09 μ	7.15 μ	0.85
5 k	35.45 μ	35.71 μ	0.73
10 k	70.91 μ	71.51 μ	0.84
50 k	354.56 μ	358.05 μ	0.98

In this case, in order to use the proposed system with the low resistance values shown by the chosen sensor (typically ranging from about 1 up to 90 kΩ) and its small variation in the presence of reduced ppm of hydrogen, the system accuracy has been improved (in terms of a reduced relative error, lower than 1%, in sensor resistance estimation), as reported in Table 4.3, in a reduced resistive variation range (from 500 Ω up to 50 kΩ with a high precision (1%) commercial sample resistors emulating R_{SENS}). More in detail, in this case the sensitivity of the realized circuit has been increased to about 7.1μs/kΩ by means of the choice of the following experimental values: $R_1 = 10\,k\Omega, R_2 = 15\,k\Omega, R_3 = 30\,k\Omega, R_4 = 30\,k\Omega, R_S = 330\,\Omega, C = 1\,nF, A = 2, V_{EXC} = 1.2\,V, V_{SAT+} = -V_{SAT-} \cong 11V$. Successively, in order to also characterize the dry air baseline sensor resistance at different operating temperatures, we have driven the sensor heater with three different current values, 38, 40 and 42 mA, obtaining a heater power consumption ranging from about 167 up to 215 mW. The sensor has been heated for 30 min for each power level and the measurements (taken once a second) have been performed during the last minute, as reported in Table 4.4. Then, we have set the heater current value at 42 mA and fluxed into a closed chamber a mixture of hydrogen and nitrogen at

Table 4.4 Heater power consumption and corresponding estimated R_{SENS}

Heater power consumption [mW]	Estimated R_{SENS} [kΩ]
167.20	59.52
191.21	45.42
215.46	36.96

Table 4.5 Gas concentration and corresponding measured period and estimated R_{SENS}

Hydrogen concentration [ppm]	Measured period [μs]	Estimated R_{SENS} [kΩ]
0 (dry air)	262.01	36.96
40	10.92	1.54
80	10.76	1.51

different concentrations for 10 min, alternating with a 60 min dry air flux. Such a cycle has been repeated in several measurement sessions. Table 4.5 shows the measured output period and the corresponding estimated sensor resistance versus the specific gas concentration, considering the last experimental setup.

References

1. G. Ferri, N. Guerrini, *Low Voltage Low Power CMOS Current Conveyors* (Kluwer Academic Publishers, Boston, 2003). ISBN 1402074867
2. A. Sedra, K.C. Smith, The current conveyor – A new circuit building basic block. IEEE Proc. **56**, 1368–1369 (1968)
3. A.S. Sedra, G.W. Roberts, Current Conveyor Theory and Practice, in *Analogue IC Design: The Current Mode Approach* (Peter Peregrinus, London, 1990)
4. C. Cantalini, G. Ferri, N. Guerrini, S. Santucci, A low voltage low power current mode gas sensor integrate interface, in *Proceedings of International Conference on Microelectronics*, Tunisia, pp. 194–197, 2004
5. A.M. Soliman, Simple sinusoidal active RC oscillators. Int. J. Electron. **39**, 455–458 (1975)
6. A.M. Soliman, A novel variable frequency sinusoidal oscillator using a single current conveyor. Proc. IEEE **66**, 800 (1978)
7. P.A. Martinez, S. Celma, I. Gutierrez, Wien-type oscillators using CCII+. Analog Integr. Circ. Signal Process. **7**, 139–147 (1995)
8. A.M. Soliman, A.S. Elwakil, Wien oscillators using current conveyors. Comput. Electr. Eng. **25**, 45–55 (1999)
9. A.M. Soliman, Synthesis of grounded capacitor and grounded resistor oscillators. J. Franklin Inst. **336**, 735–746 (1999)
10. M.T. Abuelma'Atti, H. Al-Daghrier, New single element controlled sinusoidal oscillator employing CCII+. Microelectron. J. **29**, 83–86 (1998)
11. A.M. Soliman, Current mode CCII oscillators using grounded capacitors and resistors. Int. J. Circuit Theory Appl. **26**, 431–438 (1998)
12. A.M. Soliman, New grounded-capacitor current-mode oscillators using single-output CCIIs. J. Circuit Syst. Comput. **8**(3), 363–378 (1998)
13. M.T. Abuelma'Atti, M.A. Al-Qahtani, A new current-controlled multiphase sinusoidal oscillator using translinear current conveyors. IEEE Trans. Circuits Syst. **45**(7), 881–885 (1998)

14. H. Barthelemy, G. Ferri, N. Guerrini, A 1.5 V CCII-based tunable oscillator for portable industrial applications, in *Proceedings of the IEEE International Conference on Industrial Electronics,* L'Aquila, 2002
15. V. Stornelli, G. Ferri, A. De Marcellis, CCII-based oscillator for sensor interface, in *Proceedings of AISEM (Associazione Italiana Sensori e Microsistemi) Conference,* Napoli, Feb 2007
16. S.S. Gupta, R. Senani, Grounded-capacitor current-mode SRCO: novel application of DVCCC. Electron. Lett **36**(3), 195–196 (2000)
17. C.M. Chang, B.M. Al-Hashimi, H.P. Chen, S.H. Tu, J.A. Wan, Current mode single resistance controlled oscillators using only grounded passive components. Electron. Lett **38**(19), 1071–1072 (2002)
18. S. Del Re, A. De Marcellis, G. Ferri, V. Stornelli, Low voltage integrated astable multivibrator based on a single CCII, in *Proceedings of PRIME,* Bordeaux, pp. 177–180, July 2007
19. G. Di Cataldo, G. Palumbo, S. Pennisi, A schmitt trigger by means of a CCII. Int. J. Circ. Theo. Appl. **23**, 161–165 (1995)
20. A. De Marcellis, C. Di Carlo, G. Ferri, V. Stornelli, A novel general purpose current mode oscillating circuit for the read-out of capacitive sensors, in *Proceedings of IEEE IWASI,* Trani, pp.168–172, 2009
21. A. De Marcellis, C. Di Carlo, G. Ferri, V. Stornelli, A CCII-based wide frequency range square waveform generator, accepted to Int. J. Circuit Theo. Appl., May 2011, DOI: 10.1002/cta.781
22. A. Sedra, K.C. Smith, *Microelectronic Circuits,* 5th edn. (Oxford University Press, New York, 2007). ISBN 0195142527
23. Internet resource: http://www.analog.com. Datasheet AD844
24. Internet resource: http://www.figarosensor.com. Datasheet TGS2600
25. Internet resource: http://sensing.honeywell.com. Datasheet HCH-1000
26. G. Ferri, V. Stornelli, A. De Marcellis, C. Di Carlo, A. Flammini, A. Depari, D. Marioli, Uncalibrated current-mode oscillator for resistive gas sensor integrable applications, in *Proceedings of ISOEN,* Brescia, Apr 2009
27. G. Ferri, V. Stornelli, A. De Marcellis, A. Flammini, A. Depari, D. Marioli, A novel low-voltage low-power second generation current conveyor-based front-end for high valued DC-excited resistive sensors, in *Proceedings of IEEE Sensors,* Lecce, Oct 2008
28. G. Ferri, A. De Marcellis, C. Di Carlo, V. Stornelli, A. Flammini, A. Depari, D. Marioli, E. Sisinni, A CCII-based low-voltage low-power read-out circuit for DC-excited resistive gas sensors. IEEE Sens. J. **9**(12), 2035–2041 (2009)

Chapter 5
Detection of Small and Noisy Signals in Sensor Interfacing: The Analog Lock-in Amplifier

In this chapter, firstly the main methods for the signal recovery from noise will be introduced and discussed. Then, the lock-in technique, for the detection of sensor signals embedded into noise, will be described in detail. In this sense, an analog lock-in amplifier (to be used in sensor interfaces) as a complete integrated circuit, designed at transistor level in a standard *CMOS* technology (AMS 0.35 μm), will be presented, together with some experimental results on gas sensors. This integrated solution improves sensitivity and resolution of the complete gas measurement system. Finally, we also propose the block scheme and operation of a high-precision high-accuracy fully-automatic integrable analog lock-in amplifier, also employed for the detection of small quantities of gases.

5.1 Signal Recovery Techniques Overview: The SNR Enhancement

Recovering a signal from noise allows to improve the Signal to Noise ratio (*SNR*). This can be done by reducing the noise accompanying a signal, through the following two basic techniques:

- bandwidth reduction, where the noise is decreased by reducing the system noise bandwidth (B_n). This approach works well if the frequency spectra of the noise and signal do not overlap significantly, so that reducing the noise bandwidth does not affect the signal. For a random white noise, the output noise is proportional to $\sqrt{B_n}$ (for a non-white noise, other relationships must be considered);
- averaging or integrating techniques, where consecutive samples of the signal are synchronized and added together. The signal grows as the number (n) of added samples, while, considering random white noise, the noise grows as $\sqrt{n}$. This is true if the signal characteristics are stationary for the duration of the extraction process.

A. De Marcellis and G. Ferri, *Analog Circuits and Systems for Voltage-Mode and Current-Mode Sensor Interfacing Applications*, Analog Circuits and Signal Processing, DOI 10.1007/978-90-481-9828-3_5,

The bandwidth reduction technique is best looked at from a frequency-domain point of view; on the contrary, signal averaging and correlation techniques lend themselves to time-domain analysis. Sometimes it is useful to combine both these techniques. In many applications, there is a significant overlap between the signal and noise spectra and improving a *SNR* must be done at the expense of the response time or measurement time (T); for random white noise, the output *SNR* is proportional to $\sqrt{T}$.

For further simplicity, it is assumed that all noise processes are stationary and that both signal and noise are ergodic, analog variables; in the following, digital signals or discrete-time (sampled) signals will not be taken into account, except where such signals are involved in the analog enhancement techniques. They are essential in modern application methods but it is the basic idea that drives the digital methods. Therefore, only signal recovery techniques will be considered. Further processing, such as least-squares polynomial smoothing of a waveform or Fourier transformation to obtain a frequency spectrum, are not considered here.

Let us present now a classification of the *SNR* enhancement techniques to be applied in sensor interfacing, especially when the *SNR* is < 1 or $\ll 1$. Since small size and low energy sensors often provide extremely low levels of output signals to be measured under the presence of a noisy environment, a suitable signal processing operation is required to obtain relevant information. Moreover, it is even possible that the power of the superimposed noise and interferences is larger than the power of the signal of interest. Generally, in such circumstances, a linear filtering operation is not sufficient to extract the signal information, so special techniques for enhancing the *SNR* have to be adopted [1–12].

More in detail, starting from the main basic classification previously described, it is possible to considerate the following three main electronic (analog and/or digital) systems [1]:

- waveform averagers
  - box car integrators
  - signal averagers
- correlation function calculators
  - autocorrelators
  - crosscorrelators
- lock-in amplifiers (analog or digital)

It is important to highlight the relationship between the input signal (to be recovered) and the reference one (if required). In particular, concerning waveform averages, which utilize techniques for signal averaging, the following conditions have to be satisfied: the input signal has to be repetitive, even if it is not periodic; the input signal has to be either preceded by a trigger pulse or able to provide a pulse at a certain time, before of the sampling time; the input signal and trigger pulse have to be synchronized; the input signal has to be as much as possible without "jitters" which can cause errors in some cases. As regards autocorrelators and crosscorrelators, the input signal to be recovered does not require any synchronized trigger pulse. In particular, when the same signal is compared to phase shifted copies of itself, the procedure is known as autocorrelation, while when two independent signals are compared the procedure is known as crosscorrelation. On the contrary, lock-in amplifiers require that the input signal is periodic and has a fixed and well-known frequency so a reference signal having the same frequency and a suitable phase condition has to be provided.

More in detail, a waveform averager samples the applied signal at a regular sampling rate and stores the resulting waveform. It can repeat this process so that a periodic input signal can be monitored in exactly the same way on each new cycle. Each record is added to the sum of the previous records so that a continuous summation process takes place. Any asynchronous event (i.e., noise) will be reduced in amplitude in relation to the amplitude of the synchronous events (i.e., signal) and hence the summed record represents the original signal waveform recovered from the noise. In the case of Gaussian noise, the improvement in *SNR* gained from this process is approximately equal to the square root of the number of summed cycles. Hence, for example, averaging 100 records of an identical event will improve the *SNR* by ten times.

As mentioned above, the main significant types of averagers, which can be utilized when the input signal has harmonic components in a wide frequency range, are the box-car integrators and the signal averagers.

The box-car integrator (which can be stationary, scanning mode or multichannel) typically utilizes analog electronics, supported by digital control, to monitor one discrete point in time on a repetitive signal. It is based on the sampling of the input signal, at a fixed time, in a defined time interval (the so-called "gate time"). The sampling is always controlled by a suitable trigger pulse depending on the same input signal. In particular, it builds up an average of that point over many cycles before recording it as a value. It may then move on to a different (later) point and repeat the process, averaging for the same number of cycles as for the first point, before recording a second value. In this way it can "step" across a waveform, monitoring it at discrete points to build up a complete averaged representation of the input signal.

The signal averager uses digital techniques to record all of the waveforms on each cycle. This makes it much more time efficient than box-car systems. Nonetheless the time taken to do the summation limits the maximum data throughput unless a dedicated hardware averager is included.

Therefore, box-car systems are particularly well suited to average a single point in time repetitively. As an example, the amplitude of one peak of a spectrum, derived from a repetitively swept monochromator, could be averaged easily and recorded as a function of time using a box-car system. This technology can also give a good time resolution, lower than 1 ns. Signal averagers can provide maximum time resolutions of a similar level, but are better suited to waveform recovery and to monitoring short lived phenomena due to their better time efficiency.

The correlation techniques (i.e., correlation function calculators) for the *SNR* improvement regard the existing relationship between either a signal to be revealed and an its delayed copy (autocorrelation) or two different signals (crosscorrelation). A function which is related to the correlation function, but arithmetically less complex, is the average magnitude difference function.

In particular, the autocorrelation function allows to extract an information (i.e., the amplitude of the signal), buried into noise, but the same kind of operation is not suitable to reveal also the information related to the signal phase. Moreover, this operation can provide, as a result, also a function which is completely different from that utilized as input data in the autocorrelation operation, depending on the kind of input waveform. It is important to consider that, for example, the autocorrelation of the "white" noise, having a wide bandwidth, provide as a result a correlation function which tends to zero with a decay time depending on its bandwidth; moreover, since this kind of noise is completely random and has not any temporal correlation, its resulting autocorrelation is a "delta" function. Therefore, when a sinusoidal signal is buried into noise (also with a reduced bandwidth), the autocorrelation function provides two contributes: one related to the noise, which decays to zero, and the other, associated to the sinusoidal signal, which will be extracted after a suitable time delay, being still a sinusoidal waveform from which it is possible to extract the desired information related to the amplitude value of the AC component (i.e., the peak value of the input sinusoidal signal).

The crosscorrelation function is similar to the autocorrelation one, but in this case the delayed signal, to be multiplied with the input signal, comes from another source. As a consequence, this technique provides, as a correlation result, the frequency components which are common to both the two input signals. The advantage provided by the crosscorrelation is in the very strong rejection to noise and disturbs. In fact, when there is no correlation between the input signal and the noise, the resulting signal coming from crosscorrelation, for a suitable long measuring time, tends to zero.

Autocorrelation is a method which is frequently used for the extraction of a fundamental frequency (f_0): if a copy of the signal is shifted in phase, the distance between correlation peaks is taken to be the fundamental period of the signal (directly related to f_0). The method may be combined either with the simple smoothing operations of peak and centre clipping or with other low-pass filter operations. On the contrary, crosscorrelation is the method which basically underlies implementations of the Fourier transformation: signals of varying frequency and phase are correlated with the input signal and the degree of correlation in terms of frequency and phase represents the frequency and phase spectrums of the input signal.

Finally, lock-in amplifiers (analog or digital) are extremely powerful signal recovery instruments if the signal is, or can be made to be, an amplitude-modulated AC waveform, where the envelope of the modulation is the required output. In fact, a lock-in amplifier, based on a phase sensitive detector, provides a DC output voltage signal which is proportional to the root mean square value of the AC input noisy signal (typically, slowly time variable). Generally, long time constants can increase the accuracy of the measurement system by averaging out AC noise, but if the meter itself experiences DC drifts during that time, the measurements cannot be valid and, in addition, a very long time constant occur. On the contrary, the lock-in technique provides for rejecting both AC and DC noise sources before the signal is measured. Typically, in lock-in amplifiers the measured signal can be averaged to much shorter time constants, allowing faster and more accurate results [2–12].

More in detail, also referring to sensor microsystems, the application of lock-in principle to extract, in a synchronous way, signal from noise is possible under the condition that the noisy signal (e.g., coming from the sensor) has a fixed and well-known frequency [13–18]. In particular, the lock-in technique, operating with a single reference frequency, can be utilized in electronic interfaces and optical sensor applications for recovering signal from noise or, in alternative, to operate very high-resolution measures of "clean" or "noisy" signals with different amplitudes and frequencies.

In the next Paragraphs, analog lock-in amplifiers will be described and their application to sensor interface to enhance system sensitivity and resolution is proved.

5.2 The Lock-in Amplifier

The lock-in amplifier measures the magnitude of a signal in a very narrow frequency bandwidth, while rejects all the components of the signal that are outside it. The lock-in technique has revealed to be better than a simple filtering operation, thanks to its superior performance. In fact, because of the automatic tracking, lock-in amplifiers can give effective quality factor Q values (a measure of filter selectivity) over 100,000, whereas a normal band-pass filter becomes difficult to use with a Q greater than 50.

Its main active block is the Phase-Sensitive Detector (*PSD*): it is a "special waveform rectifier", performing an AC-to-DC conversion, which increases only the useful signal while reduces the noise effect overlapped to the same signal. In order to properly work, the *PSD* has to be excited by a reference signal having the same frequency of the input noisy signal and a suitable phase delay. The use of this "locked" reference (from which the technique takes its name) assures the capability of the system to pursue the input noisy signal increasing its *SNR*. In fact, the system reduces the noise bandwidth through a synchronous operation which needs the knowledge of the useful signal frequency, giving a *SNR* improvement equal to the ratio between the *SNR* at the lock-in output and that at its input.

In order to recover a signal from noise, the lock-in amplifier must be provided with a relatively clean reference signal of the same frequency as the signal to be measured. Therefore, lock-in amplifiers can use a Phase-Locked-Loop (*PLL*) to generate the reference signal, otherwise an external reference signal must be provided. In particular, in the lock-in system, the *PLL* locks the internal reference oscillator to the external signal, resulting in a reference waveform with a proper phase shift. Since the *PLL* actively tracks the external signal, changes in the external reference frequency do not affect the measurement. Moreover, if the input noisy signal to be measured is of DC kind, it must be opportunely modulated by an AC waveform either electrically (e.g., as in exciting a strain gauge with an AC voltage) or mechanically (e.g., as in passing a light beam through an optical chopper). The signal and its modulation frequency (the reference signal) can be then easily fed to a lock-in amplifier.

Lock-in amplifiers can be of analog or digital kind. Those which use an analog signal processing channel are invariably known as analog instruments, even if sometimes they include digital output filters. The term "digital lock-in amplifier" usually refers to units which utilize a *DSP* demodulator. In fact, although there is a commercially available instrument described as a high-frequency *DSP* lock-in amplifier, it is an analog unit used as a "down converter" followed by a low frequency *DSP* final detector stage. *DSP* instruments generally give better performance than their analog counterparts and have inevitably become the first choice for the user. However, it is worth remembering that there are still some applications for which the analog instruments will offer different advantages (e.g., higher operating frequencies, since *DSP* units are currently restricted to operation at about few MHz or below, whereas analog units can operate to many MHz). Among these, we mention the use of lock-in amplifiers as first analog front-ends in sensor applications when the information coming from sensor systems can be very small and buried into noise. This allows to detect very small quantities of measurands as, for an example, few ppm (or ppb) of target gases.

In the following, the basic operation of an analog lock-in amplifier will be described more in detail so to better understand how this system works and how the choices made in their design influence its performances.

In Fig. 5.1 a possible block scheme of an analog lock-in amplifier implementation is shown. It is possible to highlight two different channels: the input and the reference signal channels. The first active block, which processes the input noisy signal, is a Low Noise Amplifier (*LNA*). It provides a high DC gain, adding noise as small as possible, to the input signal. Since the spectrum of the signal of interest

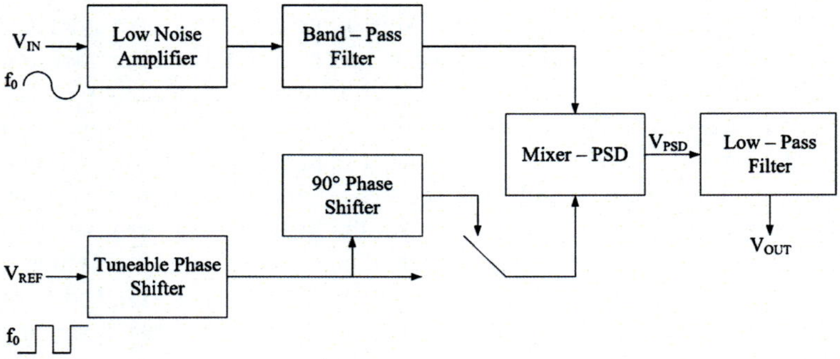

Fig. 5.1 Basic block scheme of an analog lock-in amplifier architecture. *Upper path*: input signal channel. *Lower path*: reference signal channel (V_{IN} = AC input signal, V_{REF} = AC reference signal, V_{PSD} = mixer output signal, V_{OUT} = DC output signal)

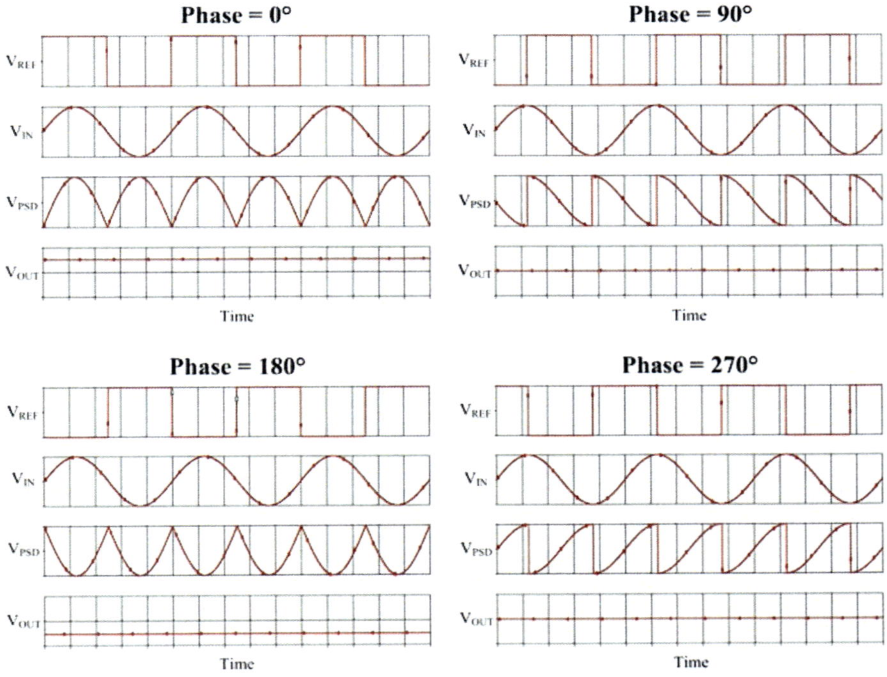

Fig. 5.2 Main signals in the lock-in amplifier for various phase differences between the input signal of interest and the reference signal

is zero for all frequencies but the signal operating frequency f_0, a suitable band-pass filter, whose center frequency must be exactly f_0, can increases the *SNR*. In the reference signal channel, different phase shifters must be used in order to both null and put "in quadrature" the phase difference between the reference signal and input signal (e.g., coming from a sensor). In particular, the relative phase of reference signal can be easily synchronized with the input signal through two active blocks: a tunable phase shifter and a 90° fixed phase shifter. In this way, the next block, a mixer or *PSD*, generates a periodic signal, whose DC component is proportional to the amplitude of AC input signal and depends from mentioned phase difference, as shown in Fig. 5.2: if the phase difference between the reference signal and the signal of interest is 0° or 180°, the output signal has a non-zero DC component which is proportional to the amplitude of the input signal. The signal generated by the mixer may easily be extracted by means of a suitable low-pass filter, which represents the final block of the complete system. In order to pull out exactly the DC component, from the periodic signal generated by the mixer, a suitable choice of the low-pass filter cut-off frequency (possibly the lowest) must be done.

Lock-in amplifiers can be seen as very narrow filters with a central frequency f_0 and a quality factor Q which, neglecting the band-pass filtering, can be expressed as following:

$$Q = \frac{f_0}{\Delta f} \tag{5.1}$$

where Δf is the band-width of the low-pass filter. Obviously, the smaller the bandwidth of the low pass filter is, the higher both the Q and the rejection of the disturbances are; on the other hand, the complete system may not be faster than the low pass filter itself, so a trade-off exists between the Q value (related to the disturbance rejection) and the speed (i.e., time response) of the lock-in amplifier (in other words, better results require long measurement times).

In order to express the lock-in amplifier *SNR* improvement quantitatively, we have that the *SNR* at the output of the lock-in amplifier is given by the *SNR* at the input multiplied by the square root of the ratio between the equivalent noise bandwidth and the bandwidth of the low pass filter, as reported in the following expression:

$$SNR_{OUT} = SNR_{IN}\sqrt{\frac{B_{EQ\_NOISE}}{B_{LPF}}} \tag{5.2}$$

Therefore, as an example, if the equivalent noise bandwidth is 10^4 Hz and the bandwidth of the low pass filter is 10^{-2} Hz, the *SNR* improvement is 10^3. On the other hand, the advantage of the lock-in amplifier, with respect to a conventional filter, is in the fact that the system bandwidth (i.e., the bandwidth of the low pass filter) can be imposed orders of magnitude lower than that related to a conventional filtering device.

Finally, it is important to highlight that the basic architecture reported in Fig. 5.1 can be conveniently integrated into a single chip, after a suitable design of the single blocks at transistor level, as demonstrated in the next Paragraph; therefore, the lock-in amplifier is also suitable for portable sensor applications [1–12].

5.3 An Integrated LV LP Analog Lock-in Amplifier for Low Concentration Detection of Gas

Commercial lock-in amplifiers typically show large dimensions, high costs and are not suitable for portable applications. In the literature, both digital and analog lock-in amplifiers for sensor applications have been presented (e.g., in [13–18]).

Recently, a low-cost integrated analog lock-in amplifier, powered by a low dual supply voltage (±1 V) and showing low power consumption (3 mW), has been proposed. It has been also used as a part of the first sensor analog front-end and has been introduced in a sensor system with the aim to detect very low quantities of dangerous gases [19, 20].

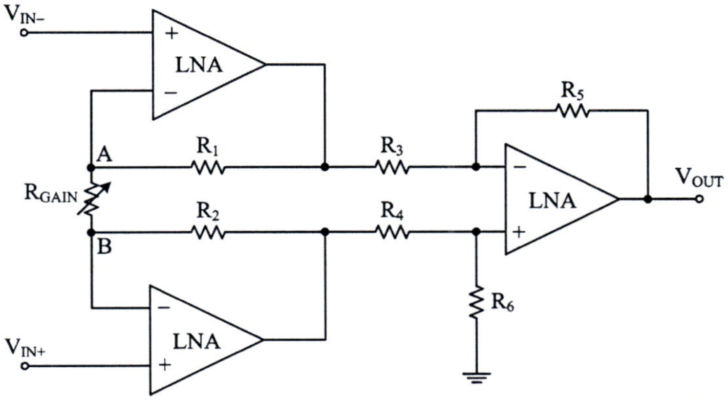

Fig. 5.3 Traditional differential input instrumentation voltage amplifier implemented by LNA

The proposed lock-in amplifier has been designed, at transistor level in a standard *CMOS* technology (AMS 0.35 μm), in order to work at a low fixed operating frequency $f_0 = 77$ Hz. This value has been chosen so to reduce, as much as possible, any kind of interference with the net supply oscillation frequency (50 Hz) and its harmonics. Moreover, this frequency is also compatible with typical characteristics of the resistive gas sensor. The presented lock-in system, totally formed by analog blocks, has been integrated in a reduced silicon area (about 5 mm^2) and is able to reveal very small signals, also thanks to a particular design of its internal blocks and of the first amplifier stage showing very low noise characteristics. In particular, the latter has been designed in the configuration of a traditional *VM OA*-based instrumentation amplifier, as reported in Fig. 5.3. In order to achieve good noise and common mode rejection performances, a *LNA*, to be utilized in the three active blocks, has been designed at transistor level, according to the internal topology shown in Fig. 5.4. Since that, in the utilized technology, the *pMOS* transistors provide a noise factor K_f lower than *nMOS* transistors, *LNA* input stage has been based on a double *pMOS* differential pair (M_4–M_7), where each transistor has $W = 800\,\mu$ and $L = 20\,\mu$ sizes, so to provide a reduced input equivalent noise (about 22 nV/$\sqrt{}$Hz at 77 Hz reference frequency). The output stage, formed by M_{10} and M_{11}, is a class-AB Push-Pull inverter topology where frequency compensation has been obtained through the capacitance C_1, set to 1.5 pF. The complete AC differential amplifier provides a very high DC tuneable gain A (from 10 to 110 dB) and a very small input equivalent noise level. In particular, its DC gain can be simply set through the external variable resistance R_{GAIN}, as shown in Fig. 5.3, so the total equivalent input noise is about 34 nV/$\sqrt{}$ Hz at 77 Hz, for a 110 dB DC gain (obtained with $R_{GAIN} = 1\,\Omega$). Then, a band-pass filter has been implemented through the block scheme reported in Fig. 5.5. It uses only a single active component having the required specific central frequency, showing unitary voltage gain at central frequency and a quality factor Q of about 2. This not high value of Q has been chosen to avoid errors due to the variation of the central frequency owed

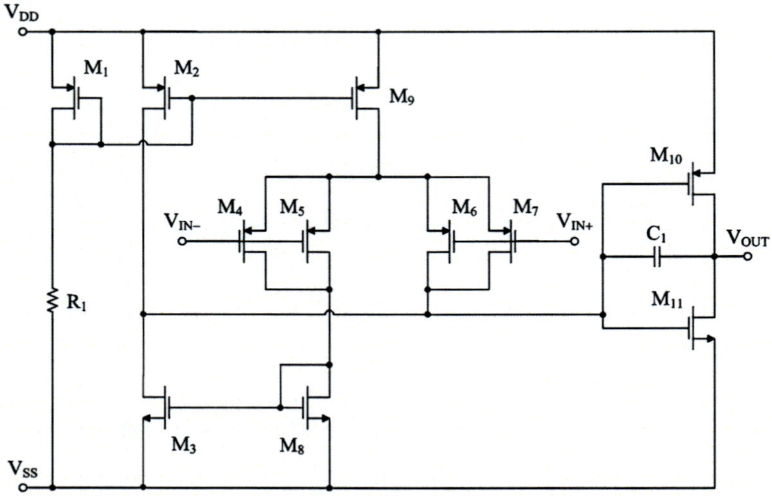

Fig. 5.4 Internal topology, at transistor level, of the implemented LNA

Fig. 5.5 Block scheme of the implemented active Bband-Ppass filter

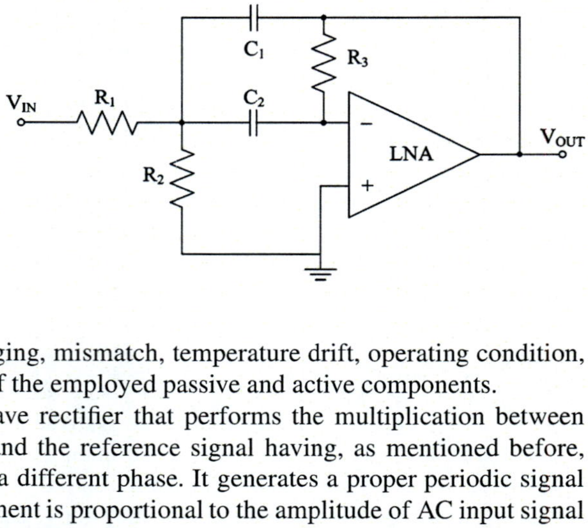

to the non-idealities (e.g., aging, mismatch, temperature drift, operating condition, technological spread, etc.) of the employed passive and active components.

The mixer block is a wave rectifier that performs the multiplication between the amplified input signal and the reference signal having, as mentioned before, the same frequency f_0, but a different phase. It generates a proper periodic signal $V_{OUT,MIX}$, whose DC component is proportional to the amplitude of AC input signal V_{IN} and depends on this phase difference (φ), according to the following expression:

$$V_{OUT,MIX} = \frac{2}{\pi} k V_{IN} [\cos\varphi - \cos(2\omega_0 t + \varphi)] \tag{5.3}$$

being k the system total amplification (which takes into account also A).

Fig. 5.6 shows the block scheme of the implemented mixer, which utilizes the same *LNA* as main active block, together with two matched resistances R and two analog switches (S_1, S_2). The latter are controlled by the square wave reference

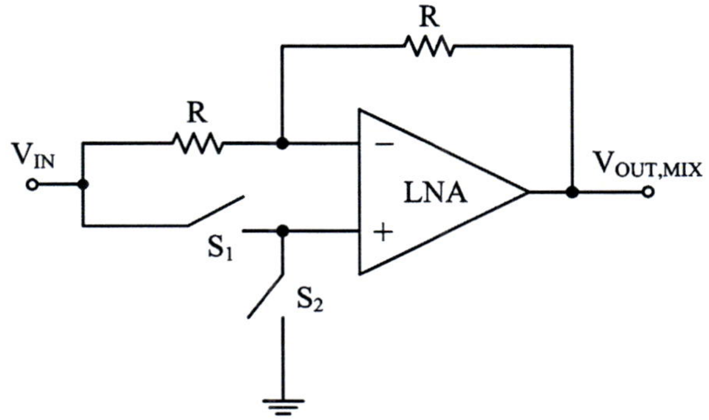

Fig. 5.6 Block scheme of the mixer or *PSD*

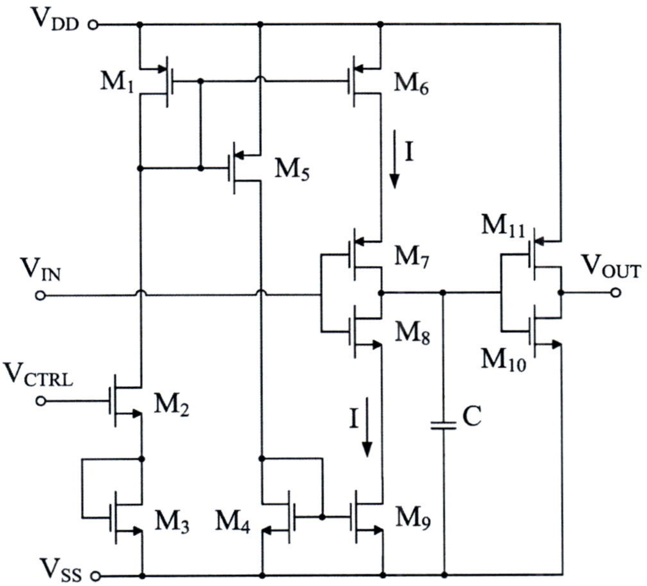

Fig. 5.7 Schematic circuit of the tunable phase shifter

signal which, through a *NOT* gate, is also inverted so to have two control lines in non overlapping phase-opposition. In this way, the switches are properly controlled since, referring to Fig. 5.6, when S_1 is open, S_2 is closed and vice versa. The gain of the mixer is ideally $+1$ or -1, depending on the relative phase between input and reference signals.

In order both to null and put in-quadrature the phase difference between the reference signal and that coming from the sensor (mixer input signals), a tuneable phase shifter and a 90° (fixed) phase shifter have been implemented, so to easily synchronize the two mixer inputs. Both of these blocks have been designed utilizing traditional internal circuit topologies, based on the cascade of two Push-Pull stages and the capacitance charge–discharge effect. Fig. 5.7 shows the schematic circuit, at

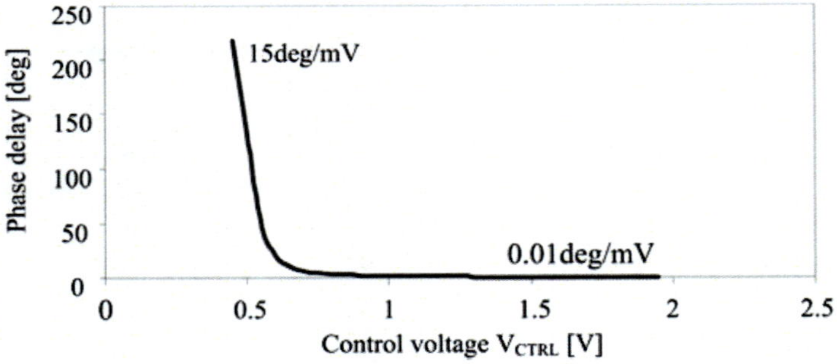

Fig. 5.8 The tuneable phase shifter: relationship between the phase delay, expressed in degree, and the applied external control voltage V_{CTRL}, showing two different sensitivity ranges

transistor level, of the designed tuneable phase shifter. In this scheme, C is charged through M_6 and M_7 and discharged through M_8 and M_9, with a constant current I. This circuit generates a time delay, thus a phase shifting, proportional to the capacitance value, according to the following expression:

$$T_{DELAY} = \frac{V_{DD} - V_{SS}}{2} \cdot \frac{C}{I} \tag{5.4}$$

being I the current shown in Fig. 5.7, whose value is determined by an external control voltage, V_{CTRL}. In order to perform a phase tuning in a large degree range, a cascade of four independent tuneable phase shifters has been implemented. In particular, the phase tuning can be easily performed, firstly, by activating the tuneable phase shifters through voltage controlled *CMOS* switches, and, then, by varying the single external control voltage V_{CTRL} (see always Fig. 5.7), so to tune the previously selected phase shifters. In this manner, it is possible both to properly regulate the current I which flows into the capacitances and to adjust the relative phase between input and reference signals.

Fig. 5.8 shows the relationship between the achieved phase delay, expressed in degree, and the applied external control voltage V_{CTRL}, highlighting two different sensitivity ranges. The 90°-phase shifter has been designed starting from the schematic circuit shown in Fig. 5.7 and adding two further Push-Pull stages and another capacitance. Through other suitable voltage controlled *CMOS* switches, it is possible to activate this shifter, so to provide exactly the required 90°.

The final block of the proposed complete architecture, which follows the mixer, is a low-pass filter that reduces the noise contribution through a DC extraction. The filter output is a DC voltage whose level is proportional to the amplitude of the input signal. An active low-pass filter based on a Transconductance active block (*Gm*), which, together with a capacitance C_{Gm}, allows to obtain a *Gm-C* integrator cell, has been designed, as it shown in Fig. 5.9. A fourth order low-pass filter has been simply obtained by cascading four *Gm-C* cells, having chosen all the capacitances C_{Gm} equal to 100 pF. Fig. 5.10 shows the internal topology of the

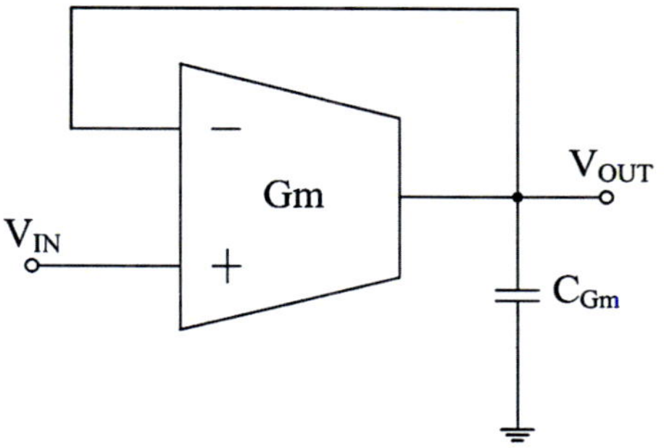

Fig. 5.9 Block scheme of a single Gm-C cell implementing an active low-pass filter

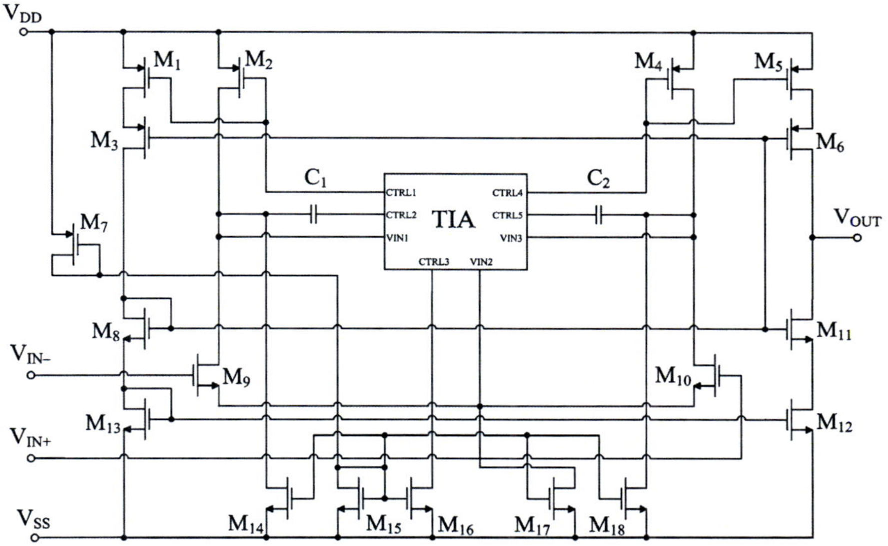

Fig. 5.10 Internal topology of the Gm block

designed Transconductance blocks, implemented through a Three Input Amplifier (*TIA*) [21], which allows to achieve a very low cut-off frequency, of about 1 mHz. Fig. 5.11 shows the circuit schematic at transistor level, of the designed *TIA*.

Fig. 5.12 shows the photo of the integrated lock-in, highlighted, on the left, by the arrow. A prototype *PCB* has been fabricated and utilized for the complete system on-chip testing. Fig. 5.13a, b depict the measured signals generated by the mixer, when a clean input signal and the reference signal are "in-quadrature" and "in-phase", respectively. Fig. 5.14a, b show the generated input noisy signal and the corresponding time response of the DC voltage signal at the system output. In particular, we detect a null response, related to "in-quadrature" mixer inputs, and a non-zero DC signal, with its transient response, achieved through the activation of the 90°-phase shifter ("in-phase" mixer inputs).

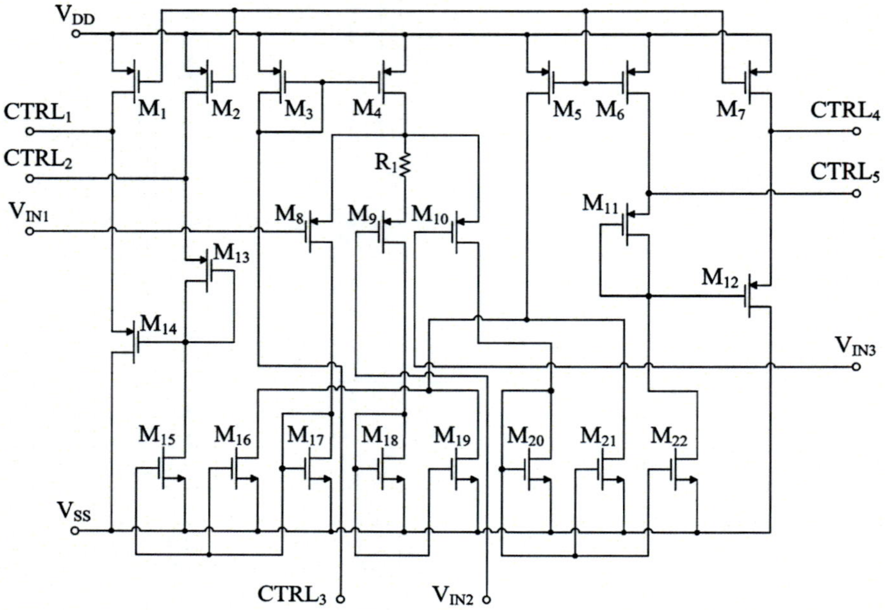

Fig. 5.11 The designed TIA circuit schematic at transistor level

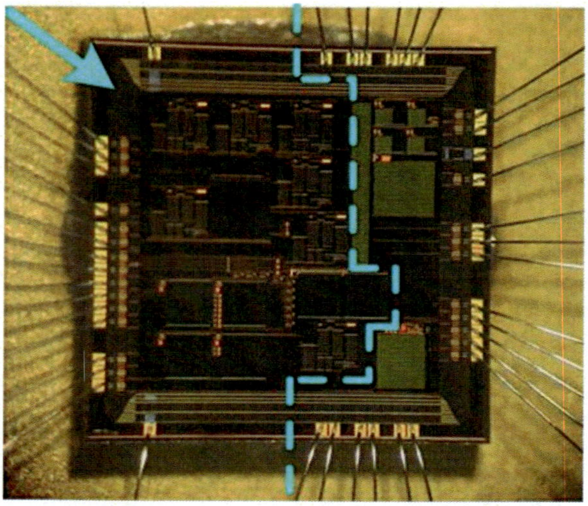

Fig. 5.12 Photo of fabricated chip: the designed lock-in is in the *left part*, delimited by the *dashed line*

This analog lock-in system is able to recover with success very small noisy signals (down to about 500 nV, with $SNR < 1$) without performance degradation. In particular, measurement results are reported in Fig. 5.15 showing a good linearity

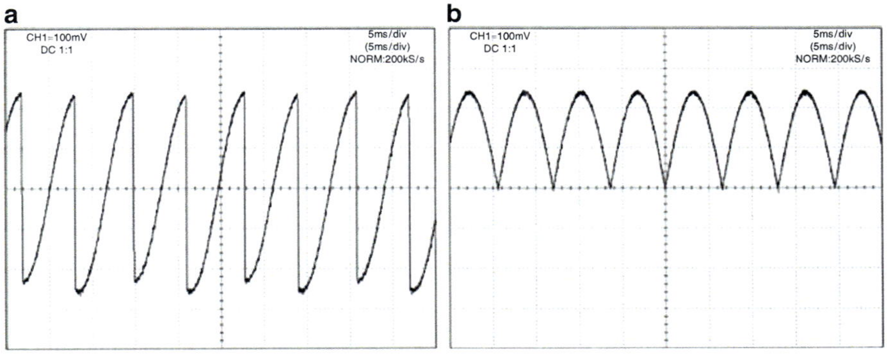

Fig. 5.13 Measurement results at the mixer output for in-quadrature (**a**) and in-phase (**b**) inputs

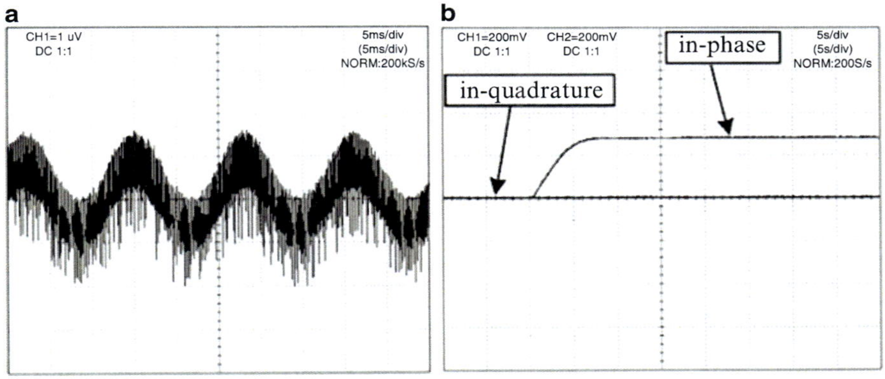

Fig. 5.14 Measured noisy input signal at the instrumentation amplifier (**a**) and the time response of the extracted DC signal at the system output with in-quadrature and in-phase input and reference signals (**b**)

between the extracted output DC voltage and the input AC noisy signal, according to the following relationship:

$$V_{OUT} = \frac{2kV_{IN}}{\pi} \tag{5.5}$$

being k the system total amplification, of about 320,000.

The fabricated system has been also tested with a suitable experimental apparatus to detect, for example, the presence of CO into a closed chamber, as shown in Fig. 5.16. In this case, for the experimental measurements, a commercial resistive gas sensor (FIGARO TGS2600, R_{SENS}) has been utilized [22], excited with a 77 Hz sinusoidal voltage signal (V_{REF} is a 77 Hz square wave signal), whose amplitude has been fixed to 40 mV, in series with a reference resistance R_{REF} valued 10 kΩ, according to the measurement block scheme depicted in Fig. 5.17.

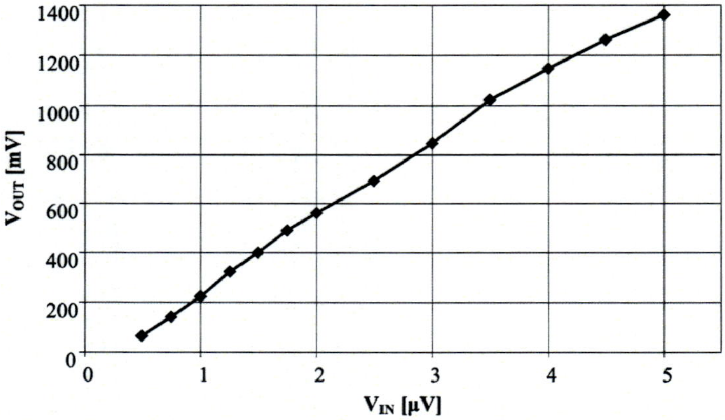

Fig. 5.15 Measured output DC voltage vs. input AC signal amplitudes (the system voltage gain A is about 110 dB)

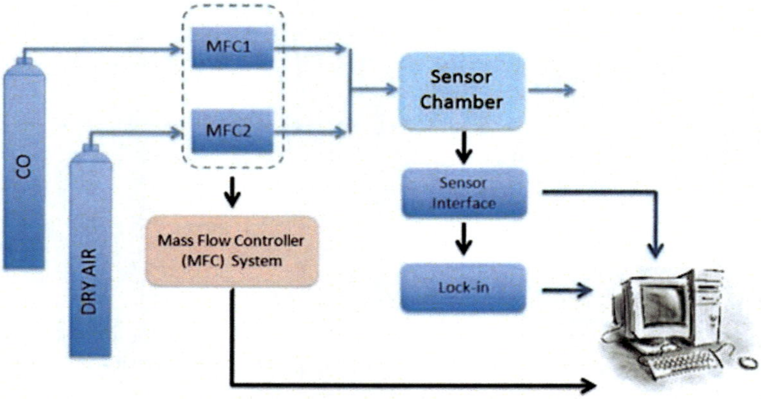

Fig. 5.16 A sketch of the experimental set-up utilized for the CO revelation

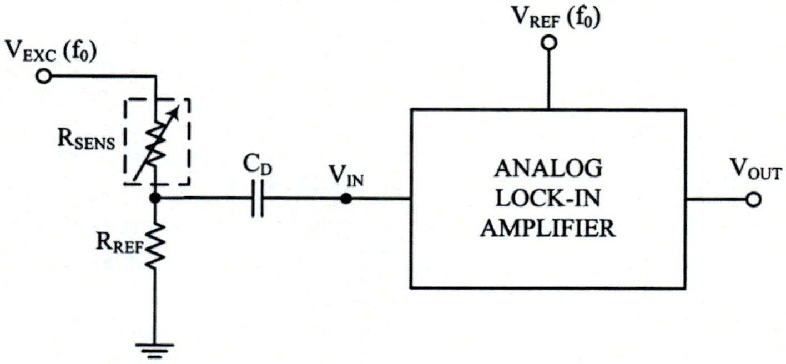

Fig. 5.17 Measurement scheme for the lock-in testing

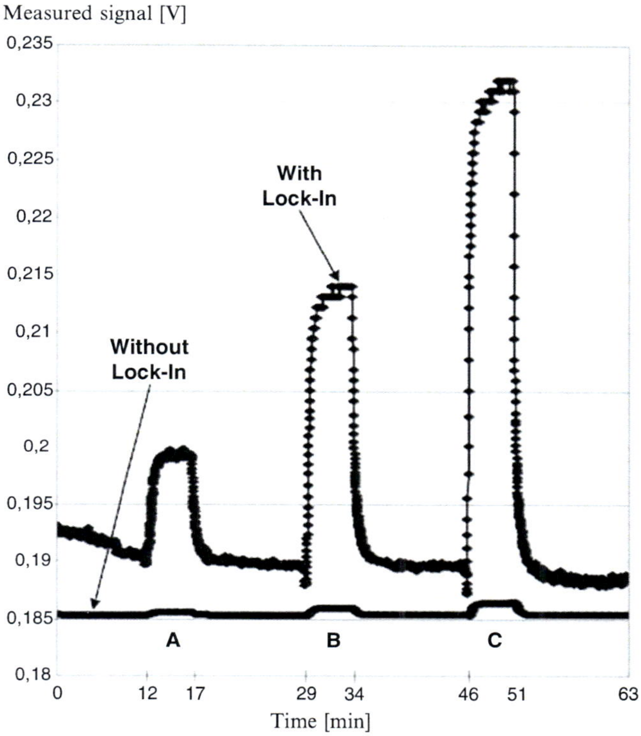

Fig. 5.18 Measured time response of the extracted DC voltage signal at the system output and voltage signal at the system input vs. time, for three different CO concentrations (A = 10 ppm, B = 20 ppm, C = 30 ppm)

In particular, in several and repetitive measurement sessions, for 5 min into a closed chamber, a mixture of dry air and *CO* at different concentrations have been fluxed, alternating it with a 12 min dry air flux. Fig. 5.18 shows the typical system time responses, considering both the input and the output voltages of the lock-in amplifier for different *CO* concentrations, as detailed in Table 5.1 (A = 10 ppm, B = 20 ppm, C = 30 ppm), where the mean values of the sensor resistance have been calculated over all the experimental measurements.

The proposed lock-in system, whose internal instrumentation amplifier operates, in this case, with a voltage gain of about 34 dB (obtained with $R_{GAIN} = 6\,k\Omega$), has improved the system sensitivity of a factor of about 40. More accurately, the system sensitivity (considered as a constant in the measured concentration range), evaluated before the lock-in amplifier application, is about 0.04 mV/ppm, while the complete analog integrated system shows an improved sensitivity of about 1.6 mV/ppm. In addition, considering the relative experimental noise levels, system resolutions before and after the lock-in amplifier are about 10 and 0.2 ppm, respectively, showing a resolution improvement, obtained through the lock-in technique, being it reduced of a factor of about 50.

Table 5.1 Experimental results achieved through the fabricated chip with related sensor resistance (R_{SENS}) estimation (see Fig. 5.18)

Measurement time [min]	CO concentration [ppm]	Mean sensor resistance $< R_{SENS} >$ [kΩ]
0–12 Initial cleaning	– (Dry air only)	61.1
12–17 (A) Dry air+CO mixture	10	57.9
17–29 Cleaning	– (Dry air only)	61.4
29–34 (B) Dry air+CO mixture	20	53.2
34–46 Cleaning	– (Dry air only)	61.5
46–51 (C) Dry air+CO mixture	30	48.3
51–63 Final cleaning	– (Dry air only)	61.7

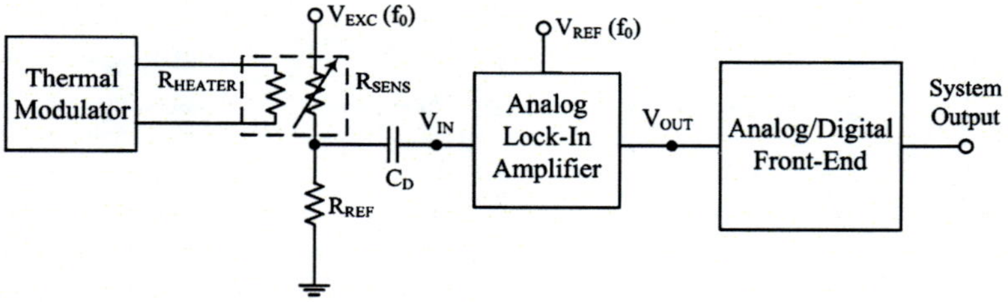

Fig. 5.19 Scheme for lock-in utilization in resistive gas sensor interface together with thermal modulation technique

Finally, the designed lock-in amplifier can also exploit the advantages of the resistive gas sensor thermal modulation (which allows to increase the sensor sensitivity), so both to further improve the whole system resolution and to detect very small quantities of gas reagent substances, very lower than 1 ppm. As an example, Fig. 5.19 shows a possible block scheme which combines the employment of these two standard techniques.

5.4 An Automatic Analog Lock-in Amplifier for Accurate Detection of Very Small Gas Quantities

In this Paragraph we want to introduce an automatic analog lock-in amplifier, together with some preliminary experimental results, which does not need the initial phase alignment and is able to recover the signal from noise also for a very low input *SNR* [23, 24].

A traditional lock-in amplifier needs the initial phase alignment of the system through the zeroing of the output signal. This means that input signals at the *PSD* block are "in quadrature". In order to reveal and measure the noisy signal at the lock-in input through the evaluation of the generated DC output signal, these two *PSD* input waveforms must be "in phase". This condition is achieved by the use of suitable control signals and switches which must be regulated and activated by manual operations. It important to highlight that an inaccurate phase alignment of the system involves measurement errors and, sometimes, also the impossibility to recover the signal.

Both in the literature and in the commercial systems, this problem has been solved through the introduction of suitable digital circuits, more complicated (*DSP* and storage elements) with respect to analog solutions, which operate a precise control of each single block employed in the system. Unfortunately, many of them, introducing a very complex architecture based on micro-processor (and/or micro-controller), for the digital signal processing, often show, consequently, large dimensions, very high costs and weight, so are not suitable for portable systems as in sensor applications [5–7, 10–12].

Starting from these considerations, a fully analog high-accuracy high-precision lock-in amplifier operating automatic phase self-alignment has been recently developed (patented system) [23, 24]. In particular, the proposed architecture allows automatically and continuously to provide the required "in phase" condition of the considered signals (automatic phase alignment of the relative phase between input and reference signals), through suitable negative feedbacks. This lock-in system allows both to perform the necessary initial phase alignment (at the power-on of the circuit) and to continuously ensure such a condition during measurement run-time (for any variation of the input noisy signal phase and amplitude during the working time), implementing a completely automatic circuit through the use of simple analog blocks. In this way, the system allows to detect, in a continuous way, the correct mean value of the input signal (buried into noise). Furthermore, no manual operations are required, avoiding errors due to input signal phase shifting, so overcoming also the problems due to temperature drifts and components aging.

Fig. 5.20 shows a possible block scheme for this automatic system. It is constituted by three main channels: "Calibration" channel, "Measure" channel and "Calibration 90°" channel. The main blocks necessary for the noisy signal recovery are the followings (standard parts of a classic lock-in): a low noise amplifier, *LNA*; a band-pass filter, *BP*; a multiplier or *PSD*, PSD_1; a low-pass filter, LP_1; a tuneable phase shifter, TPS_1 and a 90° phase shifter, TPS_2.

The operating principle of the system can be simply described as follows: the small noisy signal, introduced at the system input terminal V_{IN}, is amplified through the *LNA*, then filtered by the *BP*, so to reduce its harmonic content, and finally multiplied with the reference signal (having the same frequency of the input signal), introduced at the system input terminal V_{REF}, through the PSD_1. Finally, the filter LP_1, whose cut-off frequency has to be very small, allows to reveal the mean value of the signal generated by the PSD_1, operating an extraction of the DC component, V_{CAL}, whose value is proportional to the amplitude of the input signal and dependent

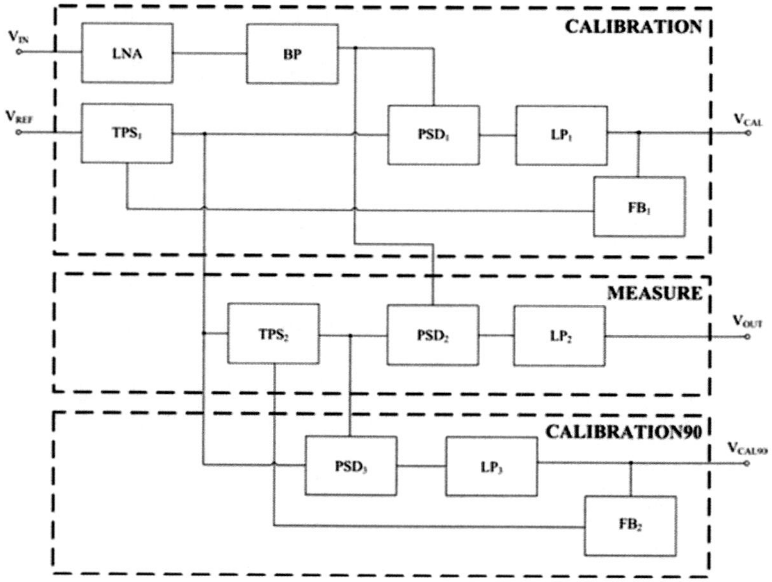

Fig. 5.20 Complete block scheme of the novel automatic analog lock-in amplifier

on its phase difference with respect to the reference signal. The precise and correct phase-alignment of the system corresponds to put the input and reference signals "in-quadrature": this is guaranteed only when the signal at LP_1 output is equal to zero. Such a condition is sustained, automatically and continuously, through suitable control blocks: the FB_1 allows to regulate opportunely the TPS_1 so that the relative phase, among input and reference signals, is always equal to 90°; at the same time, the correct measure of the DC signal V_{OUT}, whose amplitude is proportional to that of the AC input noisy signal, is achieved through the use of PSD_2 and LP_2, considering that in this case a further phase shifting of 90° between the two input signals of PSD_2 is necessary. Therefore, in order to operate and guarantee a further stable phase shifting of 90°, to be added to the relative phase among the signal coming from the PSD_1 inputs (which are "in-quadrature" by the operation of the "Calibration" channel), an additional feedback loop has been introduced. This is constituted by the blocks PSD_3, LP_3 and FB_2 which properly control the TPS_2, guaranteeing the correct relative phase delay between the input and reference signals.

This system presents two AC inputs and three DC outputs; the measured value at V_{OUT} is proportional to the input signal amplitude only when the other two calibration outputs are zero. In this sense, the final blocks of the proposed architecture are low-pass filters that reduce the noise contribution through a DC extraction.

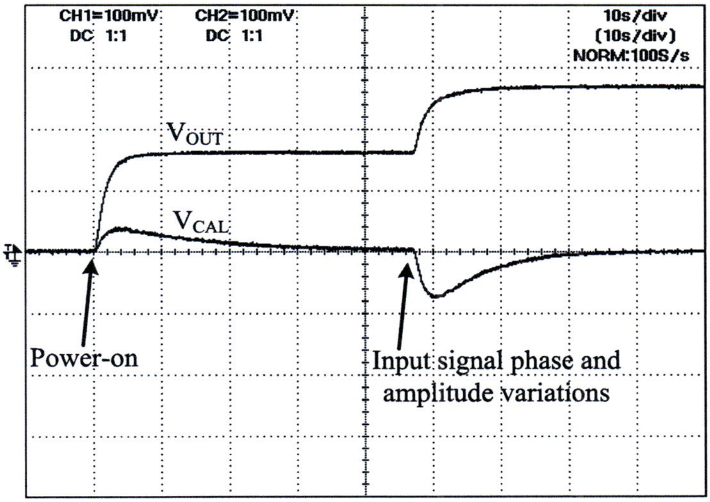

Fig. 5.21 Automatic lock-in time responses: system self-alignment at power-on followed by input signal amplitude and phase variations

The block scheme in Fig. 5.20 has been completely designed implementing the different analog blocks with suitable *VM* circuit topology, employing commercial components and precise passive elements, and then developing a discrete element prototype *PCB* (LF411 of Texas Instruments has been employed as *OA*). In particular, the *LNA* has been implemented through a well-known differential instrumentation amplifier (see Fig. 5.3), the band-pass filters have been implemented by a second-order topology (see Fig. 5.5), the low-pass filters are constituted by four *RC*-cells in cascade configuration, the *PSDs* are high-precision waveform rectifiers, while phase shifters and feedback blocks have been implemented by suitable well-known active Miller integrators. More in detail, in phase shifters, voltage controlled capacitors as electronic variable active components (i.e., AD633 of Analog Devices utilized in a suitable configuration [25]) have been utilized, achieving their tunability through the control voltage generated by relative feedback blocks.

Fig. 5.21 depicts the measured main signals V_{OUT} and V_{CAL} when an input clean sinusoidal signal has been applied, highlighting the system self-alignment at its power-on and when amplitude ($\Delta V_{IN} = 2\,\text{mV}$) and phase ($\Delta\varphi = 20°$) variations have simultaneously occurred. In particular, after a transient time due to the self-alignment operation, V_{CAL} returns at zero level, while V_{OUT} changes its DC value owing to the input signal amplitude variation.

The complete designed system has been tested by a suitable experimental apparatus (see Fig. 5.16) to detect the presence of different *CO* concentrations (10, 20 and 30 ppm), into a closed chamber, using FIGARO TGS2600 as resistive gas sensor [22]. More in detail, the commercial sensor has been excited with a 77 Hz sinusoidal voltage signal whose maximum amplitude has been fixed to 30 mV (with

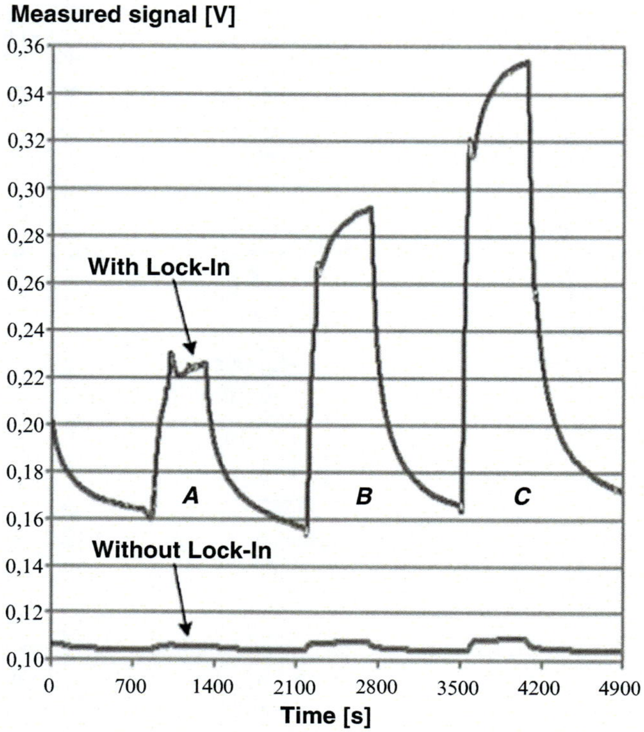

Fig. 5.22 Measured time response of the extracted DC voltage signal at the proposed lock-in output and voltage signal at the system input vs. time for different CO concentrations (CO concentrations: A = 10 ppm, B = 20 ppm, C = 30 ppm)

a DC level of 5 V), in series with a reference load resistance (see R_{REF} in Fig. 5.17) valued 10 kΩ. The sensor heater resistance has been powered with a DC voltage level equal to 5 V. In several measurement sessions, we have fluxed, for 9 min into a closed chamber, a mixture of dry air and *CO* at different concentrations, alternating it with a 14 min dry air flux. Fig. 5.22 shows the typical system time responses, considering both input and output DC lock-in amplifier signals for different *CO* concentrations, as detailed in Table 5.2 (A = 10 ppm, B = 20 ppm, C = 30 ppm), where the mean values of the sensor resistance have been reported. These voltage signals have been revealed and acquired through a *DAQ* board (NI USB-6353 by National Instruments), with a sampling rate equal to 1 s, allowing to estimate both the gas sensor resistance value and its variation, under the presence of different *CO* concentrations (see Table 5.2). Through a straightforward analysis of the experimental results and with respect to the simple resistive gas sensor interface implemented by a resistive voltage divider (as suggested by the gas sensor datasheet [22]), the sensitivity improvement given by the proposed lock-in amplifier is of a factor of about 80 (circuit input sensitivity $\approx$ 0.08 mV/ppm;

Table 5.2 Experimental results achieved through the fabricated *PCB* prototype with related sensor resistance R_{SENS} estimation (see Fig. 5.22)

Measurement time [min]	CO concentration [ppm]	Mean sensor resistance $<R_{SENS}>$ [kΩ]
0–14 Initial cleaning	(Dry air only)	128
14–23 (A) Dry air+CO mixture	10	91
23–37 Cleaning	(Dry air only)	130
37–46 (B) Dry air+CO mixture	20	69
46–60 Cleaning	(Dry air only)	129
60–69 (C) Dry air+CO mixture	30	56
69–83 Final cleaning	(Dry air only)	129

circuit output sensitivity ≈ 6.50 mV/ppm), while, the resolution, starting from about 5 ppm (system input resolution), has been enhanced to a calculated theoretical value of about 0.05 ppm (system output resolution), achieving an improvement factor of about 100 for a measured noise level of about 0.30 mV.

References

1. A. D'Amico, M. Faccio, G. Ferri, F. Mancini, Tecniche di rivelazione di segnale in condizioni di rapporto segnale-rumore molto minore di uno (S/N≪1), in *School of Sensors for Industrial Applications*, Portici, July 1989, pp. 467–511
2. W. Kester, *Mixed-signal and DSP Design Techniques* (Engineering Staff of Analog Devices Inc./Newnes, London, 2002). ISBN 0750676116
3. L.A. Wainshtein, *Extraction of Signals from Noise,* reprinted from (Dover Publications, Wokingham, 1970). ISBN 0486626253
4. R. Burdett, Signals in the Presence of Noise, Signal Recovery, in *Handbook of Measuring System Design* (Wiley, Wokingham, 2005). ISBN 9780470021439
5. M.L. Meade, *Lock-in Amplifiers: Principles and Applications* (Peter Peregrinus Ltd, London, 1983). ISBN 090604894X
6. Lock-in amplifiers and pre-amplifiers, Princeton Appl. Res. Corp., Datasheet, 1971
7. Lock-in amplifiers, appl. notes, Stanford Res. Sys., Datasheet, 1999
8. U. Marschner, H. Grätz, B. Jettkant, D. Ruwisch, G. Woldt, W.J. Fischer, B. Clasbrummel, Integration of a wireless lock-in measurement of hip prosthesis vibrations for loosening detection, in Proceedings of Eurosensors, Dresden, Sept 2008, pp. 789–792
9. M.O. Sonnaillon, F.J. Bonetto, A low-cost, high-performance, digital signal processor-based lock-in amplifier capable of measuring multiple frequency sweeps simultaneously. Review of Scientific Instr **76**, 024703-1-024703-7 (2005)
10. M.L. Meade, Advances in lock-in amplifiers. J. Phys. Sci. Instrum. **15**, 395–403 (1982)
11. Internet resource: http://www.signalrecovery.com. What is a Lock-in Amplifier, PerkinElmer, T.N. 1000

12. Internet resource: http://www.signalrecovery.com. PerkinElmer – The analog Lock in Amplifier, T.N. 1002
13. G. Ferri, P. De Laurentiis, C. Di Natale, A. D'Amico, A low voltage integrated CMOS lock in amplifier prototype for LAPS applications. Sensors Actuators A. **92**, 263–272 (2001)
14. G. Ferri, V. Stornelli, A. De Marcellis, M. Patrizi, A. D'Amico, C. Di Natale, E. Martinelli, A. Alimelli, R. Paolesse, An integrated analog lock-in amplifier for low-voltage low-frequency sensor interface,in *Proceedings of IWASI*, Bari, June 2007
15. A. Gnudi, L. Colalongo, G. Baccarani, Integrated lock-in amplifier for sensor applications, in *Proceedings of IEEE ESSCIRC,* Duisburg, Sept 1999, pp. 58–61
16. C. Azzolini, A. Magnanini, M. Tonelli, G. Chiorboli, C. Morandi, Integrated lock-in amplifier for contact-less interface to magnetically stimulated mechanical resonators, in *Proceedings IEEE International Conference on Design & Technology of Integrated Systems in Nanoscale Era,* 2008
17. C. Falconi, E. Martinelli, C. Di Natale, A. D'Amico, F. Maloberti, P. Malcovati, A. Baschirotto, V. Stornelli, G. Ferri, Electronic interfaces. Sensors Actuators B. **121**, 295–329 (2007)
18. M. Tavakoli, R. Sarpeshkar, An offset-canceling low-noise lock-in architecture for capacitive sensing. IEEE J. Solid-St Circ. **38**(2), 244–253 (2004)
19. A. De Marcellis, G. Ferri, V. Stornelli, E. Martinelli, C. Di Natale, A. D'Amico, Low-voltage low-power integrated CMOS analog lock-in amplifier for thermally modulated sensors,in *Proceedings of Eurosensors,* Dresden, Sept 2008
20. A. D'Amico, A. De Marcellis, C. Di Carlo, C. Di Natale, G. Ferri, E. Martinelli, R. Paolesse, V. Stornelli, Low-voltage low-power integrated analog lock-in amplifier for gas sensor applications. Sensors Actuators B. **144**(2), 400–406 (2010)
21. M. Schipani, F. Sebastiano, N. Nizza, P. Bruschi, A fully integrated very low frequency single-ended Gm-C filter based on a novel transconductor, in *Proceedings of IEEE PRIME,* Otranto, 2006, pp. 25–28
22. Internet resource: http://www.figarosensor.com. Datasheet TGS2600
23. A. De Marcellis, A. Di Giansante, C. Di Natale, G. Ferri, E. Martinelli, A. D'Amico, Analog automatic lock-in amplifier for very low gas concentration detection, in *Proceedings of Eurosensors,* vol 5, Linz, Sept 2010, pp. 200–203
24. A. De Marcellis, G. Ferri, V. Stornelli, A. D'Amico, C. Di Natale, E. Martinelli, C. Falconi, Analog system based on a lock-in amplifier for signal from noise detection showing a continuous and automatic phase alignment and frequency tuning, Patent n. RM-2008-A000194, 2008
25. G.Q. Zhong, R. Bargar, K.S. Halle, Circuits for voltage tuning the parameters of chuas circuit: experimental application for musical signal generation. J. Franklin Inst. **331 B**(6), 743–784 (1994). Elsevier

Appendix

Appendix 1: The Second Generation Current-Conveyor (CCII)

The Second Generation Current Conveyor (CCII): A Basic Building Block

The *CM* approach, which considers the information flowing on time-varying currents, proposes a new way to "see" integrated circuits. The Second Generation Current Conveyor (*CCII*) is considered the main *CM* basic block [1–4].

Sedra and Smith introduced the first Current-Conveyor, which actually represents for designers a possible alternative to *OA*, in 1968 [5] but its advantages and innovative impact were not immediately clear. In fact, at the same time, electronic companies started to put their main efforts in the design and fabrication of monolithic *OAs*. Consequently, the relevant value of the new invention was partially overshadowed. Only in recent years, with the growing diffusion of the *CM* approach as a way to design *LV LP* circuits, Current Conveyors have gained an increased popularity [6]. A basic well-known *CM* circuit is the Current-Feedback Operational Amplifier (*CFOA*) [7–11]. This circuit, if compared to the traditional voltage *OA*, shows a constant bandwidth with respect to the closed-loop gain and a very high slew-rate. This makes it of primary importance in the design of modern *LV LP ICs*; in addition, the first stage of *CFOA* is exactly a Current Conveyor.

The original example presented by Sedra and Smith in 1968 was generically named by the authors "Current Conveyor". The first block was identified as "First Generation Current Conveyor", or *CCI*, only when its evolved topology was called "Second Generation Current Conveyor", or *CCII*, in 1970 [6, 12]. When *CCI* was firstly introduced, it was employed as a new building block in the design of simple analog signal processing circuits (i.e., it is possible to easily implement few simple circuit topologies such as Voltage-to-Current (*V-I*) and Current-to-Voltage (*I-V*) converters, Negative Impedance Converter (*NIC*), etc.) [1], but it did not show an input voltage terminal so the new version of the conveyor (i.e., the *CCII*) was developed.

A. De Marcellis and G. Ferri, *Analog Circuits and Systems for Voltage-Mode and Current-Mode Sensor Interfacing Applications*, Analog Circuits and Signal Processing, DOI 10.1007/978-90-481-9828-3, © Springer Science+Business Media B.V. 2011

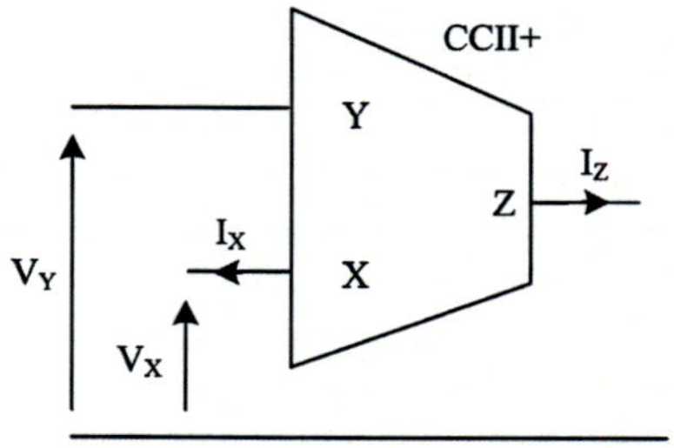

Fig. A1.1 The *CCII* schematic symbol representation

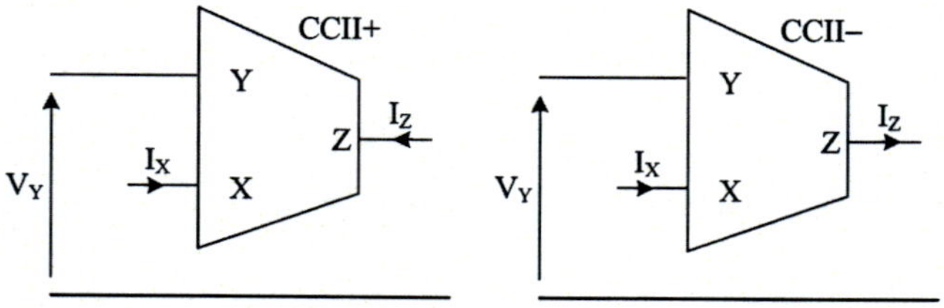

Fig. A1.2 Positive and negative *CCII* basic building block

The *CCII*, whose schematic symbol is reported in Fig. A1.1, shows the main ideal electrical characteristics (from the *CCII* theory [1,6,12]) as follows:

$$\begin{bmatrix} I_Y \\ V_X \\ I_Z \end{bmatrix} = \begin{bmatrix} 0 & 0 & 0 \\ 1 & 0 & 0 \\ 0 & \pm 1 & 0 \end{bmatrix} \cdot \begin{bmatrix} V_Y \\ I_X \\ V_Z \end{bmatrix} \tag{A1.1}$$

It has a low impedance (ideally zero) current input (X node, which can be also a voltage output). On the contrary, the other voltage input terminal (Y node) shows a high impedance (ideally infinite), while Z node shows also a high impedance level (ideally infinite), so it is an output current terminal. Moreover, currents flowing at X and Z nodes are always equal in magnitude (the current flowing at X node is "conveyed" to the current output Z node), while if a voltage is applied to Y node, the same voltage will appear at X node. Current at Z node (I_Z) can flow either in the same direction of I_X or in the opposite one. In the matrix description reported in Eq. A1.1, we assume that sign "+" stays for both these currents flowing in the same direction, while sign "–" stays for the opposite situation, considering the *CCII* as reference. In the first case we have a "positive *CCII*" (*CCII+*), in the second case a "negative *CCII*" (*CCII-*), as also shown in Fig. A1.2 [1].

Parasitic impedances are the main drawback that affects the *CCII* ideal behaviour and, sometimes, their utilization in typical analog applications. Their kind and value depend on the *CCII* internal topology, developed at transistor level. In Fig. A1.3 an equivalent model of a non-ideal *CCII*, showing its main typical parasitic impedances and non-idealities at its terminals, has been reported.

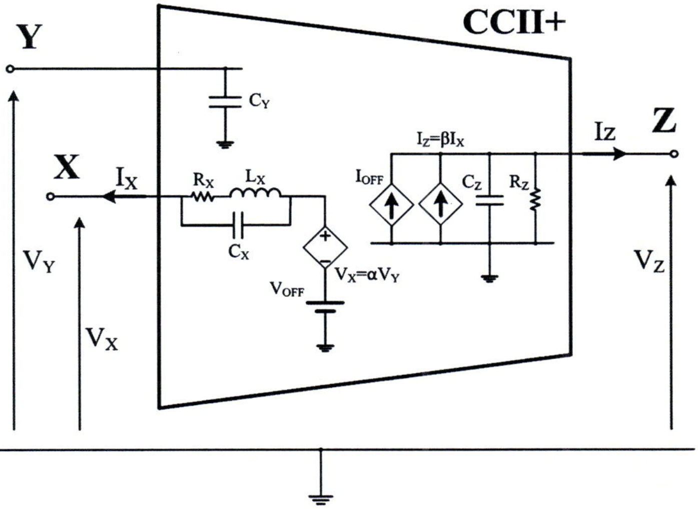

Fig. A1.3 A complete equivalent non-ideal *CCII+* representation showing typical parasitic impedances and non-idealities at each terminals (equivalent *CCII* macromodel)

Moreover, due to non-ideal behaviour of *CCII*, V_X is not exactly equal to V_Y as well as I_Z can be slightly different from $\pm I_X$. In particular, we can have $\alpha = V_Y / V_X$ and $\beta = I_Z / I_X$ parameters which, for a non-ideal *CCII*, can be non-unitary. Then, also an offset between X and Y node voltages can be present. In particular, always referring to Fig. A1.3, the voltage offset V_{OFF} causes a current offset I_{OFF} because I_X current is mirrored to high impedance Z node, so to obtain the output current I_Z, whose value is dependent on the load connected to X node. Therefore, also when $V_Y = 0$, an offset current, dependent on the external load impedance value, flows into X node (due to the voltage offset), so also into Z node (due to the current buffer operation).

In the literature, several possible implementations, at transistor level, for integrated *CCIIs*, have been presented, but if there is not a particular need of *LV LP* topologies (or it is not possible to design and fabricate a chip dedicated to a *CCII*-based application), a commercially available component can be employed. The only available commercial *CCII* is the AD844 by Analog Devices [13], whose simplified internal block scheme is shown in Fig. A1.4. It has been heavily utilized in discrete component prototype *PCB* implementations of *CCII*-based circuits, among which also sensor interface topologies.

At the output we have two nodes: a high impedance node (I_{OUT}), implementing Z node of a *CCII*, and a low impedance one (V_{OUT}). This is why this device can be regarded as a Current Feedback Operational Amplifier (*CFOA*), that can be viewed as a *CCII* followed by a voltage buffer. In fact, considering in Fig. A1.4 only the terminals V_{IN+}, V_{IN-} and I_{OUT}, the AD844 operates as a $CCII+$, allowing a quick implementation of different *CCII* based circuit solutions. The main characteristic values of the AD844 are reported in Table A1.1. More detailed information can be found in the datasheet of the component [13].

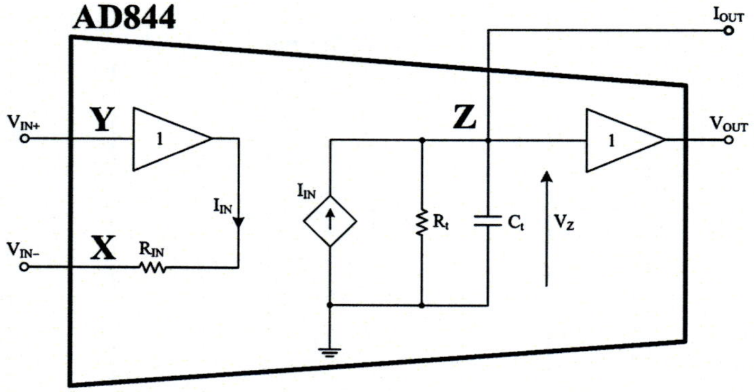

Fig. A1.4 Analog Devices AD844 simplified internal block scheme (V_{IN+} = terminal 3, V_{IN-} = terminal 2, I_{OUT} = terminal 5, V_{OUT} = terminal 6)

Table A1.1 AD844 main characteristics (nominal values @±15 V)

AD844 parameter	Typical value
Supply voltage	±15 V
R_X	50 Ω
L_X	10 nH
C_X	2 pF
α	≈1
β	≈1
R_Y	10 MΩ
C_Y	2 pF
R_Z	3 MΩ
C_Z	5.5 pF

CCII Transistor Level CMOS Implementations

In this paragraph we will show a couple of transistor level integrated solutions for *CCIIs*, developed in a standard *CMOS* technology (AMS 0.35 μm) and utilized in the *CCII*-based sensor interfaces shown in Chap. 4.

A first example of a *CCII* showing negligible parasitic impedances and unitary voltage and current gains for a very large bandwidth (quasi-ideal characteristics) is shown in Fig. A1.5 [14]. This circuit can be supplied at ±0.75 V and shows a 118 μW static power consumption. It is formed by a differential input stage ($M_1 - M_7$; R_3), a class AB output stage ($M_8 - M_{11}$; R_1, R_2; $M_{16} - M_{17}$) and a *LV* cascode Wilson current mirror ($M_{12} - M_{15}$; $M_{18} - M_{21}$). The class AB output stage allows to decrease the X parasitic impedance, whereas the cascode current mirror increases the Z impedance. Its bandwidth is about 10 MHz, while parasitic components have been minimized, while voltage and current gains are very close to one. Table A1.2 summarises the main characteristics of this *CCII*.

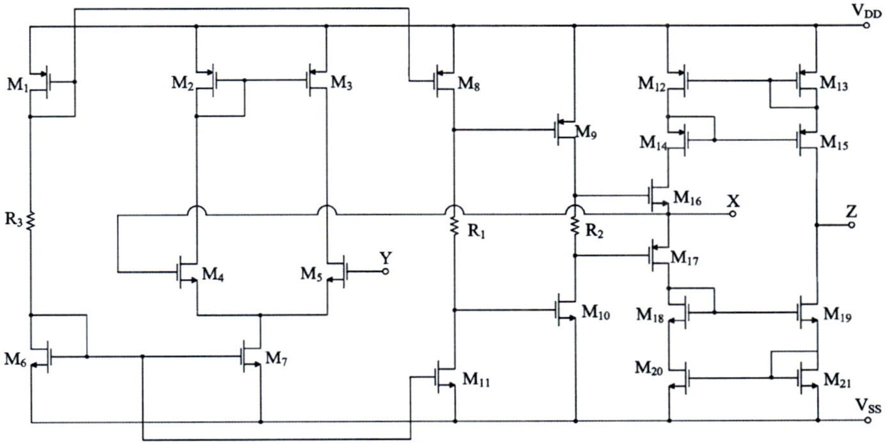

Fig. A1.5 Quasi-ideal *CCII* schematic at transistor level

Table A1.2 Quasi-ideal implemented *CCII* main performances

CCII parameter	Value
Supply voltage	±0.75 V
Power consumption	118 μW
3 dB bandwidth	10.5 MHz
Biasing currents	6 μA
Voltage gain (α)	1.00
Current gain (β)	1.00 (Rx = Rz = 10 kΩ)
X parasitic resistance R_X	13 Ω
X parasitic inductance L_X	0.4 μH
X parasitic capacitance C_X	0.1 pF
Z parasitic resistance R_Z	2.6 MΩ
Z parasitic capacitance C_Z	0.03 pF
Y parasitic capacitance C_Y	0.1 pF

Another *CCII* topology solution, showing negligible parasitic impedances and quasi-ideal characteristics is shown in Fig. A1.6 [1]. The topology is based on two complementary differential pairs in parallel as input stage ($M_1 - M_2$ and $M_3 - M_4$) and on inverter output stages ($M_5 - M_6$ and $M_7 - M_8$). Compared to the previous solution, this topology shows a complete input and output dynamic range (rail-to-rail operation). Table A1.3 reports its main characteristics.

Although *CCII* block is powerful and simple at the same time, the wide spread of possible applications has led to the development of evolutions and improvements of the basic *CCII* topology [1]. For example, the simplest modification of the basic *CCII* topology is represented by its dual output version, that shows, simultaneously available, both the output currents, inverted and non-inverted: in this case the device is named Dual Output Current Conveyor (*DOCCII*). This device can be employed

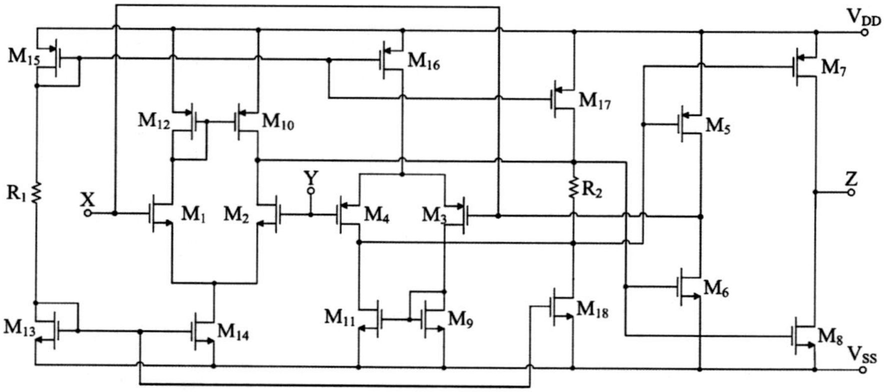

Fig. A1.6 Quasi-ideal rail-to-rail *CCII* schematic at transistor level

Table A1.3 *CCII* main characteristics

CCII parameter	Value
Supply voltage	±0.75 V
Power consumption	750 μW
3 dB bandwidth	17.7 MHz
Biasing currents	10 μA
Voltage gain (α)	1.00
Current gain (β)	1.00 (Rx = Rz = 10 kΩ)
X parasitic resistance R_X	150 Ω
X parasitic inductance L_X	1.7 μH
X parasitic capacitance C_X	1 pF
Z parasitic resistance R_Z	1.39 MΩ
Z parasitic capacitance C_Z	0.03 pF
Y parasitic capacitance C_Y	1 pF

with success in applications that require a feedback between input and output terminals, such as impedance converters or simulators, multifunction active filters, sensor interfaces, signal conditioning and signal processing, etc.

Finally, we want to mention that all the evolutions of the basic *CCII*, towards other building blocks with multiple input and output terminals (e.g., *VGCCII* and *FDCCII*), can be implemented starting from the basic block and can be obtained either modifying the internal *CCII* topology or, more simply, through suitable connections of more basic *CCII* blocks and some passive components [1]. In this case, it is important to underline that, often, the employed passive components have to be perfectly matched, otherwise a current transfer error between X and Z nodes and/or a voltage transfer error between Y and X nodes will be introduced.

Appendix 2: Noise and Offset Compensation Techniques

Noise and Noisy Signals

Noise represents an undesired signal. There are two main types of noise, especially in laboratory experimentals, that are intrinsic noise (i.e., internal noise) and external interferences (i.e., external noise). Intrinsic noise sources like thermal (Johnson), shot and flicker noises are due to all the physical fluctuations and processes inherent to devices. Though we cannot get rid of intrinsic noise sources, by being aware of their nature, we can minimize their effects. In particular, the intrinsic noise is more predictable and can be reduced already by a proper design of transistor sizes, choosing also a more suitable technology. On the contrary, external noise sources (e.g., power line noise, broadcast stations, that coming from supply, wires, electromagnetic fields, etc.) depends on the environment. Their effects can be minimized by a careful attention to grounding, shielding and other aspects of experimental design such as an accurate wiring and layout.

Several electronic devices show a noise component strongly affected by frequency. In some situations, noise shows a component inversely proportional to the frequency, that is its level increases when frequency is lowered. In standard *CMOS* technology, in particular in *MOS* transistors, the main noise contributions are given by thermal and flicker noises. The first is also called "white" noise, because its spectral density is constant over a given frequency and the thermal motion of electrons is reputed to be its main source. On the contrary, flicker noise, discovered in thermoionic valves by Schottky and Johnson in 1925, is also known as $1/f$ noise, because its spectral density is inversely proportional to frequency, so it is dominant at low frequencies (other sources of $1/f$ noise include noise found in vacuum tubes and semiconductors). It has many different origins and is not clearly understood but exhibits a $1/f^n$ power spectrum with n usually in the range 0.9–1.35 (note that DC drift can be considered a very low frequency form of flicker noise). The intersection between flicker and white noise is called $1/f$ noise corner.

Moreover, it is known how, in *MOS* devices, thermal noise is inversely proportional to its transconductance (and, consequently, to its aspect ratio W/L), while flicker noise is inversely proportional to $W \cdot L$ product [15]. As a consequence, the choice of transistor sizes is important to minimize noise and has to be done according to the working frequency of the circuit. In addition, the shot noise has been observed especially in all those situations where a current is controlled by a voltage, as, for example, the gate of a *MOS* transistors. As a consequence, in *MOS* devices, where the gate current is extremely low, the shot noise contribution can be easily neglected in a first order evaluation. On the contrary, $1/f$ noise can be modelled by a noise current or voltage generator whose level is related to a noise factor K_f which depends numerically on the employed technology. Hence, as for the offset, an evaluation of the noise is more important at reduced supplies, where signals have reduced amplitudes.

It is well known that a resistor generates a noise voltage across its terminals, caused by random motion of thermally agitated electrons; its mean-square noise voltage is given by:

$$v_n^2 = 4kTR\Delta f \tag{A2.1}$$

where k is Boltzmann's constant (1.381×10^{-23}JK^{-1}), T is the absolute temperature (in K) and R is the resistance (Ω) and Δf is the bandwidth of measurement (Hz). Alternatively, from Ohm's law, the mean-square noise current related to a resistance R is given by:

$$i_n^2 = \left(\frac{v_n}{R}\right)^2 = \frac{4kT\Delta f}{R}. \tag{A2.2}$$

The shot noise is caused by the random movements of electrons, for example, at the electrodes of electron tubes or transistor junctions (there is always some non-uniformity in the electron flow which generates noise in the current). A DC current I has a noise-current component (shot noise) i_n given by:

$$i_n^2 = 2qI\Delta f, \tag{A2.3}$$

where q is the charge of one electron (about 1.6×10^{-19}C), I is the *RMS* AC current or DC current depending upon the circuit and Δf is the bandwidth.

More in general, noise becomes of interest when buries the signal to be detected. Fig. A2.1 shows the power spectral density (power/unit bandwidth) of the most commonly encountered types of noise. Deterministic noise can range from simple discrete-frequency components such as power-line buzz at harmonics of 50 or 60 Hz, to Radio Frequency Interference (*RFI*) caused by narrow, high-energy pulses from power-line switching spikes, pulsed lasers, radar transmitters, etc. [16, 17].

Stochastic or random noise is found in many systems both as white noise and also as flicker noise. It is known that for an *RMS* voltage of v (Volts) and a frequency range of Δf (Hz), the power spectral density S is given by:

$$S = \frac{v^2}{\Delta f} = \left(\frac{v}{\sqrt{\Delta f}}\right)^2. \tag{A2.4}$$

The quantity $v/\sqrt{\Delta f}$ is usually referred to as voltage spectral density and is measured in *RMS* V/$\sqrt{\ }$ Hz. As concerns the bandwidth, in the simple low-pass filter shown in Fig. A2.2, for example, the signal bandwidth is usually defined, see Fig. A2.3, as the cut-off frequency f_c where $V_{OUT}/V_{IN} = 1/\sqrt{2} = 70.7\%$ (i.e., -3 dB) or $(V_{OUT}/V_{IN})^2 = 50\%$. In this case:

$$f_c = \frac{1}{2\pi RC}. \tag{A2.5}$$

In addition, the Equivalent Noise BandWidth (*ENBW*) B_n is defined as:

$$B_n = \frac{1}{G^2}\int_0^\infty |H(j2\pi f)|^2 \, df, \tag{A2.6}$$

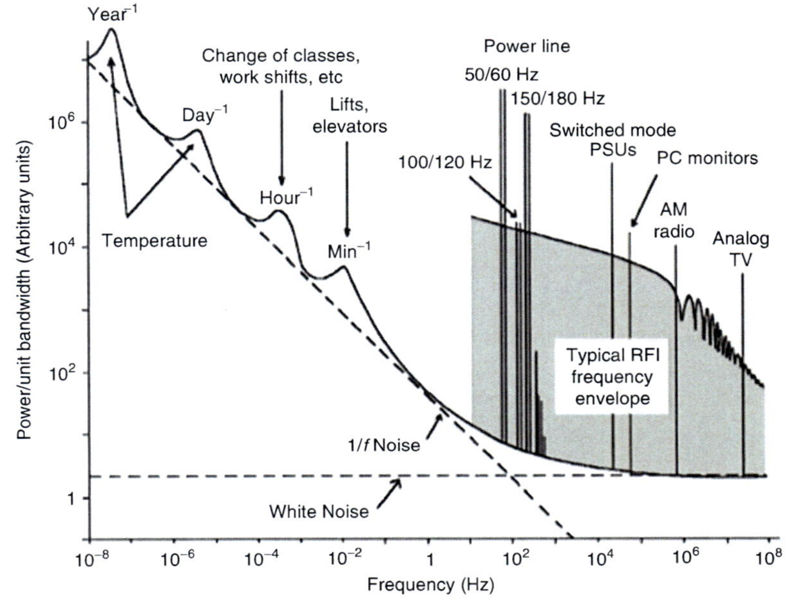

Fig. A2.1 Environmental and type of noises versus frequency

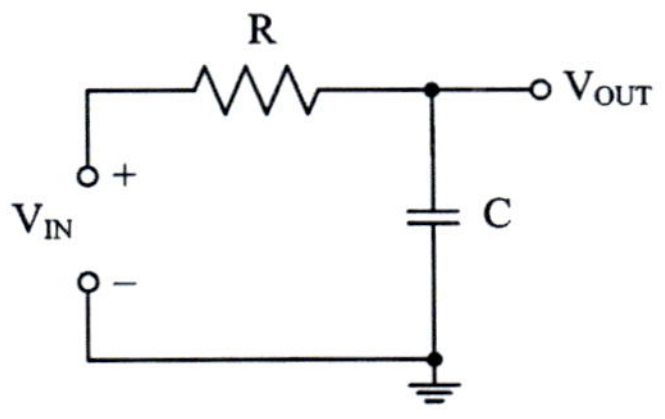

Fig. A2.2 Low-pass filter circuit

being $H(j2\pi f)$ the frequency transfer function and G the DC gain of the system. For the simple RC filter, shown in Fig. A2.2, we have:

$$B_n = \frac{1}{4RC} \tag{A2.7}$$

and this value is slightly higher than f_c, as shown in Fig. A2.3.

Many of these noise sources can be also minimized with good laboratory practice and experiment design or through suitable compensation techniques, described in the next Paragraphs [15–17].

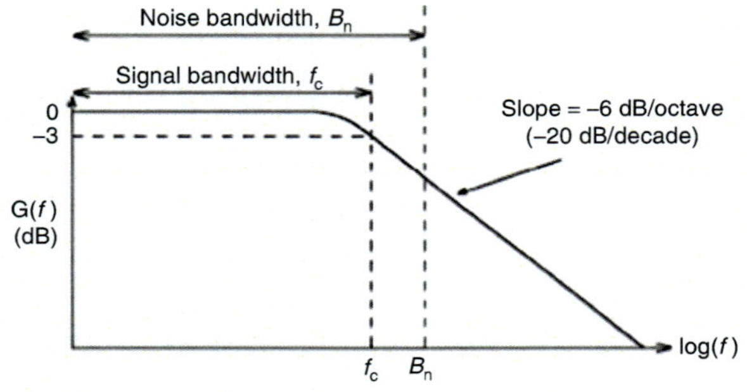

Fig. A2.3 Low-pass filter transfer characteristic

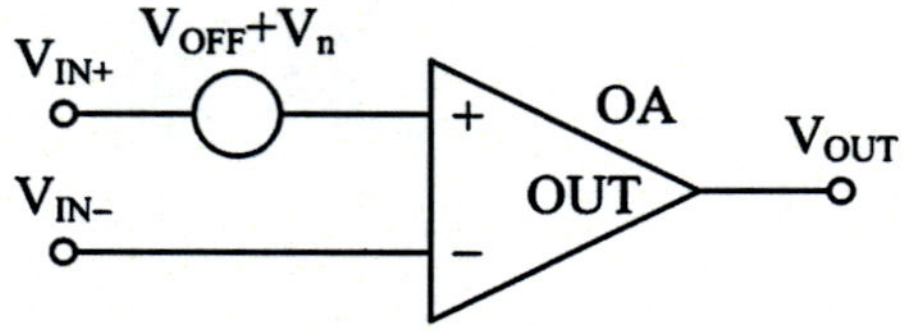

Fig. A2.4 The equivalent model for "noisy" *OA*

Noise and Offset Compensation Techniques: An Overview

Noise and offset, especially in *CMOS* circuits like amplifiers, can be reduced through suitable compensation techniques, which can be divided into two categories: static and dynamic [18]. Static techniques can be applied to balance the input stage, concerning the biasing conditions, when no input signal is applied. Dynamic compensation techniques, which need a clock for temporization, can be categorized into three main classes: autozero, chopper and Dynamic Element Matching *(DEM)* techniques [18–26].

Noise and offset can be both modelled through an input voltage source, as shown in Fig. A2.4, where the (voltage) amplifier (for example, an operational amplifier, *OA*) is now considered free from these disturbs (ideal *OA*).

Concerning to *CM* approach, a feasible model for the "noisy" *CCII* can be drawn as in the Fig. A2.5. The two external equivalent noise sources depend on the designed *CCII* internal topology, at transistor level (in particular, the noise contribution is directly affected by both transistor operating points and transistor sizes).

In the following, we will describe in a deeper detail the main dynamic techniques which can be implemented so to reduce the errors in the circuits due to the presence of both noise and offset (in particular, they can be applied to both *OA* and *CCII*) [18–20].

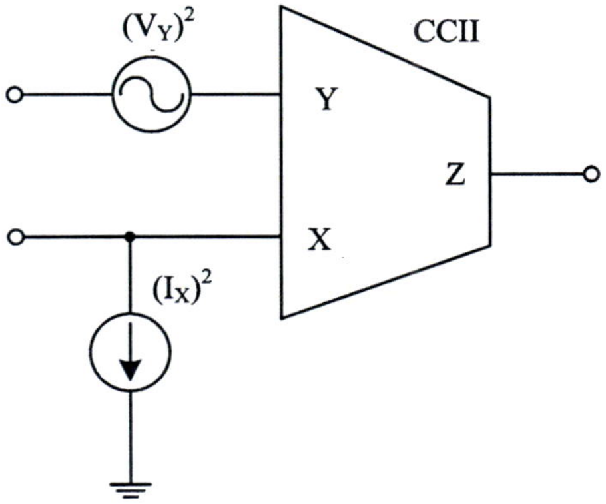

Fig. A2.5 The complete equivalent model for "noisy" *CCII*

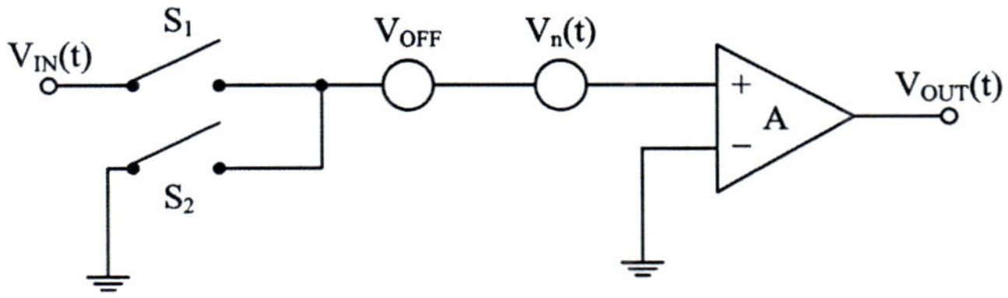

Fig. A2.6 Example of autozero amplifier (basic circuit)

The Autozero Technique

The autozero technique (see Fig. A2.6, related to a generic *OA*) utilizes two switches that active themselves in opposite way. In particular, we consider two non-overlapping phases: a sampling phase (S_1 open, S_2 closed), during which only the offset voltage V_{OFF} and the noise voltage V_n are amplified, and a signal processing phase (S_1 closed, S_2 open), when also the input signal $V_{IN}(t)$ is amplified. The two outputs must be somehow stored and, then, a post-processing operation must perform the difference between these two values, so to extract the information of input signal free from noise and offset [18–20].

The Chopper Technique

The chopper technique is shown in Fig. A2.7 at schematic block level. Also in this case, switches and temporization must be employed. In Fig. A2.8 the chopper blocks (CH_1 and CH_2) are better specified, while Fig. A2.9 shows a *CMOS* chopper differential amplifier, where the chopper technique is applied also at transistor level. Referring to Fig. A2.8, through the φ_1 and φ_2 switches, input signal V_{IN} can be

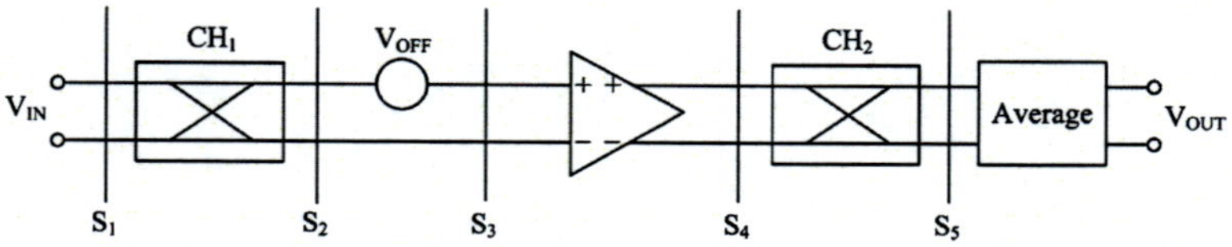

Fig. A2.7 Example of a chopper amplifier

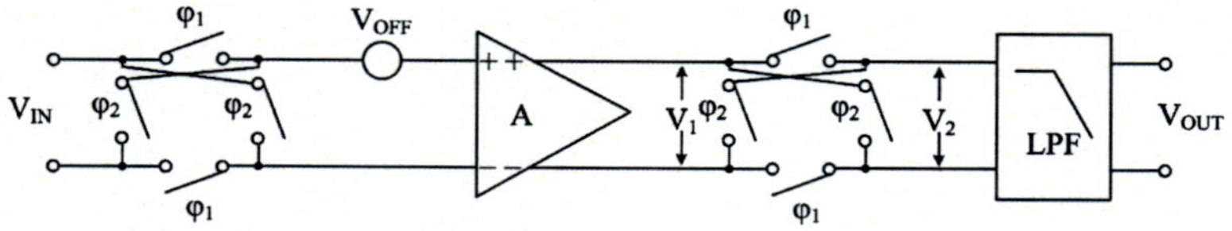

Fig. A2.8 Detailed block scheme of the chopper circuit shown in Fig. A2.7

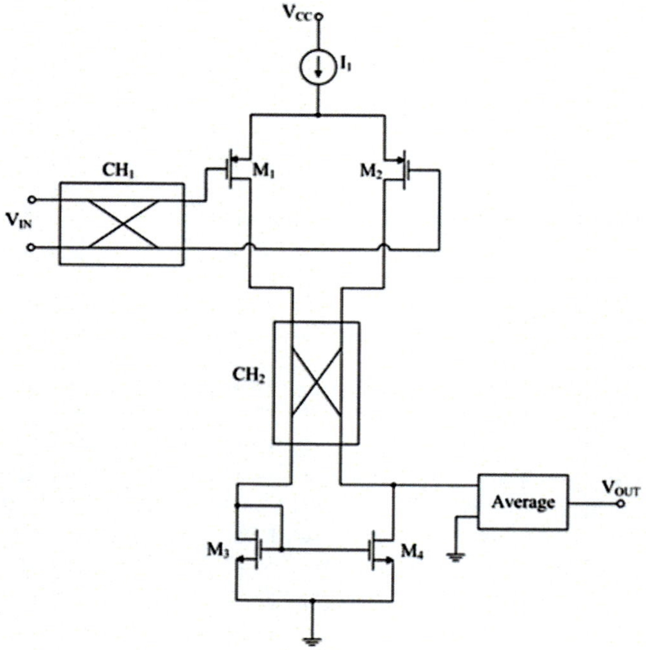

Fig. A2.9 Schematic circuit of a simple *CMOS* chopper differential amplifier at transistor level

carried at the amplifier input terminals either as is (φ_1 closed and φ_2 open) or with a opposite phase (φ_1 open and φ_2 closed). Then, a low-pass filter (*LPF*) performs the averaging of the output voltage that, in this way, results to be independent from voltage offset and noise [18–20].

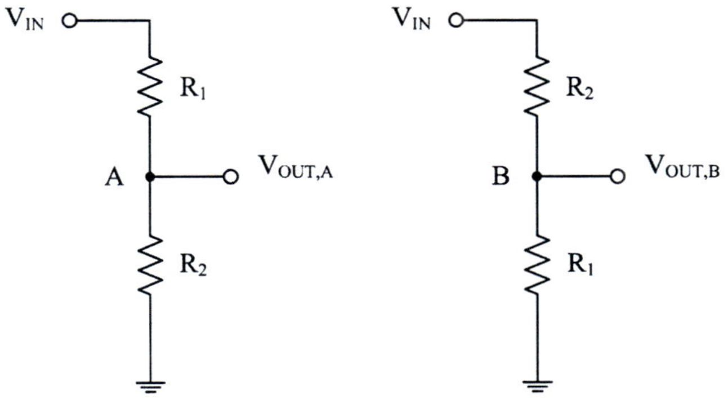

Fig. A2.10 Resistance-based voltage dividers

The Dynamic Element Matching Technique

DEM technique performs a "dynamical matching" of the circuit "mismatched" elements, that is it dynamically interchanges them and, after, takes the average of the two outputs, opportunely stored. In Fig. A2.10, the simple resistive voltage divider (based on two in series resistances), as an example, is considered. If the two resistances R_1 and R_2 have the same value, the output voltages, for both cases illustrated in Fig. A2.10, are also the same, as follows:

$$V_{OUT,A} = V_{OUT,B} = V_{OUT} = \frac{V_{IN}}{2}. \tag{A2.8}$$

Since the two resistances are not never perfectly equal (in fact, they are "mismatched"), we have for example:

$$R_2 = R_1 (1 + \delta). \tag{A2.9}$$

Calling with $V_{OUT,A}$ and $V_{OUT,B}$ the output voltages (see Fig. A2.10), after an averaging calculation, we can write (neglecting switch non-idealities):

$$\frac{1}{2}(V_{OUT,A} + V_{OUT,B}) = \overline{V_{OUT}} = \frac{V_{IN}}{2}. \tag{A2.10}$$

This technique can be employed in *VM* and *CM* circuit design and applied both at transistor level and at whole active block (exchanging different *OAs* or *CCIIs* in the schematic circuit). The dynamic technique in both the last cases is named Dynamic *OA* Matching (*DOAM*) and Dynamic *CCII* Matching (*DCCIIM*), respectively [18–20].

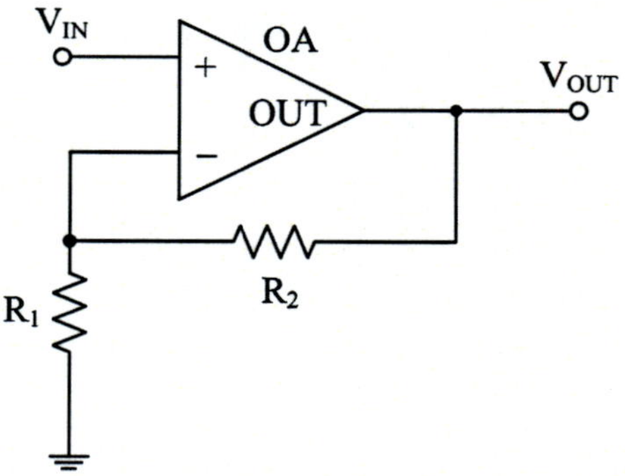

Fig. A2.11 *OA*-based non-inverting voltage amplifier

Voltage Gain Error in Voltage Amplifiers: OA vs. CCII

In the following, a comparison of the voltage gain errors (caused by non ideal active devices) for both the traditional *OA*-based voltage amplifier and the *CCII*-based is reported [27,28]. In particular, in Fig. A2.11, a traditional *OA*-based voltage amplifier is pictured (non-inverting configuration).

If the *OA* is an ideal block, we have the ideal output $V_{OUT,I}$ as follows:

$$V_{OUT,I} = A \cdot V_{IN} = \left(1 + \frac{R_2}{R_1}\right) \cdot V_{IN}. \tag{A2.11}$$

Considering a finite open loop voltage gain G for the *OA*, Eq. A2.11 becomes:

$$V_{OUT,NI} = A \cdot V_{IN} = \frac{\left(1 + \frac{R_2}{R_1}\right)}{1 + \frac{\left(1+\frac{R_2}{R_1}\right)}{G}} \cdot V_{IN}. \tag{A2.12}$$

In particular, the Voltage Gain Error (*VGE*), considering the errors introduced by this non-infinite gain G, can be expressed by:

$$VGE = \frac{\left(1 + \frac{R_2}{R_1}\right)}{G + \left(1 + \frac{R_2}{R_1}\right)}, \tag{A2.13}$$

considering that *VGE* is calculated as $(V_{OUT,NI} - V_{OUT,I})/V_{OUT,I}$. Since G depends also on frequency, it comes that, for an example, for $R_2/R_1 = 10$ and DC *OA* gain equal to 60 dB, at low frequencies *VGE* is about 0.1%, but this error increases dramatically for higher frequencies. For this reason, the *CCII*-based solution of the voltage amplifier, presented in Fig. A2.12, can be used to overcome these drawbacks. In fact, α parameter is close to its ideal unitary value for a frequency range much larger with respect to that where *OAs* present a very high voltage gain.

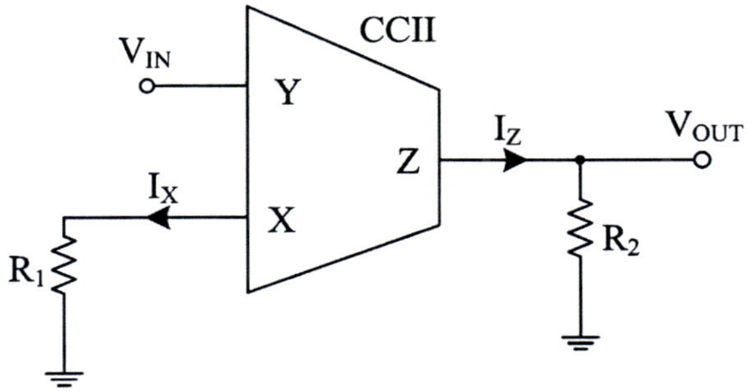

Fig. A2.12 *CCII*-based non-inverting voltage amplifier

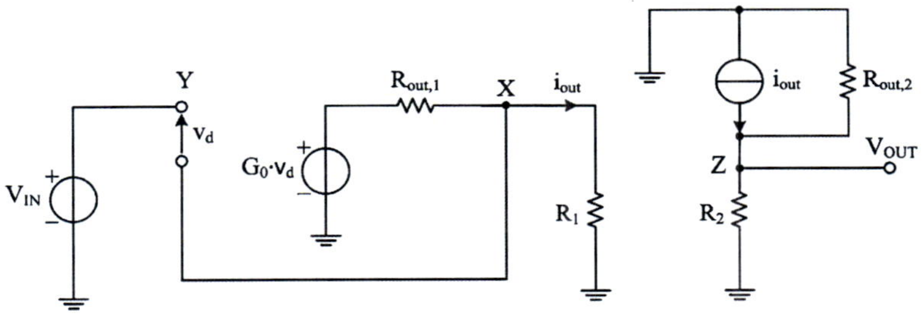

Fig. A2.13 *CCII*-based voltage amplifier with a Thevenin equivalent circuit for the unity-feedback *OA*

More in detail, the *CCII*-based voltage amplifier, reported in Fig. A2.12, shows the following voltage gain (that considers also α and β parameters):

$$V_{OUT} = R_2 I_Z = \beta R_2 I_X = \beta R_2 \frac{V_X}{R_1} = \alpha\beta \frac{R_2}{R_1} V_{IN}. \tag{A2.14}$$

Generally, α and β parameters are very close to unit, so they do not directly affect the voltage error but the latter, unfortunately, can depend on other *CCII* non idealities, in particular its parasitic impedances.

In Fig. A2.13, in the *CCII*-based voltage amplifier, we have evidenced the open-loop gain G_0 and equivalent parasitic resistances $R_{out,1}$ and $R_{out,2}$.

A straightforward analysis [27, 28] has given the following result for the voltage gain error:

$$VGE_{CCII} = \frac{A_V - A_{V,ideal}}{A_{V,ideal}} = \delta_1 + \delta_2, \tag{A2.15}$$

being:

$$\delta_1 = -\frac{(R_1 + R_{out,1})}{R_1 + R_{out,1} + G_0 R_1}, \tag{A2.16}$$

$$\delta_2 = -\frac{R_2}{R_2 + R_{out,2}}. \tag{A2.17}$$

Then, the relation that allows to reduce *VGE* is the following:

$$(1 + \delta_1)(1 + \delta_2) \cong 1. \tag{A2.18}$$

Imposing $|\delta_1| \ll 1$ and $|\delta_2| \ll 1$ corresponds to the following design conditions for the *CCII*:

$$\begin{cases} G_0 \cdot R_1 \gg R_1 + R_{out,1} \\ R_{out,2} \gg R_2. \end{cases} \tag{A2.19}$$

In comparison, for the *OA* based amplifier with negligible load, it is possible to find the following equations:

$$VGE_{OA} = -\frac{R_1 + R_2 + R_{out}}{(1 + G_0)R_1 + R_2 + R_{out}}, \tag{A2.20}$$

where R_{out} is the *OA* output resistance and G_0 its open-loop voltage gain.

Comparing the two expressions of *VGE* for the *CCII* (Eq. A2.15) and the *OA* (Eq. A2.20), it is evident that the design conditions are the following: high gain G_0 (for both), low $R_{out,1}$ (for the *CCII*) and low R_{out} (for the *OA*). The design of a low output impedance is clearly more difficult for *OAs*, unless a reduction of output swing must be accepted, while in the *CCII* amplifier $R_{out,1}$ is the dynamic resistance of an internal node and, in principle, can be made low without reducing the output swing.

Moreover, it is also possible to design a suitable *CCII*, internally at transistor level, so to fulfil the constraints of Eq. A2.19. Recently, it has been proved that a suitable designed *CCII* gives, in the basic voltage amplifier, a voltage gain error about 10 times lower than that given by the use of *AD844* as *CCII*.

An Offset-Compensated CMOS CCII

In order to further improve the *VGE*, a compensation technique has been also applied to an integrated *CCII* topology, so to implement an offset-compensated *CCII*, shown in Fig. A2.14 [27, 28]. The circuit is based on a two stages Miller-compensated *OTA* with a class AB output stage and a *LV* cascode current mirror for injecting the output current of the buffer-connected *OTA* into the *Z* node. Its main characteristics are summarized in Table A2.1.

A CCII-Based Instrumentation Amplifier for Sensor Interfaces

Instrumentation amplifiers (*IA*) are widely used in sensor interfaces and, in some cases, *CMIAs* based on *CCIIs* may offer important advantages over conventional *IA*

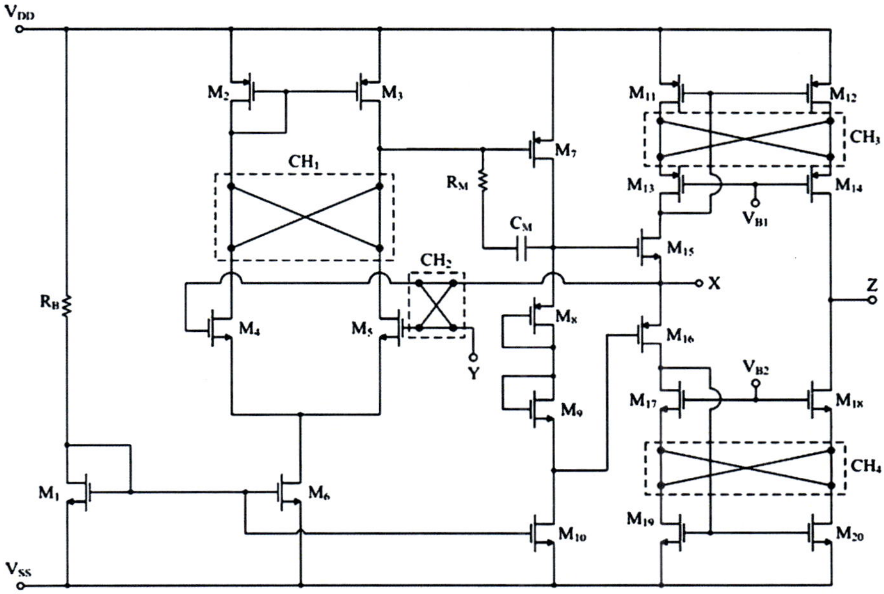

Fig. A2.14 Proposed Miller offset compensated *CCII* with cascode output current mirror

Table A2.1 Main characteristics of the offset compensated *CCII* reported in Fig. A2.14

Parameter	Value
Voltage supply	±1.65 V
Power consumption	309 μW
Open loop voltage gain (*OA*)	3990
Slew rate	13.5 V/μs
Z node resistance	5.4 MΩ
X node resistance	107 mΩ ($R_{out,1}/G_0$)
Systematic input offset voltage	43 μV
Chopper clock frequency	100 kHz
Output swing	up: +1.19 V; down: −1.14 V

architectures based on *OA*, with particular reference to Common-Mode Rejection Ratio (*CMRR*) especially if resistor matching is not too good, bandwidth and lower *VGE* [29–31]. These characteristics make the *CCII*-based *IA* a really attractive and useful block for smart sensor systems which include both the sensing device and the electronic interface integrated in the same chip allowing, in many practical cases, to reduce the cost of the sensor system, the errors of the electronic interface, the power consumption, the susceptibility to interferences, the size and the weight of the whole system. The use of compensation techniques to reduce input offset is particularly useful in these *IA* configurations.

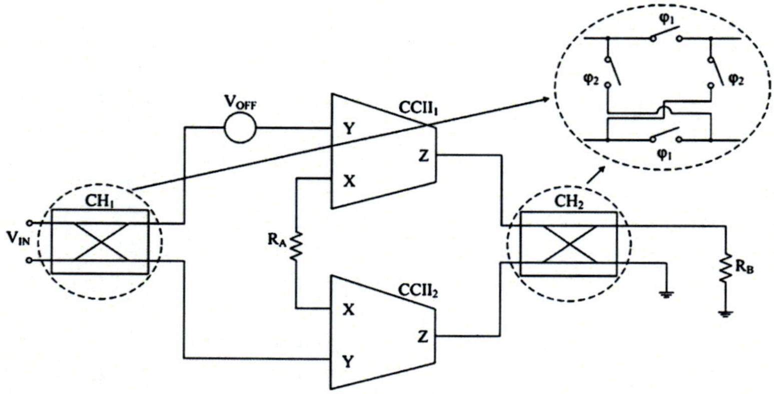

Fig. A2.15 Complete schematic block of the proposed offset compensated *CMIA*

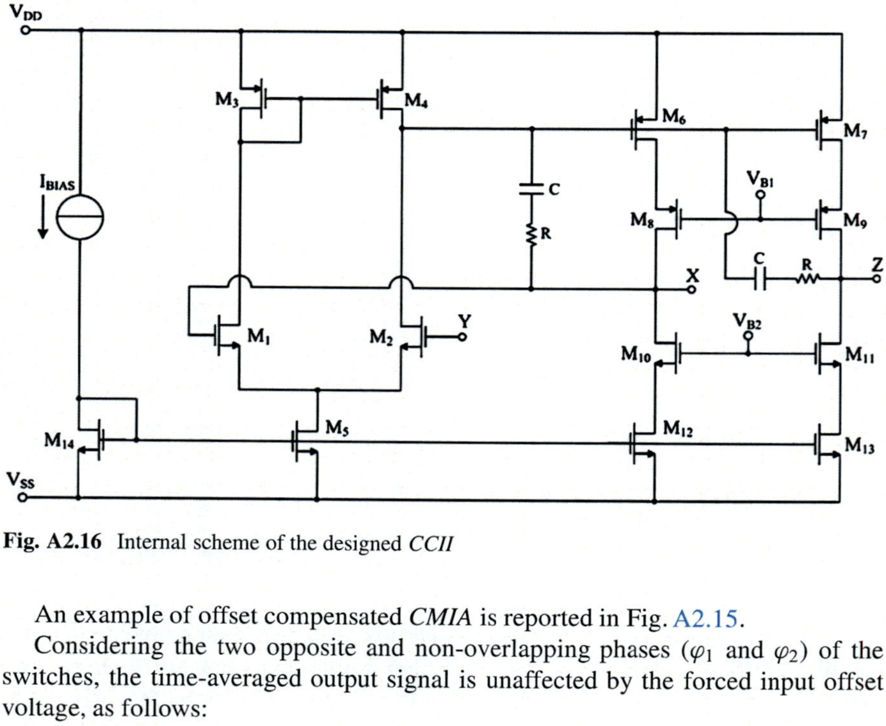

Fig. A2.16 Internal scheme of the designed *CCII*

An example of offset compensated *CMIA* is reported in Fig. A2.15.

Considering the two opposite and non-overlapping phases (φ_1 and φ_2) of the switches, the time-averaged output signal is unaffected by the forced input offset voltage, as follows:

$$V_{OUT} = \frac{R_B}{R_A}\left(\frac{V_{IN} + V_{OFF} + V_{IN} - V_{OFF}}{2}\right) = \frac{R_B}{R_A} V_{IN} = A \cdot V_{IN}. \tag{A2.21}$$

Implementing the *CCII* with the circuit shown in Fig. A2.16, the circuit has shown good *VGE* properties (of few percent), as reported in Fig. A2.17 that shows the

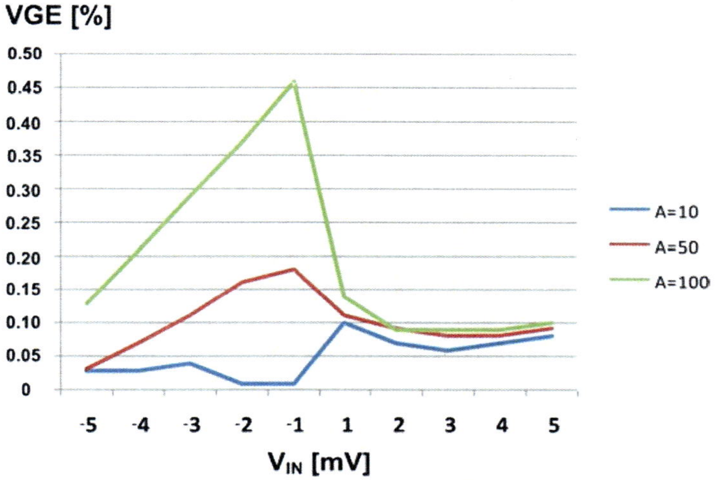

Fig. A2.17 *VGE* of the offset compensated *CMIA* vs. input voltage V_{IN} for different ideal gains A and with an auxiliary DC voltage signal equal to 500 μV added to the inputs (forced voltage offset)

simulated *VGEs*, with the activation of the switches, for three different gains ($A = R_B/R_A$) as a function of the input voltage when auxiliary DC voltage sources equal to 500 μV have been added in series to one of the input of the *CMIA*.

References

1. G. Ferri, N. Guerrini, Low Voltage Low Power CMOS Current Conveyors (Kluwer Academic Publishers, Boston, 2003). ISBN 1402074867
2. C. Toumazou, A. Payne, D. Haigh, Analogue IC Design: The Current Mode Approach (Peregrinus, London, 1990). ISBN 9780863412974
3. C. Toumazou, J. Lidgey, Universal Current Mode Analogue Amplifiers, in Analogue IC Design: The Current Mode Approach (Peregrinus, London, 1990)
4. G. Palumbo, S. Palmisano, S. Pennisi, CMOS Current Amplifiers (Kluwer Academic Publishers, Boston, 1999). ISBN 9780792384694
5. A. Sedra, K.C. Smith, The current conveyor – A new circuit building basic block. IEEE Proc. **56**, 1368–1369 (1968)
6. A.S. Sedra, G.W. Roberts, Current Conveyor Theory and Practice, in Analogue IC Design: The Current Mode Approach (Peregrinus, London, 1990)
7. S. Franco, Analytical foundation of current feedback amplifiers, in Proceedings of the IEEE International Symposium on Circuits and Systems, Chicago, vol. 2, 1993 pp. 1050–1053
8. D.F. Bowers, Applying Current Feedback to Voltage Amplifiers, in Analogue IC Design: The Current Mode Approach (Peregrinus, London, 1990)
9. A. Soliman, Applications of the current feedback operational amplifier. Analog Integr. Circ. Signal Process. **11**, 265–302 (1996)

10. C. Toumazou, A. Payne, J. Lidgey, Current-feedback versus voltage amplifiers: Hystory, Insight and Relationships, in Proceedings of the IEEE International Symposium on Circuits and Systems, Chicago, 1993
11. G. Palumbo, S. Pennisi, Current feedback amplifiers versus voltage operational amplifiers. IEEE Trans. Circuit Syst. I **5**, 617–623 (2001)
12. A. Sedra, K.C. Smith, A second generation current conveyor and its applications. IEEE Trans. Circuit Theory **CT-17**, 132–134 (1970)
13. Internet resource: http://www.analog.com. Datasheet AD844
14. G. Ferri, V. Stornelli, M. Fragnoli, An integrated improved CCII topology for resistive sensor application. Analog Integr. Circ. Signal Process. **48**(3), 247–250 (2006)
15. F. Maloberti, *Analog Design For CMOS VLSI Systems* (Kluwer Academic Publishers, Dordrecht, 2001). ISBN 1441949194
16. Internet resource: http://www.signalrecovery.com. "Signal Recovery", part of AMETEK (Advanced Measurement Technology)
17. R.C. Dorf, The Electrical Engineering Handbook (CRC Press LLC, Boca Raton, 2000). ISBN 0849385741
18. C. Falconi, Principles and circuits for integrated thermal sensors, Ph.D. Thesis, University of Roma Tor Vergata, 2001
19. C. Falconi, C. Di Natale, A. D'Amico, M. Faccio, Electronic interface for the accurate readout of resistive sensors in low voltage-low power integrated systems. Sensors Actuators A **117**, 121–126 (2005)
20. C. Enz, G. Temes, Circuit techniques for reducing the effects of op-amp imperfections: autozeroing, correlated double sampling and chopper stabilization, Proceedings of IEEE, vol. 84(11) (Nov 1996), pp. 1584–1613
21. C. Falconi, A. D'Amico, M. Faccio, Design of accurate analog circuits for low voltage low power CMOS systems, Proceedings of IEEE ISCAS, vol. 1 (2003) pp. 429–432
22. C. Falconi, C. Di Natale, A. D'Amico, Dynamic Op amp matching: a new approach to the design of accurate electronic interfaces for low voltage/low power integrated sensors systems, Proceedings of Eurosensors, Prague (2002)
23. R.J. van de Plassche, Dynamic element matching for high accuracy monolithic D/A converters. IEEE J. Solid-State Circuits **SC-11**, 795–800 (1976)
24. J.F. Witte, K.A. Makinwa, J.H. Huijsing, A CMOS Chopper Offset-Stabilized Opamp. Proc. IEEE ESSCIRC **1**, 360–363 (2006)
25. C. Falconi, D. Mazzieri, A. D'Amico, V. Stornelli, A. De Marcellis, G. Ferri, Dynamic element matched CCII for high accuracy electronic interfaces in deep sub-micron CMOS microsystems, Proceedings of Eurosensors, Goteborg (2006)
26. C. Falconi, E. Martinelli, C. Di Natale, A. D'Amico, F. Maloberti, P. Malcovati, A. Baschirotto, V. Stornelli, G. Ferri, Electronic interfaces. Sensors Actuators B **121**, 295–329 (2007)
27. V. Stornelli, G. Ferri, A. De Marcellis, C. Falconi, D. Mazzieri, A. D'Amico, High accuracy, high precision DEM-CCII amplifiers, Proceedings of ISCAS 2007, New Orleans (2007) pp. 2196–2199
28. C. Falconi, G. Ferri, V. Stornelli, A. De Marcellis, D. Mazzieri, A. D'Amico, Current mode high accuracy high precision CMOS amplifiers. IEEE Trans. Circ. Syst. II **55**(5), 394–398 (2008)
29. Y.H. Ghallab, W. Badawy, K.V.I.S. Kaler, B.J. Maundy, A novel current-mode instrumentation amplifier based on operational floating current conveyor. IEEE Trans. Instrum. Meas. **54**, 1941–1949 (October 2005)
30. K. Koli, K.A.I. Halonen, CMRR enhancement techniques for current-mode instrumentation amplifiers. IEEE Trans. Circ. Syst. I **47**, 622–632 (May 2000)
31. S.J. Azhari, H. Fazlalipoor, CMRR in Voltage-op-amp-based Current-Mode Instrumentation Amplifiers (CMIA). IEEE Trans. Instrum. Meas. **58**(3), 563–569 (2009)

Book Overview

The book describes novel circuit and system solutions for the design of analog electronic interfaces for resistive, capacitive and temperature sensors, also showing a wide variation range, with the intent to give a complete overview of the first analog front-ends. After a description of the main kinds of sensors and their definitions, the monograph presents novel electronic circuits, most of which do not require any initial calibration, also designed with analog microelectronic techniques, at transistor level in a standard *CMOS* integrated technology. These solutions utilize both AC and DC excitation voltages for the employed sensor and are developed both in Voltage-Mode approach (which considers the use of Operational Amplifiers or Operational Transconductance Amplifiers as the main active blocks) and in Current-Mode approach (being the Second Generation Current Conveyor the main active device) as well as with Low Voltage Low Power characteristics when designed for portable applications and instrumentations. The presented interfaces can be easily fabricated both as prototype boards, for a fast characterization (in this sense, they can be simply implemented by students and technicians) and as integrated circuits, also using modern design techniques (well known to specialist analog microelectronic students and designers). The book can give a practical help to analog electronic circuit designers, sensor companies and Ph.D. students, as well as in advanced graduate courses, in the implementation of analog electronic front-ends, also for the detection of small signals coming from sensors.

The main aspects of the book are the following:

- A deep introduction and description on sensors and basic resistive, capacitive and temperature sensor interfacing together with their main characteristics and parameter definitions.
- A complete overview of the first analog front-ends, in particular for wide-range resistive and capacitive physical and chemical sensors, easy to understand and also to implement both with discrete components for *PCB* fabrication and in a standard *CMOS* integrated technology.

Author Biographies

Andrea De Marcellis was born in Giulianova, Italy, in 1980. He graduated in Electronic Engineering in 2005 and received the Ph.D. degree in Electronic and Information Engineering in 2009, at the University of L'Aquila, Italy. He currently works on signal conditioning design for portable integrated applications, in particular, on analog design of integrated circuits for Voltage-Mode and Current-Mode sensor interfacing and signal processing. He is co-author of a patent and more than 70 scientific publications on international journals and talks at national and international conferences.

Giuseppe Ferri was born in L'Aquila, Italy, in 1965. He is a Professor of Analog Electronics and Microelectronics at University of L'Aquila. His research activity is centered on the analog design of integrated circuits for portable applications and circuit theory. He is author and co-author of two patents, five books and about 300 publications on scientific journals and international conferences. He is an IEEE senior member and Associate Editor of Journal of Circuits, Computers and Systems. Actually he serves also as the director of the Ph.D. School in Electrical and Information Engineering at University of L'Aquila.

Index

H

I

L

M

N

O

P

R

S